Heavy Metal Removal from Water

This book embarks on a thorough exploration of metal ion removal processes and cutting-edge technologies, with a particular focus on membranes, photocatalysts, and a diverse array of adsorbents. Going beyond the fundamentals, the authors illuminate the principles and practical intricacies underlying membrane-based separation processes, offering readers a comprehensive understanding of their applications and nuances. In addition to membrane separation techniques, this book examines alternative approaches utilizing adsorbents and photocatalysts, comparing their advantages and drawbacks against conventional membrane-based methods.

Features:

- Includes the latest technological advancements, such as the utilization of innovative materials like MXenes and graphene for heavy ion removal from water.
- Features a dedicated chapter addressing the critical aspects of extraction, recovery, and disposal of heavy metals.
- Examines the current landscape of heavy metal contamination in both developed and developing countries, providing valuable insights into the challenges at hand.
- Offers practical guidance on effective management strategies spanning the entire lifecycle of heavy metal pollutants, from extraction through to safe disposal.

This text is aimed at various stakeholders, including researchers, students, practitioners, and policymakers, committed to tackling the pressing challenges posed by heavy metal contamination.

Umapada Pal is Distinguished Professor at the Institute of Physics, Autonomous University of Puebla, Mexico. He is a pioneer in nanomaterials research in the country, having established Mexico's inaugural Laboratory on Nanomaterials Research.

Jose Luis Ortiz-Quiñonez is Visiting Professor in the Chemistry Department at the Metropolitan Autonomous University, Mexico. His research centers on synthesizing nanomaterials and their application in purifying water by removing pollutants.

Heavy Metal Removal from Water

Materials and Technologies

Edited by
Umapada Pal and Jose Luis Ortiz-Quiñonez

CRC Press
Taylor & Francis Group
Boca Raton London New York

CRC Press is an imprint of the
Taylor & Francis Group, an informa business

Designed cover image: © Shutterstock

First edition published 2027
by CRC Press
2385 NW Executive Center Drive, Suite 320, Boca Raton FL 33431

and by CRC Press
4 Park Square, Milton Park, Abingdon, Oxon, OX14 4RN

CRC Press is an imprint of Taylor & Francis Group, LLC

ISBN: 978-1-032-98519-0 (hbk)
ISBN: 978-1-032-98521-3 (pbk)
ISBN: 978-1-003-59909-8 (ebk)

DOI: 10.1201/9781003599098

Typeset in Times
by Apex CoVantage, LLC

Contents

Preface

Water is one of the most precious resources on our planet, essential for sustaining life, supporting ecosystems, and driving industrial and agricultural processes. However, with increasing population growth, urbanization, industrialization, and environmental degradation, the quality of water bodies worldwide is under threat. Contamination from chemicals, expired antibiotics, heavy metals, pathogens, and excess nutrients poses significant risks to human health and the environment. In the face of this mounting crisis, the imperative for water remediation has never been more urgent.

Cleaning up water, known as water remediation, encompasses a range of techniques and technologies aimed at restoring its quality to acceptable standards for intended use, whether for drinking, agriculture, industrial processes, or ecosystem support. This process is crucial not only for safeguarding public health but also for preserving the integrity of aquatic ecosystems and ensuring sustainable water resources for future generations. Among the multitude of pollutants jeopardizing water quality, heavy metals stand out as particularly insidious, demanding specialized attention amid the currents of industrialization and the surge in electronic waste, a characteristic of modern lifestyle.

This book extensively explores the pressing need to eliminate heavy metals from water, which is crucial for protecting human health, maintaining ecological harmony, stimulating economic progress, and promoting social equity. Through a comprehensive exploration of cutting-edge technologies and methodologies, we highlight the pathways toward mitigating heavy metal pollution in water sources. Drawing upon the latest advancements in materials science and engineering, we showcase innovative approaches poised to revolutionize the field of water remediation. We also discuss some key challenges and opportunities associated with water remediation efforts, highlighting the importance of interdisciplinary collaboration.

We trust that this book will serve as a valuable resource for graduate and post-graduate students, researchers, and professionals, empowering them not only to grasp the pivotal significance of water remediation in ensuring a pure, secure, and sustainable water reservoir for all inhabitants but also to cultivate the expertise required to transform polluted water into the pristine, and precious lifeblood of our planet.

Umapada Pal
Institute of Physics, Autonomous University of Puebla, Mexico

Jose Luis Ortiz-Quiñonez
Department of Chemistry, Metropolitan Autonomous University, Iztapalapa, Mexico

Contributors

Joyeeta Bose
National Institute of Technology Durgapur
Durgapur, West Bengal, India

Sudip Chakraborty
Department of Computer Engineering, Modeling, Electronics and Systems
University of Calabria
Cosenza, Italy

Ana María Herrera González
Autonomous University of the State of Hidalgo
Mineral de la Reforma, Hidalgo, México

Sherif Hegazy
University of Oulu
Oulu, Finland

Sidra Iftekhar
University of Eastern Finland
Kuopio, Finland

Arpita Jash
National Institute of Technology Durgapur
Durgapur, West Bengal, India

Ulla Lassi
University of Oulu
Oulu, Finland

Jose Luis Ortiz-Quiñonez
Metropolitan Autonomous University-Iztapalapa
México City, México

Umapada Pal
Autonomous University of Puebla
Puebla, México

Susmita Pandit
National Intitute of Technology Durgapur
Durgapur, West Bengal, India

Hanna Runtti
University of Oulu
Oulu, Finland

Jesús García Serrano
Autonomous University of the State of Hidalgo
Mineral de la Reforma, Hidalgo, México

Jaya Sikder
National Institute of Technology Durgapur
Durgapur, West Bengal, India

Varsha Srivastava
University of Oulu
Oulu, Finland

Amina Tariq
University of Eastern Finland
Kuopio, Finland

1 Heavy Metals and Their Toxicity

Umapada Pal and Jose Luis Ortiz-Quiñonez

1.1 INTRODUCTION TO HEAVY METALS

1.2 Definition and Distinctive Properties of Heavy Metal

Heavy metals are a group of elements that have attracted considerable attention due to their persistence, tendency to bioaccumulate, and potential toxicity in environmental and biological systems. They naturally occur in the Earth's crust, but their levels in air, soil, and water have significantly risen due to human activities, such as industrialization, mining, and farming. Because they are non-biodegradable and can build up in living organisms, heavy metals pose serious risks to ecosystems and human health. Understanding their properties, sources, and toxic effects is crucial for evaluating their environmental behavior and associated dangers. In this chapter, we first define heavy metals and highlight their key characteristics, then provide an overview of common metals and their natural abundance. Next, we explore their physicochemical properties and toxicity, along with permissible limits in drinking water. Finally, the importance of removing heavy metals from water systems is emphasized to underline their role in environmental cleanup and safeguarding public health.

Heavy metals are broadly defined as metals and metalloids with atomic numbers greater than 20 and densities exceeding 5 g/cm^3 (Chaemiso 2019). There is no universally accepted definition of heavy metals. Depending on the criteria, different scientific disciplines and regulatory bodies classify them differently. The term "heavy metal" is commonly associated with their potential for environmental contamination, toxicity, and ecotoxicity. In this context, heavy metals are elements that pose significant health risks to humans, animals, and ecosystems (Raychaudhuri et al. 2021). In another approach, heavy metals are categorized only by atomic number. Any element with an atomic number greater than 20 is designated as a heavy metal. Some classifications emphasize density, identifying elements with densities in between 4 and 6 g/cm^3 as heavy metals (Raychaudhuri et al. 2021).

Apart from density and atomic number, toxicity is a crucial factor in determining whether an element is a heavy metal or not. Elements such as cadmium (Cd), chromium (Cr), copper (Cu), lead (Pb), mercury (Hg), nickel (Ni), selenium (Se), and zinc (Zn) are recognized as heavy metals due to their industrial use and harmful effects on biological systems. However, the term "heavy metals" can be misleading, as certain dense metals, such as gold (Au) and platinum (Pt), with relative densities

DOI: 10.1201/9781003599098-1

of approximately 20, are not considered toxic and, for the same reason, do not fall under the heavy metal category (Smith and Scott 2005).

Despite the lack of a standardized definition, all heavy metals share several common properties, such as high thermal and electrical conductivity, corrosion resistance, and the ability to form stable complexes.

1.3 COMMON HEAVY METALS

The most commonly identified heavy metals are Cd, Cr, Cu, Pb, Hg, Ni, Se, and Zn. These elements are frequently used in industrial processes, such as metal plating, battery production, chemical synthesis, and electronic manufacturing. Due to their widespread use and inadequate disposal, they are among the most prevalent pollutants in soil, water, and air.

While many elements in the periodic table have densities higher than $5 g/cm^3$, not all of them are considered toxic; certain elements, such as antimony (Sb), niobium (Nb), tantalum (Ta), strontium (Sr), cobalt (Co), and cesium (Cs), are also considered heavy metals due to their high density and toxicity (Wang and Guo 2023).

The toxicity of heavy metals varies depending on their chemical form, bioavailability, and environmental interactions. Some heavy metals, such as zinc and copper, are essential trace elements required for biological functions in small quantities but become toxic at higher concentrations. Others, such as mercury and lead, have no known beneficial role in biological systems and pose serious health risks even at low exposure levels.

To develop effective strategies for wastewater treatment, environmental remediation, and pollution control, we need to understand the classification, properties, and environmental impacts of heavy metals. Their extensive industrial use necessitates stringent regulatory measures to minimize adverse effects on human health and the ecosystem. The toxicity of heavy metals remains a global concern due to extensive industrialization, which generates wastewater and sludge containing heavy metals. The extraction and processing of heavy metals generate CO_2, which contributes to global warming. At higher temperatures, the mobility of heavy metals in soil and aqueous environments increases, thereby increasing water contamination.

1.4 NATURAL ABUNDANCE OF HEAVY METALS

1.4.1 Distribution of Heavy Metals in the Earth's Crust

Geochemists such as A.P. Vinogradov, Taylor, A.B. Ronov, A.A. Yaroshevsky, and A.A. Migdisov have extensively investigated the distribution of chemical elements within the Earth's crust.[5] They utilized the principles of geochemical balance, sedimentary process analysis, and the compositional attributes of primary rock types (e.g., magmatic, metamorphic, and sedimentary) to estimate elemental abundances. Table 1.1 summarizes the average concentrations of various elements in stony meteorites, different rock types (ultrabasic, basic, intermediate, and acidic), and the overall composition of the upper continental crust. This comparative dataset provides valuable insights into the distribution of heavy metals and trace elements in geological materials. The rock types considered in Table 1.1 are (i) Ultrabasic rocks (e.g.,

TABLE 1.1
Average Elemental Abundances (10^{-4} wt%) in the Major Rock Types and the Upper Continental Crust

Elements*	Stony meteorites	Ultrabasic rocks	Basic rocks	Intermediate rocks	Acid rocks	Earth's crust
1 H	–	–	–	–	–	–
3 Li	3.0	0.5	15	20	40	32
4 Be	3.6	0.2	0.4	1.8	5.5	3.8
5 B	2.0	1.0	5	15	15	12
6 C	400	100	100	200	300	230
7 N	1.0	6.0	18	22	20	19
8 O	350000	425000	435000	460000	487000	470000
9 F	28	100	370	500	800	660
11 Na	7000	5700	19400	30000	27700	25000
12 Mg	140000	259000	45000	21800	5600	18700
13 Al	13000	4500	87600	88500	77000	80500
14 Si	180000	190000	240000	260000	323000	295000
15 P	500	170	1400	1600	700	930
16 S	20000	100	300	200	400	470
17 Cl	70	50	50	1100	240	170
19 K	850	300	8300	23000	33400	25000
20 Ca	14000	7000	67200	46500	15800	29600
21 Sc	6	5	24	2.5	3	10
22 Ti	500	300	9000	8000	2300	4500
23 V	70	40	200	100	40	90
24 Cr	250	2000	200	50	25	83
25 Mn	2000	1500	2000	1200	600	1000
26 Fe	250000	98500	85600	58500	27000	46500
27 Co	800	200	45	10	5	18
28 Ni	13500	2000	160	55	8	58
29 Cu	100	20	100	35	20	47
30 Zn	50	30	130	73	60	83
31 Ga	3	2	18	20	20	19
32 Ge	10	1	1.5	1.5	1.4	1.4
33 As	0.3	0.5	2	2.4	1.5	1.7
34 Se	10	0.05	0.05	0.05	0.05	0.05
35 Br	0.5	0.5	3	4.5	1.7	2.1
37 Rb	5	2	45	100	200	150
38 Sr	10	10	440	800	300	340
39 Y	0.8	–	20	–	34	29
40 Zr	30	30	100	260	170	170
41 Nb	0.3	1	20	20	20	20
42 Mo	0.6	0.2	1.4	0.9	1	1.1
44 Ru	1	–	–	–	–	–
45 Rh	0.19	–	–	–	–	–

(*Continued*)

TABLE 1.1 (*Continued*)
Average Elemental Abundances (10^{-4} wt%) in the Major Rock Types and the Upper Continental Crust

Elements*	Stony meteorites	Ultrabasic rocks	Basic rocks	Intermediate rocks	Acid rocks	Earth's crust
46 Pd	1	0.12	0.019	–	–	–
47 Ag	0.094	0.05	0.1	0.07	0.05	0.07
48 Cd	0.1	0.05	0.19	–	0.1	0.13
49 In	0.001	0.013	0.022	–	0.25	0.25
50 Sn	1	0.5	1.5	–	3	2.5
51 Sb	0.1	0.1	1	0.2	0.26	0.5
52 Te	0.5	0.001	0.001	0.001	0.001	0.001
53 I	0.04	0.01	0.5	0.3	0.4	0.4
55 Cs	0.1	0.1	1	–	5	3.7
56 Ba	6	1	300	650	830	650
57 La	0.3	–	27	–	60	29
58 Ce	0.5	–	45	–	100	70
59 Pr	0.1	–	4	–	12	9
60 Nd	0.6	–	20	–	46	37
62 Sm	0.2	–	5	–	9	8
63 Eu	0.08	0.01	1	–	1.5	1.3
64 Gd	0.4	–	5	–	9	8
65 Tb	0.05	–	0.8	–	2.5	4.3
66 Dy	0.35	0.05	2		6.7	5
67 Ho	0.07	–	1	–	2	1.7
68 Er	0.2	–	2	–	4	3.3
69 Tm	0.04	–	0.2	2—	0.3	0.27
70 Yb	0.2	–	2	–	4	3.3
71 Lu	0.035	–	0.6	–	1	0.08
72 Hf	0.5	0.1	1	1	1	1
73 Ta	0.02	0.018	0.48	0.7	3.5	2.5
74 W	0.15	0.1	1	1	1.5	1.3
75 Re	0.0008	–	0.00071	–	0.00067	0.0007
76 Os	0.5	–	–	–	–	–
77 Ir	0.48	–	–	–	0.0063	–
78 Pt	2	0.2	0.1	–	–	–
79 Au	0.17	0.005	0.004	–	0.0045	0.0043
80 Hg	3	0.01	0.01	0.09	0.08	0.083
81 Tl	0.001	0.01	0.2	0.5	1.5	1
82 Pb	0.2	0.1	8	15	20	16
83 Bi	0.003	0.001	0.007	0.01	0.01	0.009
90 Th	0.04	0.005	3	7	18	13
92 U	0.015	0.003	0.05	1.8	3.5	2.5

* Corresponding atomic number at the left.
Source: Yaroshevsky (2006).

peridotite), which are low in silica but rich in iron and magnesium. These rocks are typically found in the Earth's mantle and are a significant source of elements like Ni, Cr, and Mg; and (ii) Basic rocks (e.g., basalt), which are relatively low in silica content, but rich in calcium, magnesium, and iron. They are frequently present in oceanic crust and contain elevated levels of Co, Cu, and Zn; (iii) Intermediate rocks (e.g., andesite), having moderate silica content and present a compositional bridge between basic and acidic rocks; (iv) Acid rocks (e.g., granite) are rich in silica and aluminum and have higher concentrations of Rb and K; (v) Stony meteorites provide insights into early solar system materials and act as references for Earth's geochemical composition. These classifications are significant because the type of parent rock directly influences the natural occurrence and concentration of heavy metals in Earth's crust.

1.4.2 Natural Occurrence of Heavy Metals in Minerals, Ores, and Soil

Heavy metals are naturally present in the environment due to various geological processes. Weathering of rocks, volcanic eruptions, crustal degassing, and even natural forest fires significantly contribute to their release. While these natural processes account for a certain level, human activities, such as industrialization, mining, and urban development, have significantly elevated heavy metal concentrations in several regions. The Earth's crust comprises approximately 95% igneous and 5% sedimentary rocks (Angon et al. 2024). Igneous rocks, especially basaltic types, typically contain higher concentrations of heavy metals such as Co, Cd, Zn, Cu, and Ni. On the other hand, sedimentary rocks such as shales, especially black shales with high organic content, also accumulate significant amounts of heavy metals such as Zn, Cu, Mn, Cd, and Pb. The concentration range (in ppm) of selected heavy metals in various rock types, including igneous (basaltic and granitic) and sedimentary (clays, shales, black shales, and sandstone), is summarized in Table 1.2.

In addition to heavy metals with considerable concentration, several trace elements are present in specific rock-forming minerals. These elements are released gradually during weathering and soil formation, altering soil chemistry and the environment.

Table 1.3 highlights common rock-forming minerals, the trace elements they contain, and their relative resistance to weathering. The weathering behavior of these minerals plays a critical role in controlling the release and bioavailability of heavy metals in surrounding soils and water bodies.

TABLE 1.2
Heavy Metal Concentrations in Igneous and Sedimentary Rocks in ppm

Element	Basaltic igneous	Granite igneous	Clays and shales	Black Shales	Sandstone
Cu	48–240	5–140	18–180	34–1500	2–41
Zn	2–18	6–30	16–50	7–150	<1–31
Pb	30–160	4–30	18–120	20–200	–
Cd	0.006–0.6	0.003–0.18	0–11	<0.3–8.4	–

Source: Angon et al. (2024).

Igneous or magmatic rocks are formed when magma (molten rock derived from Earth's mantle) cools and crystallizes. These rocks are transported to the surface via geological processes such as volcanic eruptions and tectonic activity (Saravanan et al. 2024). Along with associated minerals and soils, these rocks constitute the primary natural reservoirs of heavy metals in the environment. The weathering behavior of these minerals controls the release rate of heavy metals into surrounding soils and water bodies. Heavy metals in rocks, minerals, and ores may be released into the soil by natural processes, such as erosion, leaching, volcanic eruptions, biological activity, terrestrial processes, and surface winds (Angon et al. 2024). Anthropogenic activities, such as mining, smelting, burning fossil fuels, and improper waste disposal, further increase heavy metal levels in the environment (Angon et al. 2024).

1.4.3 Presence of Heavy Metals in Rivers, Lakes, and Oceans

Heavy metal pollution in aquatic systems originates from a variety of sources, including mining waste, landfill leachates, municipal and industrial wastewater, urban runoff, and natural events such as volcanism, rock weathering, and rock abrasion (Saravanan et al. 2024). Although weathering naturally introduces heavy metals into rivers, lakes, and oceans, industrial activities significantly amplify their levels. Table 1.4 outlines some key industrial processes that release heavy metals into

TABLE 1.3
Trace Elements in Common Rock-Forming Minerals

Mineral	Trace element	Resistance to weathering
Olivine	Ni, Co, Mn, Li, Zn, Cu, Mo	weathers easily
Hornblende	Ni, Co, Mn, Se, Li, V, Zn, Cu, Ga	–
Augite	Ni, Co, Mn, Se, Li, V, Zn, Pb, Cu, Ga	–
Biotite	Rb, Ba, Ni, Co, Mn, Se, Li, V, Zn, Cu, Ga	–
Apatite	Rare earth elements, Pb, Sr	–
Anorthite	Sr, Cu, Ga, Mn	–
Andesine	Sr, Cu, Ga, Mn	–
Oligoclase	Cu, Ga	–
Albite	Cu, Ga	–
Garnet	Mn, Cr, Ga	intermediate stability
Orthoclase	Rb, Ga, Sr, Cu, Ga	–
Muscovite	F, Rb, Ba, Sr, Cu, Ga, V	–
Titanite	Rare earth elements, V, Sn	–
Ilmenite	Co, Ni, Cr, V	–
Magnetite	Zn, Co, Ni, Cr, V	–
Turmaline	Li, F, Ga	–
Zircon	Hf, U	–
Quartz	–	resistant

Source: Bradl (2005).

TABLE 1.4
Some Common Industrial Sources of Heavy Metals

Element	Sources	References
Lead	Paint, industrial emissions, electronic waste, herbal products, toys, cosmetics, manufacturing of lead-acid batteries, ceramics, industrial dust and fumes, fertilizers, steelwork, pesticides, pharmaceuticals, combustion of fossil fuels and pigments, gasoline additives, solid waste incineration, explosives, dishware, mining, metallurgy, urban runoff.	Obeng-Gyasi (2019), Shrestha et al. (2021), Muhammad et al. (2016)
Nickel	Stainless steel production, paper mills, fertilizers, petroleum refining, steelwork, aerospace industry, engine alloys, pharmaceuticals, fine chemicals, dyes, pigments, pesticides, electroplating, food processing, hydrogenation catalysts, chemical industries, waste incineration, surgical waste, automobile industries, industrial aerosols, mining, batteries, and coal combustion.	Shrestha et al. (2021), Johnson et al. (2006), "Nickel," Chen et al. (2017)
Chromium	Mining, metallurgy, corrosion-resistant alloys, paper mills, fertilizers, petroleum refinery, aircraft, glass and cement production, textile industry, leather tanning, metal plating, power plants, cooling systems, electroplating, pigments, wood preservation, rubber, photography, nuclear power plants, chemical industries.	Shrestha et al. (2021), Sharma et al. (2021), Kerur et al. (2020), Johnson et al. (2006), Muhammad et al. (2016)
Mercury	Coal-fired plants, oil combustion, biomass burning, municipal waste incineration, non-ferrous metal (Zn, Pb, Cu) smelting, cement production, metal (Au, Al, Fe and Hg) and steel industries, chlor-alkali plants, fluorescent lamps, thermometers, batteries, pesticides, fertilizers, fungicides, electronic materials, dental amalgams, textiles, photographic and pharmaceutical industries.	Bhardwaj et al. (2020), Zhang et al. (2015), Mark Wilhelm (2001), Liang et al. (2014), Cai et al. (2020), Sharma et al. (2021), Muhammad et al. (2016)
Copper	Electrical wiring, switches, transformers, telecommunications equipment, heat exchangers, copper tubing, PCB manufacturing, pesticides, fertilizers, dyes, fine chemicals, and copper mining.	Izydorczyk et al. (2021), "Copper Uses in Industry," Zhan and Xu (2014), Tang et al. (2013), Shrestha et al. (2021)
Arsenic	Pharmaceuticals (e.g., Trisenox and Melarsoprol), herbicides, pesticides, lead alloys, semiconductors, LEDs, mining industry, smelting, animal feed, glass, and electrical industries.	Seidl et al. (2019), Mere et al. (2022)
Cadmium	Ni-Cd battery electrodes, stabilizers for polyvinyl chloride (PVC) against heat and light, pigments for plastics, glasses and ceramics, coatings on steel and some nonferrous metals, photography, photocopying, dyeing, vacuum tube manufacturing, electroplating, lubricants, CdTe-based solar cells, welding, nuclear plants, fertilizers.	"Toxicological Profile for Cadmium," Muhammad et al. (2016)

(*Continued*)

TABLE 1.4 (*Continued*)

Some Common Industrial Sources of Heavy Metals

Element	Sources	References
Zinc	Rubber, paint, chemical and agricultural industries, dry-cell batteries, galvanization, mining, iron and steel plating, alloys like bronze and brass, photocopiers, sunscreens, cathode ray tubes, tires, fertilizers.	Angon et al. (2024), "Zinc Statistics and Information," Rhodes et al. (2012), Angon et al. (2024)
Cobalt	Lithium-ion batteries, cutting and grinding tools, electroplating, medical implants, pigments, paints, and colored glasses.	"IARC Monographs" (2025)
Antimony	Flame retardants, lead alloys, lead–acid batteries, plastics, ceramics, primer in explosives, brake pads, clutch discs, glass.	"IARC Monographs" (2025)
Thallium	Coal combustion and smelting activities	Kellerhals et al. (2010)
Rhodium	Automobile catalysts	Soyol-Erdene et al. (2011)
Iridium	Catalytic converters	Ravindra et al. (2004)

water bodies. When wastewater from these processes is not adequately treated, some heavy metals eventually reach surface water bodies (e.g., lakes, rivers, and oceans).

1.4.4 Heavy Metals from Volcanic Eruptions, Erosion, and Weathering

Volcanic eruptions release a range of pollutants, including ash, carbon dioxide, sulfur dioxide, hydrogen sulfide, hydrogen chloride, hydrogen fluoride, radon, and various metals. During volcanic eruptions, magma degassing releases volatile metals such as cadmium, lead, bismuth, and thallium as soluble aerosols. These aerosols, with particle sizes around 0.2 μm, can travel thousands of kilometers and deposit on soils, ice cores, and water systems (Mason et al. 2022).

Other metals such as mercury, gold, and copper are also emitted during volcanic eruptions. The metal-rich gases emitted by the volcano rapidly condense into particulate matter (aerosols), forming sulfates, chlorates, and oxides. The presence of trace elements near a volcano can be used to track its past activities. For example, thallium, found in nearby ice cores, serves as a reliable tracer of volcanic activity due to its low abundance in Earth's crust (~0.52 mg kg^{-1}) (Kellerhals et al. 2010). Elevated concentrations of sulfate and fluoride ions have also been exploited as other indicators. In Antarctic snow, spikes in platinum, iridium, and rhodium concentrations at different depths correspond to volcanic eruptions at specific periods (Soyol-Erdene et al. 2011).

Volcanic activity significantly increases heavy metal content in surrounding agricultural soils. As shown in Table 1.5, soils in volcanic regions of Hainan Province, China, contain elevated levels of Cd, Cr, Hg, Pb, Ni, Zn, and Mg compared to non-volcanic soils.

Soil erosion caused by wind or water transports dust particles containing heavy metals into water systems. Over time, these particles dissolve, increasing the concentration of heavy metals in aquatic bodies. Also, some heavy metals are absorbed by plants and vegetables (Carrera-Beltrán et al. 2024), posing potential health risks.

TABLE 1.5
Heavy Metal Concentrations (mg/kg) Reported in Soils from the Volcanic Regions of Hainan Province, China

Sample	As	Cd	Cr	Cu	Hg	Ni	Pb	Zn	Mn
Volcanic agricultural soil, Hainan	7.06	67.51	156.88	33.43	49.09	72.47	19.48	65.57	749.85
Background soil, Hainan	1.34	0.04	27.5	6.1	0.02	7.24	24.4	44.4	163.5
Qiongzhong County, Hainan	3.30	0.063	49.9	9.93	0.035	13.4	33.4	60.1	-
Chinese topsoil	11	0.097	61	23	0.065	27	26	74	585
Gancheng Town, Hainan	22.7	0.128	33.4	14.5	0.032	9.32	32.5	43.3	366.2

Source: Yang et al. (2022).

Several researchers have demonstrated that sediments formed by water erosion often contain high concentrations of heavy metals (Quinton and Catt 2007). Several phosphate fertilizers contain toxic elements such as As, Cd, Cr, Hg, Ni, and V. Cu and As elements can also be present in pesticides (Quinton and Catt 2007). Therefore, erosion of agricultural soils can increase the concentration of heavy metals in adjacent sediments and water bodies. Because heavy metals bind strongly to silt, clay, and organic matter, they are transported with fine soil particles during erosion (Quinton and Catt 2007).

Heavy metals entrapped in rocks are gradually released through physical or chemical weathering. Physical weathering, such as temperature fluctuations, breaks down rocks into smaller fragments. In deserts and coastal regions, the high concentration of dissolved salts percolates into the pores and cracks of rocks. Due to high temperatures in such regions, water evaporates from salt crystals, and these crystals grow larger over time, promoting rock fragmentation (Moosdorf et al. 2014).

In chemical weathering, the minerals in rocks suffer chemical changes. A typical process involves the oxidation of iron ions to form iron oxides. Other mechanisms include reactions of minerals with H^+ ions in acidic conditions or with CO_2, as shown in reaction (R1.1) (Moosdorf et al. 2014). These processes contribute to the long-term release of heavy metals into the environment.

$$Mg_2SiO_2 + 4CO_2 + 4H_2O \rightarrow 2Mg^{2+} + 4HCO_3^- + H_4SiO_4 \qquad \text{(R1.1)}$$

1.5 CHARACTERISTICS AND TOXICITY OF HEAVY METALS

1.5.1 Physical and Chemical Characteristics of Heavy Metals

An important physical characteristic of heavy metals is their high specific gravity. In this section, we outline the fundamental physical and chemical characteristics of selected heavy metals, which are highly relevant to environmental and toxicological contexts.

Lead (Pb): Lead is a soft, malleable, ductile, silvery-gray metal with a bright luster, which crystallizes in a face-centered cubic (fcc) structure. It has a high density of 11.35 g/cm^3, and commonly exists in the +2 and +4 oxidation states. Galena (PbS) is the primary natural source of lead (Patnaik 2003, Hammond 2010). Galena is frequently associated with sulfides of Fe, Cu, Zn, Ag, Bi, As, Sb, and Sn. Lead reacts with water and oxygen to form $Pb(OH)_2$ and with nitric acid to produce lead nitrate, following the reaction (R1.2):

$$3Pb + 8HNO_3 \rightarrow 3Pb(NO_3)_2 + 2NO + 4H_2O \qquad \text{(R1.2)}$$

Nickel (Ni): Nickel is a silvery-white, lustrous metal with a face-centered cubic crystal structure. It is ductile, ferromagnetic, and has a density of 8.908 g/cm^3. Naturally occurring nickel is found in minerals such as pentlandite [$(Ni, Fe)_9S_{16}$], limonite [$(Fe,Ni)O(OH)\cdot nH_2O$], and garnierite [$(Ni,Mg)_6Si_4O_{10}(OH)_8$]. Pentlandite is frequently found in nature alongside other sulfide minerals, such as Fe_7S_8 and $CuFeS_2$. Nickel typically occurs in the +2 oxidation state. Finely divided nickel reacts with water, oxygen, or carbon monoxide (CO), and dissolves readily in dilute nitric acid (Patnaik 2003, Hammond 2010).

Chromium (Cr): Chromium is a blue-white metal with a body-centered cubic structure and a density of 7.19 g/cm^3. It naturally occurs in chromite [$FeO\cdot Cr_2O_3$] and crocoite [$PbCrO_4$]. Metallic Cr readily reacts with oxygen to form chromium(III) oxide, Cr_2O_3. The Cr^{2+} ion oxidizes in acidic media following reaction (R1.3). In basic media, Cr^{3+} ions form chromium hydroxide following reaction (R1.4).

$$4Cr^{2+}{}_{(aq)} + O_{2(g)} + 4H^{+}{}_{(aq)} \rightarrow 4Cr^{3+}{}_{(aq)} + 2H_2O \quad (R1.3)$$

$$Cr(H_2O)_6{}^{3+}{}_{(aq)} + 3OH^{-}{}_{(aq)} \rightarrow Cr(H_2O)_3(OH)_{3(s)} + H_2O \quad (R1.4)$$

Chromium also forms chromate (CrO_4^-) and dichromate ($Cr_2O_7^{2-}$) ions in aqueous environments (Patnaik 2003, Hammond 2010).

Mercury (Hg): Mercury is the only metal that is liquid at room temperature, appearing silver-white with a density of 13.534 g/cm^3. It melts at –38.83 °C and boils at 356.73 °C. Common oxidation states of mercury are 0, +1, and +2. Mercury is found in nature mainly as HgS (cinnabar). Upon heating cinnabar at high temperature (600–700 °C) in the presence of O_2, elemental Hg and SO_2 are produced (R1.5). Mercury reacts with halogens and H_2S to form corresponding halides and HgS, as described in (R1.6) and (R1.7).

$$2Hg + O_2 \xrightarrow{(350^\circ C)} 2HgO \quad (R1.5)$$

$$Hg + X_2\ (X = Cl, Br, I) \rightarrow HgX_2 \quad (R1.6)$$

$$Hg + H_2S \rightarrow HgS + H_2 \quad (R1.7)$$

Additionally, it reacts with concentrated nitric and sulfuric acids to generate mercury salts (R1.8, R1.9) (Patnaik 2003, Hammond 2010).

$$Hg + 4HNO_3 \rightarrow Hg(NO_3)_2 + 2NO_2 + 2H_2O \quad (R1.8)$$

$$Hg + 2H_2SO_4 \rightarrow HgSO_4 + 2H_2O + SO_2 \quad (R1.9)$$

Copper (Cu): Copper is a reddish-brown metal with a face-centered cubic (fcc) structure and a density of 8.92 g/cm^3. The Mohs hardness of Cu is 2.5 to 3.0. Copper is highly conductive, with low electrical resistivity (~ 1.71×10^{-8} Ω-cm at 25°C). It melts at 1,083 °C and boils at 2,567 °C. While insoluble in water, copper dissolves in nitric acid, hot sulfuric acid, ammonium hydroxide, and cyanide solutions. The common valence states of copper are 0, +1, and +2 (Patnaik 2003, Hammond 2010).

Arsenic (As): Arsenic is a brittle, steel-gray crystalline metal with a hexagonal crystal structure. It exists in three allotropes: α–metallic, β–arsenic (also named black amorphous), and yellow. It sublimes at 613°C (normal atmospheric pressure) and has densities of 5.72 g/cm^3 (β-metallic form) and 4.70 g/cm^3 (β-amorphous form). Arsenic is insoluble in water and reacts with concentrated hydrochloric acid in the presence of an oxidant. In HCl solution, it reacts with H_2S to form arsenic sulfide, as shown in (R1.10). In concentrated nitric acid, arsenic forms orthoarsenic acid (R1.11). Arsenic also reacts with halogen to form corresponding trihalides (e.g., $AsCl_3$), as shown in (R1.12), and with fluorine (F_2) forms AsF_5. Reaction of As with sulfur generates a mixture of As_2S_3, As_2S_2, and As_2S_5 (Patnaik 2003, Hammond 2010).

$$2\text{As} + 3\text{H}_2S + \text{Conc.HCl} \rightarrow \text{As}_2S_3 + 3\text{H}_2 \quad \text{(R1.10)}$$

$$\text{As} + 5\text{HNO}_3 \rightarrow H_2\text{AsO}_4 + 5\text{NO}_2 + H_2O \quad \text{(R1.11)}$$

$$2\text{As} + 3\text{X}_2 \left(\text{X} = \text{Cl, Br, I}\right) \rightarrow 2\text{AsX}_3 \quad \text{(R1.12)}$$

Cadmium (Cd): Cadmium is a soft, bluish–white metal with a closed–packed hexagonal crystal system. It has a density of 8.69 g/cm^3, a Brinell hardness of 21, and melts at 321.1°C. It is insoluble in water but reacts with nonmetals and halogens when heated, forming compounds such as Cd_3P_2, CdS, and CdTe. Cadmium dissolves in hydrochloric and sulfuric acids, releasing hydrogen gas (Patnaik 2003, Hammond 2010).

Zinc (Zn): Zinc is a bluish-white, lustrous metal with a hexagonal close-packed structure and a density of 7.14 g/cm^3. It is brittle at room temperature but becomes malleable between 100 and 150°C. Principal zinc minerals include ZnS, $ZnCO_3$, ZnO, $ZnFe_2O_4$, and $ZnAl_2O_4$. Typically, zinc exhibits a +2 oxidation state and reacts with sulfuric and hydrochloric acids, producing hydrogen. Reacting with nitric acid, it produces NO_x gases (Patnaik 2003, Hammond 2010). In alkali solutions, zinc forms zincates (e.g., Na_2ZnO_2), releasing H_2 gas (R1.13):

$$\text{Zn} + 2\text{NaOH} \rightarrow \text{Na}_2\text{ZnO}_2 + \text{H}_2 \quad \text{(R1.13)}$$

Cobalt (Co): Cobalt is a silvery-white metal, with a density of 8.86 g/cm^3 occurring in alpha (hexagonal close-packed) and gamma (fcc) allotropic modifications. Cobalt is commonly found as sulfide and oxide minerals. On reacting with mineral acids, it forms Co^{2+} salts, releasing H_2 gas. Reactions with halogens produce cobalt halides (Patnaik 2003, Hammond 2010).

Antimony (Sb): Antimony is a brittle, silvery-white metal with a hexagonal crystal structure and a density of 6.697 g/cm^3. It occurs in two unstable allotropes: yellow and dark gray. Antimony can exist in multiple oxidation states, including +5, +3, 0, and –3. It forms oxides (Sb_2O_3, Sb_2O_4, and Sb_2O_5) and sulfides (Sb_2S_3 and Sb_2S_5). Reacting with hot sulfuric and hydrochloric acids, it produces oxysulfate and chloride salts, respectively. On reacting with nitric and hydrofluoric (HF) acids, it produces Sb_2O_5 and SbF_3/SbF_5, respectively (Patnaik 2003, Hammond 2010).

Thallium (Tl): With metallic luster in the freshly cut state, thallium turns bluish-gray upon air exposure. With a density of 11.85 g/cm^3, it forms tetragonal crystals. The common oxidation states of thallium are +1 and +3. It reacts with halogens and acids to form a diverse range of thallium salts (Patnaik 2003, Hammond 2010).

1.5.2 Health and Environmental Impacts of Heavy Metal Toxicity

Heavy metals cause numerous adverse health effects, primarily from the sources outlined in Table 1.4. These metals can enter the food chain through absorption by plants such as rice and fruit trees, or aquatic organisms like fish, shellfish, and marine mammals (Kang et al. 2024, Basu et al. 2023). When humans consume contaminated plant-based foods or seafood, heavy metals enter and accumulate in human tissues and bones.

Some heavy metals induce the formation of free radicals and reactive oxygen species (ROS; Kang et al. 2024) including superoxide anion (O_2•–), singlet oxygen ($^1O_2$), hydrogen peroxide (H_2O_2), peroxyl radical (ROO•), nitric oxide (NO•), dimethyl arsenic peroxyl radicals [$(CH_3)_2$AsOO•], and dimethyl arsenic radical [$(CH_3)_2$As•] (Flora et al. 2022). These ROS can alter or disrupt vital cellular functions, leading to serious health hazards. Figure 1.1 highlights the organs most affected by heavy metal exposure.

ROS can deactivate important regulatory molecules, such as the tumor suppressor protein P53 and other transcription factors, which may lead to carcinogenesis. Additional health effects include: Hepatotoxicity (liver damage), DNA damage, such as strand breaks and DNA adduct formation, oxidative stress, immunotoxicity, involving inflammation pathways like NF-κB (nephrotoxicity), TNF-α (tumor necrosis factor alpha), and various interleukins (IL-1, IL-6, IL-8), potentially resulting in abnormal T Cell proliferation. Heavy metals can also induce neurotoxicity due to neurotransmitter disruption, skin toxicity, and reproductive toxicity, which include reduced testosterone levels, impaired spermatogenesis, and increased infertility due to oxidative stress (Mitra et al. 2022). Table 1.6 summarizes the potential health risks associated with long-term exposure to heavy metals above their maximum contaminant levels (MCLs) in drinking water, as well as their classifications by the International Agency for Research on Cancer (IARC).

Exposure to emerging contaminants like heavy metal nanoparticles and mixed metals, commonly found in e-waste and urban runoff, is an ongoing concern due to their potential synergistic toxicity and importance in environmental risk assessments. These nanoparticles are taken up by cells, where they release heavy metal ions, which induce oxidative stress either directly or through mitochondrial dysfunction (Zhang et al. 2022). Mixed metals may also exhibit combined toxicity that exceeds the effects of individual metals (Su et al. 2017). For example, it has been

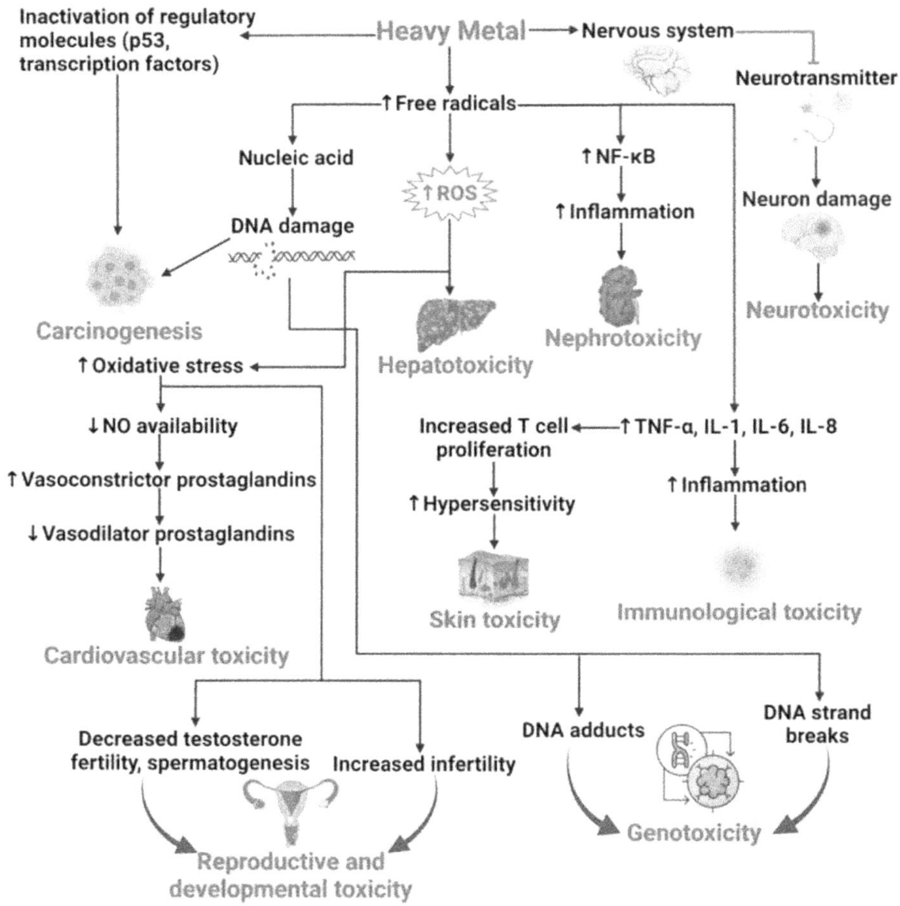

FIGURE 1.1 Mechanism of heavy metal toxicity in humans. Reproduced with permission from Mitra et al. (2022).

reported that the combined toxicity of Cd-Ni is greater than that of the single elements (Zhang et al. 2022).

1.6 TOXICITY INDEX AND PERMISSIBLE LIMITS OF COMMON HEAVY METAL IONS IN POTABLE WATER

1.6.1 Definition of Toxicity Index and Its Relevance in Assessing Heavy Metal Contamination

Various International agencies, including the IARC, the Environmental Protection Agency (EPA), and regulatory bodies from the European Union, use toxicity indices to evaluate and classify the risks associated with heavy metal exposure.

TABLE 1.6
Potential Health Effects of Long-Term Exposure to Heavy Metals above MCL and IARC Classification

Heavy metal	Potential health effects	IARC Classification	Reference
Lead	Infants and children: Development (physical and mental) delays, reduced attention span, and learning difficulties. Adults: Kidney damage, high blood pressure.	Inorganic lead compounds: *probably carcinogenic to humans.*	("EPAs National Primary Drinking Water Regulation")
Nickel	Dermatitis, asthma, cardiovascular disease, lung fibrosis, and lung and nasal cancers.	Nickel compounds: *Carcinogenic*; Metallic nickel: *possibly carcinogenic.*	Chen et al. (2017)
Chromium	Allergic dermatitis, lung cancer, respiratory issues, nephrotoxicity, intestinal disorders, teratogenic effects.	Chromium(VI): *Carcinogenic.*	("EPAs National Primary Drinking Water Regulation")
Mercury	Inhalation of the vapor of Hg causes pneumonitis, pulmonary edema. Ingestion of $HgCl_2$ causes severe damage to the gastrointestinal tract and lesions in the kidneys. Prenatal CH_3Hg exposure causes brain and cerebellum malfunction (general atrophy and hypoplasia of the brain cortex, hypoplasia of the granular cell layer in the cerebellum and deranged cortical organization of the cerebrum).	Methylmercury compounds: *possibly carcinogenic.*	Kang et al. (2024)
Copper	Short-term exposure: Gastrointestinal distress. Long-term exposure: Liver and kidney damage.	———	("EPAs National Primary Drinking Water Regulation")
Arsenic	Skin damage, circulatory issues. Long-term exposure via inhalation causes lung cancer. Ingestion causes cancer of the skin, lung, bladder, liver, kidney, prostate, and ovary.	Arsenic and inorganic arsenic compounds: *carcinogenic to humans.*	States (2017), "EPAs National Primary Drinking Water Regulation")
Cadmium	Cancers of the lung, bladder, kidney, prostate, pancreas, ovary, and endometrium.	Cadmium and cadmium compounds: *Carcinogenic.*	"EPAs National Primary Drinking Water Regulation"; Mudipalli (2017)

(Continued)

TABLE 1.6 (*Continued*)
Potential Health Effects of Long-Term Exposure to Heavy Metals above MCL and IARC Classification

Heavy metal	Potential health effects	IARC Classification	Reference
Zinc	Vomiting blood, kidney inflammation, liver necrosis, blood clotting disorders, and irritation of the respiratory system.	———	("Zinc Toxicity")
Cobalt	Oxidative damage to DNA, lipids, and proteins; enzyme inhibition; tumors, cardiomyopathy, adverse respiratory effects, and bone resorption.	Metallic cobalt and soluble cobalt(II) salts: *probably carcinogenic to humans.*	Bellouard et al. (2022), Leyssens et al. (2017)
Antimony	Increase in blood cholesterol, decrease in blood sugar.	Trivalent antimony: *probably carcinogenic to humans.*	("EPAs National Primary Drinking Water Regulation")
Selenium	Hair and nail loss, numbness in fingers or toes, and circulatory problems.	———	("EPAs National Primary Drinking Water Regulation")
Thallium	Hair loss, blood abnormalities, kidney, intestinal, and liver problems.	———	("EPAs National Primary Drinking Water Regulation")

The EPA defines the MCL as the highest permissible concentration of a contaminant in drinking water. This limit is established on the basis that there is no known or expected risk to health when the contaminant level in drinking water is below this limit ("EPA's National Primary Drinking Water Regulation"). The IARC classified several substances as carcinogenic to humans (group 1), probably carcinogenic to humans (group 2A), possibly carcinogenic to humans (group 2B), and not classifiable as to its carcinogenicity to humans (group 3)("IARC Monographs" 2025).

1.6.2 Regulatory Standards and Permissible Limits of Heavy Metal Ions in Drinking Water

Table 1.7 presents the established MCLs for various heavy metals, along with their primary sources in drinking water. These values serve as essential benchmarks for water quality monitoring and management.

1.7 NECESSITY OF REMOVING HEAVY METALS FROM WATER

The harmful effects of heavy metals on human health have been extensively documented. Therefore, it is necessary to prevent their presence in water sources or eliminate their presence altogether. As the demand for clean water continues to rise, ensuring the removal of heavy metals from water bodies is becoming critical. Plants can absorb heavy metals in contaminated irrigation water and accumulate them in the edible parts of stems and crops. Moreover, heavy metal-contaminated soils may transfer these pollutants to plants over time. The harmful effects of heavy metals have also affected aquatic ecosystems, as dissolved heavy metals can accumulate in marine organisms, including fish and shellfish. These contaminants then enter the food chain, ultimately reaching humans and posing serious health risks on long-term exposure. A quantitative comparison of typical heavy metal concentrations in wastewater from industrial effluents with international regulatory limits (CPCB and EPA) is presented in Table 1.8. Development and implementation of effective water treatment technologies, alongside stricter regulatory measures, are essential to minimize heavy metal contamination and protect both the environment and public health.

1.8 CONCLUSION

Heavy metal contamination of drinking and agricultural water sources poses a significant threat to human health and the ecosystem. Due to their non-degradable nature and ability to bioaccumulate, heavy metals persist in the environment and can cause several toxic effects, including carcinogenicity, neurotoxicity, immunotoxicity, and reproductive disorders. The use of toxicity indices and regulatory standards, such as MCL, provides a scientific basis for monitoring and controlling heavy metal concentration in water bodies. Eliminating heavy metal pollution from water not only protects our environment but also safeguards food security and public health. Future research should prioritize emerging remediation technologies that do not generate excessive sludge and produce water with heavy metal concentrations below the regulatory-prescribed limits.

TABLE 1.7
Maximum Contaminant Level and Sources of Heavy Metals in Drinking Water Established by the EPA, WHO, and European Union

Heavy metal	MCL, (mg L^{-1})			Sources of contamination	Reference
	EPA	WHO	EU		
Lead	0.010	0.01	0.005	Corrosion of household plumbing systems; erosion of natural deposits.	("EPA's National Primary Drinking Water Regulation," Ministry of Environment, European Union)
Nickel	–	0.07	0.02	Corrosion of pipes; erosion of natural deposits.	Alkhadra et al. (2022)
Chromium	0.1	0.05	0.025	Discharge from steel and pulp mills; erosion of natural deposits.	("EPA's National Primary Drinking Water Regulation")
Mercury	0.002	0.006	0.001	Industrial discharge; runoff from landfills and croplands, artisanal and small-scale gold mining.	"EPAs National Primary Drinking Water Regulation, Basu et al. (2023)
Copper	1.3	2.0	2.0	Corrosion of household plumbing systems; erosion of natural deposits.	("EPA's National Primary Drinking Water Regulation")
Arsenic	0.010	0.01	0.01	Erosion of natural deposits; runoff from orchards, industrial waste, and runoff from glass and electronics production wastes.	("EPAs National Primary Drinking Water Regulation," Ahmad and Bhattacharya (2019)
Cadmium	0.005	0.003	0.005	Corrosion of galvanized pipes; erosion of natural deposits; discharge from metal refineries; runoff from waste batteries and paints; water supplies close to mines; battery recycling facilities, landfills, hazardous waste sites; the burning of fossil fuels; fertilizers; waste incineration; and cadmium-using industrial activities.	"EPAs National Primary Drinking Water Regulation," Jomova et al. (2025)
inc	–	—	—	Refineries, electroplating, and metal processing.	Alkhadra et al. (2022)
Cobalt	–	—	—	Coal-fired power plants, mining, and erosion of natural deposits.	Alkhadra et al. (2022)
Antimony	0.006	0.02	0.010	Discharge from refineries, electronics, solder, and fire retardants.	("EPAs National Primary Drinking Water Regulation")
Selenium	0.05	0.01	0.020	Discharge from petroleum refineries, mining, and erosion of natural deposits.	("EPAs National Primary Drinking Water Regulation")
Thallium	0.002	—	—	Leaching from ore-processing sites; discharge from electronic and pharmaceutical wastes.	("EPAs National Primary Drinking Water Regulation")

Note: EPA, US Environmental Protection Agency; EU, European Union; MCL, Maximum Contaminant Level; WHO, World Health Organization.

TABLE 1.8
Metal Concentrations (mg L^{-1}) in Three Industrial Effluents (Anpara, Kannoria, Bina) in India before Being Discharged into a Water Reservoir

Metal examined	Anpara	Kannoria	Bina	Standard (CPCB)	Standard (US EPA) 2002
Cu	39 ± 3.1	5.9 ± 1.0	17.5 ± 1.4	3.0	5.0
Cr	44 ± 4.6	7.9 ± 1.1	3.9 ± 0.7	0.1	2.0
Fe	94 ± 9.5	16.5 ± 1.4	18 ± 1.3	3.0	100
Mn	98 ± 11.0	18.9 ± 1.6	15 ± 1.7	5.0	5.0
Ni	54 ± 8.9	3.7 ± 0.8	8.6 ± 1.4	3.0	5.0
Pb	19.8 ± 3.4	5.9 ± 0.7	3.0 ± 0.4	0.10	0.10
Zn	38 ± 8.4	4.9 ± 0.5	12.0 ± 1.0	5.0	5.0
Hg	6.5 ± 1.4	9.8 ± 2.5	0.4 ± 0.01	0.01	0.01
Cd	6.0 ± 1.0	3.1 ± 0.3	3.0 ± 0.7	2.0	1.0

Note: CPCB, Central Pollution Control Board, Ministry of Environment, Forest and Climate Change, Government of India; US EPA, US Environment Protection Agency.
Source: Rai (2012).

REFERENCES

Ahmad, Arslan, and Prosun Bhattacharya. 2019. "Arsenic in Drinking Water: Is 10 Mg/L a Safe Limit?" *Current Pollution Reports* 5: 1–3. https://doi.org/10.1007/s40726-019-0102-7.

Alkhadra, Mohammad A., Matthew L. Jordan, Huanhuan Tian, Christopher G. Arges, and Martin Z. Bazant. 2022. "Selective and Chemical-Free Removal of Toxic Heavy Metal Cations from Water Using Shock Ion Extraction." *Environmental Science and Technology* 56: 14091–14098. https://doi.org/10.1021/acs.est.2c05042.

Angon, Prodipto Bishnu, Md Shafiul Islam, Shreejana KC, Arpan Das, Nafisa Anjum, Amrit Poudel, and Shaharia Akter Suchi. 2024. "Sources, Effects and Present Perspectives of Heavy Metals Contamination: Soil, Plants and Human Food Chain." *Heliyon* 10: e28357. https://doi.org/10.1016/j.heliyon.2024.e28357.

Basu, Niladri, Ashley Bastiansz, José G. Dórea, Masatake Fujimura, Milena Horvat, Emelyn Shroff, Pál Weihe, and Irina Zastenskaya. 2023. "Our Evolved Understanding of the Human Health Risks of Mercury." *Ambio* 52: 877–896. https://doi.org/10.1007/s13280-023-01831-6.

Bellouard, Marie, Marie Gasser, Sébastien Lenglet, Federica Gilardi, Nasim Bararpour, Marc Augsburger, Aurélien Thomas, and Jean Claude Alvarez. 2022. "Toxicity and Metabolomic Impact of Cobalt, Chromium, and Nickel Exposure on HepaRG Hepatocytes." *Chemical Research in Toxicology* 35: 807–816. https://doi.org/10.1021/acs.chemrestox.1c00429.

Bhardwaj, Shefali, Richa Soni, Sharad Kumar Gupta, and Dericks P. Shukla. 2020. "Mercury, Arsenic, Lead and Cadmium in Waters of the Singrauli Coal Mining and Power Plants Industrial Zone, Central East India." *Environmental Monitoring and Assessment* 192: 251. https://doi.org/10.1007/s10661-020-8225-2.

Bradl, H. B. 2005. "Sources and Origins of Heavy Metals." In *Heavy Metals in the Environment: Origin, Interaction and Remediation*, pp 1–27. Elsevier. https://doi.org/10.1016/S1573-4285(05)80020-1.

Cai, Xingrui, Bofeng Cai, Haoran Zhang, Long Chen, Chaoyue Zheng, Peifeng Tong, Huiming Lin, et al. 2020. "Establishment of High-Resolution Atmospheric Mercury Emission Inventories for Chinese Cement Plants Based on the Mass Balance Method." *Environmental Science and Technology* 54: 13399–13408. https://doi.org/10.1021/acs.est.0c02963.

Carrera-Beltrán, Lourdes, Irene Gavilanes-Terán, Julio Idrovo-Novillo, Victor Hugo Valverde, Adrián Rodríguez-Pinos, Concepción Paredes, Antonio J. Signes-Pastor, and Ángel A. Carbonell-Barrachina. 2024. "Environmental Pollution by Heavy Metals within the Area Influenced by the Tungurahua Volcano Eruption—Ecuador." *Ecotoxicology and Environmental Safety* 270:115919. https://doi.org/10.1016/j.ecoenv.2023.115919.

Chaemiso, Tamirat Dula. 2019. "Removal Methods of Heavy Metals from Laboratory Wastewater." *Journal of Natural Sciences Research* 9: 36–42. https://doi.org/10.7176/jnsr/9-2-04.

Chen, Q. Y., J. Brocato, F. Laulicht, and M. Mechanisms Costa. 2017. "Mechanisms of Nickel Carcinogenesis." In *In Essential and Non-Essential Metals: Carcino- Genesis, Prevention and Cancer Therapeutics*, edited by A. Mudipalli and J. T. Zelikoff, pp. 181–197. Springer International Publishing, Cham.

"Copper Uses in Industry." Accessed November 23, 2025. https://industrialmetalservice.com/metal-university/how-copper-is-used-in-industry/.

"EPAs National Primary Drinking Water Regulation." Accessed November 30, 2025. https://www.epa.gov/ground-water-and-drinking-water/national-primary-drinking-water-regulations#Inorganic.

European Union. "Eur-Lex." Accessed November 30, 2025. https://eur-lex.europa.eu/EN/legal-content/summary/drinking-water-essential-quality-standards.html.

Flora, Swaran J. S., Keerti Jain, Archna Panghal, and Jayant Patwa. 2022. "Chemistry, Pharmacology, and Toxicology of Monoisoamyl Dimercaptosuccinic Acid: A Chelating Agent for Chronic Metal Poisoning." *Chemical Research in Toxicology* 35: 1701–1719. https://doi.org/10.1021/acs.chemrestox.2c00129.

Hammond, C. R. 2010. "The Elements." In *CRC Handbook of Chemistry and Physics*, edited by David R. Lide and W. M. "Mickey" Haynes, 90th ed., pp. 4–7, 4–10, 4–20,4–22, 4–23, 4–32, 4–36,. CRC Press/Taylor and Francis, Boca Raton, FL.

"IARC Monographs." 2025. https://monographs.iarc.who.int/agents-classified-by-the-iarc/.

Izydorczyk, Grzegorz, Katarzyna Mikula, Dawid Skrzypczak, Konstantinos Moustakas, Anna Witek-Krowiak, and Katarzyna Chojnacka. 2021. "Potential Environmental Pollution from Copper Metallurgy and Methods of Management." *Environmental Research* 197: 111050. https://doi.org/10.1016/j.envres.2021.111050.

Johnson, Jeremiah, Laura Schewel, and T. E. Graedel. 2006. "The Contemporary Anthropogenic Chromium Cycle." *Environmental Science and Technology* 40: 7060–7069. https://doi.org/10.1021/es060061i.

Jomova, Klaudia, Suliman Y. Alomar, Eugenie Nepovimova, Kamil Kuca, and Marian Valko. 2025. *Heavy Metals: Toxicity and Human Health Effects. Archives of Toxicology* 99: 153–209. https://doi.org/10.1007/s00204-024-03903-2.

Kang, Bolun, Jinghan Wang, Shaojuan Guo, and Lixin Yang. 2024. "Mercury-Induced Toxicity: Mechanisms, Molecular Pathways, and Gene Regulation." *Science of the Total Environment* 943: 173577. https://doi.org/10.1016/j.scitotenv.2024.173577.

Kellerhals, Thomas, Leon Hard Tobler, Sabina Brütsch, Michael Sigl, Lukas Wacker, Heinz W. Gäggeler, and Margit Schwikowski. 2010. "Thallium as a Tracer for Preindustrial Volcanic Eruptions in an Ice Core Record from Illimani, Bolivia." *Environmental Science and Technology* 44: 888–893. https://doi.org/10.1021/es902492n.

Kerur, S. S., Sneha Bandekar, Manjunath S. Hanagadakar, Santosh S. Nandi, G. M. Ratnamala, and Prasad G. Hegde. 2020. "Removal of Hexavalent Chromium-Industry Treated Water and Wastewater: A Review." *Materials Today: Proceedings* 42: 1112–1121. https://doi.org/10.1016/j.matpr.2020.12.492.

Leyssens, Laura, Bart Vinck, Catherine Van Der Straeten, Floris Wuyts, and Leen Maes. 2017. "Cobalt Toxicity in Humans—A Review of the Potential Sources and Systemic Health Effects." *Toxicology* 387: 43–56. https://doi.org/10.1016/j.tox.2017.05.015.

Liang, Sai, Chao Zhang, Yafei Wang, Ming Xu, and Weidong Liu. 2014. "Virtual Atmospheric Mercury Emission Network in China." *Environmental Science and Technology* 48: 2807–2815. https://doi.org/10.1021/es500310t.

Mark Wilhelm, S. 2001. "Estimate of Mercury Emissions to the Atmosphere from Petroleum." *Environmental Science and Technology* 35: 4704–4710. https://doi.org/10.1021/es001804h.

Mason, Emily, Marie Edmonds, and Joseph R. McConnell. 2022. "Volatile Trace Metals Deposited in Ice as Soluble Volcanic Aerosols during the 17.7.Ka Eruptions of Mt Takahe, West Antarctic Rift." *Frontiers in Earth Science* 10: 1–11. https://doi.org/10.3389/feart.2022.1002366.

Mere, Aurore, Maxime Enrico, Honggang Zhou, Emmanuel Tessier, and Brice Bouyssiere. 2022. "Arsenic Analysis in the Petroleum Industry: A Review." *ACS Omega* 7: 38150–38157. https://doi.org/10.1021/acsomega.2c03708.

Ministry of Environment, Forest and climate change of government of India. "Central Pollution Control Board." https://cpcb.nic.in/who-guidelines-for-drinking-water-quality/.

Mitra, Saikat, Arka Jyoti Chakraborty, Abu Montakim Tareq, Talha Bin Emran, Firzan Nainu, Ameer Khusro, Abubakr M. Idris, et al. 2022. "Impact of Heavy Metals on the Environment and Human Health: Novel Therapeutic Insights to Counter the Toxicity." *Journal of King Saud University—Science* 34: 101865. https://doi.org/10.1016/j.jksus.2022.101865.

Moosdorf, Nils, Phil Renforth, and Jens Hartmann. 2014. "Carbon Dioxide Efficiency of Terrestrial Enhanced Weathering." *Environmental Science and Technology* 48: 4809–4816. https://doi.org/10.1021/es4052022.

Mudipalli, Anuradha. 2017. "Cadmium Carcinogenesis and Mechanistic Insights." In *Essentials and Non-Essentials Metals: Carcinogenesis, Prevention and Cancer Therapeutics*, edited by Anuradha Mudipalli and Judith T. Zelikoff, 1st ed., pp. 115–133. Springer International Publishing AG, Cham, Switzerland.

Muhammad, Habibun Nabi, Ekramul Mahmud, A. K. Obidul Huq, and Rosiyah Binti Yahya. 2016. "The Removal of Heavy Metal Ions from Wastewater/Aqueous Solution Using Polypyrrole-Based Adsorbents: A Review." *RSC Advances* 6: 14778–14791. https://doi.org/10.1039/c5ra24358k.

"Nickel." Accessed November 24, 2025. https://periodic-table.rsc.org/element/28/nickel.

Obeng-Gyasi, Emmanuel. 2019. "Sources of Lead Exposure in Various Countries." *Reviews on Environmental Health* 34: 25–34. https://doi.org/10.1515/reveh-2018-0037.

Patnaik, P. 2003. *Handbook of Inorganic Chemicals*. McGraw-Hill, New York.

Quinton, John N., and John A. Catt. 2007. "Enrichment of Heavy Metals in Sediment Resulting from Soil Erosion on Agricultural Fields." *Environmental Science and Technology* 41: 3495–3500. https://doi.org/10.1021/es062147h.

Rai, Prabhat Kumar. 2012. "An Eco-Sustainable Green Approach for Heavy Metals Management: Two Case Studies of Developing Industrial Region." *Environmental Monitoring and Assessment* 184: 421–448. https://doi.org/10.1007/s10661-011-1978-x.

Ravindra, Khaiwal, László Bencs, and René Van Grieken. 2004. "Platinum Group Elements in the Environment and Their Health Risk." *Science of the Total Environment* 318: 1–43. https://doi.org/10.1016/S0048-9697(03)00372-3.

Raychaudhuri, Sarmistha Sen, Paulami Pramanick, Pratik Talukder, and Apaala Basak. 2021. *Polyamines, Metallothioneins, and Phytochelatins—Natural Defense of Plants to*

Mitigate Heavy Metals. Studies in Natural Products Chemistry, 1st ed. Vol. 69. Elsevier B.V. https://doi.org/10.1016/B978-0-12-819487-4.00006-9.

Rhodes, Emily P., Zhiyong Ren, and David C. Mays. 2012. "Zinc Leaching from Tire Crumb Rubber." *Environmental Science and Technology* 46: 12856–12863. https://doi.org/10.1021/es3024379.

Saravanan, Panchamoorthy, V. Saravanan, R. Rajeshkannan, G. Arnica, M. Rajasimman, Gurunathan Baskar, and Arivalagan Pugazhendhi. 2024. "Comprehensive Review on Toxic Heavy Metals in the Aquatic System: Sources, Identification, Treatment Strategies, and Health Risk Assessment." *Environmental Research* 258: 119440. https://doi.org/10.1016/j.envres.2024.119440.

Seidl, Michael, Gábor Balázs, and Manfred Scheer. 2019. "The Chemistry of Yellow Arsenic." *Chemical Reviews* 119: 8406–8434. https://doi.org/10.1021/acs.chemrev.8b00713.

Sharma, Nitika, Kushneet Kaur Sodhi, Mohit Kumar, and Dileep Kumar Singh. 2021. "Heavy Metals Eco-Toxicity with Major Concern to Chromium and Recent Advancement in Remediation Technologies." *Environmental Nanotechnology, Monitoring and Management* 15: 100388. https://doi.org/10.1016/j.enmm.2020.100388.

Shrestha, Rakesh, Sagar Ban, Sijan Devkota, Sudip Sharma, Rajendra Joshi, Arjun Prasad Tiwari, Hak Yong Kim, and Mahesh Kumar Joshi. 2021. "Technological Trends in Heavy Metals Removal from Industrial Wastewater: A Review." *Journal of Environmental Chemical Engineering* 9: 105688. https://doi.org/10.1016/j.jece.2021.105688.

Smith. P. G., and J. S. Scott. 2005. *Dictionary of Waste and Water Treatment*, 2nd ed. Elsevier, Butterworths, London.

Soyol-Erdene, Tseren Ochir, Youngsook Huh, Sungmin Hong, and Soon Do Hur. 2011. "A 50-Year Record of Platinum, Iridium, and Rhodium in Antarctic Snow: Volcanic and Anthropogenic Sources." *Environmental Science and Technology* 45: 5929–5935. https://doi.org/10.1021/es2005732.

States, J. Christopher. 2017. "Arsenic Carcinogenesis." In *Essentials and Non-Essentials Metals: Carcinogenesis, Prevention and Cancer Therapeutics*, edited by Anuradha Mudipalli and Judith T. Zelikoff, pp. 95–98. Springer International Publishing.

Su, Hong, Zhou Li, Samuel Selorm, Fiati Kenston, Hongbo Shi, Yafei Wang, Xin Song, et al. 2017. "Joint Toxicity of Different Heavy Metal Mixtures after a Short-Term Oral Repeated-Administration in Rats." *Environmental Research and Public Health* 14: 1164. https://doi.org/10.3390/ijerph14101164.

Tang, Yuanyuan, Po Heng Lee, and Kaimin Shih. 2013. "Copper Sludge from Printed Circuit Board Production/Recycling for Ceramic Materials: A Quantitative Analysis of Copper Transformation and Immobilization." *Environmental Science and Technology* 47: 8609–8615. https://doi.org/10.1021/es400404x.

"Toxicological Profile for Cadmium." Accessed November 24, 2025. https://www.ncbi.nlm.nih.gov/books/NBK158841/.

Wang, Jianlong, and Xuan Guo. 2023. "Adsorption Kinetics and Isotherm Models of Heavy Metals by Various Adsorbents: An Overview." *Critical Reviews in Environmental Science and Technology* 53: 1837–1865. https://doi.org/10.1080/10643389.2023.2221157.

Yang, Jianzhou, Yanling Sun, Zhenliang Wang, Jingjing Gong, Jianweng Gao, Shixin Tang, Shengming Ma, and Zhuang Duan. 2022. "Heavy Metal Pollution in Agricultural Soils of a Typical Volcanic Area: Risk Assessment and Source Appointment." *Chemosphere* 304: 135340. https://doi.org/10.1016/j.chemosphere.2022.135340.

Yaroshevsky, A. A. 2006. "Abundances of Chemical Elements in the Earth's Crust." *Geochemistry International* 44: 48–55. https://doi.org/10.1134/S001670290601006X.

Zhan, Lu, and Zhenming Xu. 2014. "State-of-the-Art of Recycling E-Wastes by Vacuum Metallurgy Separation." *Environmental Science and Technology* 48: 14092–14102. https://doi.org/10.1021/es5030383.

Zhang, Lei, Shuxiao Wang, Long Wang, Ye Wu, Lei Duan, Qingru Wu, Fengyang Wang, et al. 2015. "Updated Emission Inventories for Speciated Atmospheric Mercury from Anthropogenic Sources in China." *Environmental Science and Technology* 49: 3185–3194. https://doi.org/10.1021/es504840m.

Zhang, Naiding, Guiya Xiong, and Zhenjie Liu. 2022. "Toxicity of Metal-Based Nanoparticles: Challenges in the Nano Era." *Frontiers in Bioengineering and Biotechnology*, November: 1–16. https://doi.org/10.3389/fbioe.2022.1001572.

"Zinc Statistics and Information." Accessed April 24, 2025. https://www.usgs.gov/centers/national-minerals-information-center/zinc-statistics-and-information.

"Zinc Toxicity." https://www.ncbi.nlm.nih.gov/books/NBK554548/.

2 Metal Ion Capture Technologies

Jose Luis Ortiz-Quiñonez and Umapada Pal

2.1 INTRODUCTION

Water pollution remains one of the major global challenges, contributing not only to illness, diseases, and mortality but also to the progressive reduction of drinkable water resources (Bolisetty et al. 2019). Among the various pollutants of concern, heavy metals are particularly problematic due to their persistence, non-biodegradability, and tendency to bioaccumulate through the food chain. Their entry into the environment is mainly attributed to human activities such as agriculture, mining, electroplating, alloy production, leather processing, household waste discharge, bleaching operations, and the inadequate disposal of electronic waste (Raouf and Raheim 2016). In addition to these human-derived sources, certain heavy metals, such as arsenic, can originate naturally; groundwater contamination by arsenic is a notable example of geogenic pollution affecting several regions worldwide.

A wide variety of technologies is currently used to remove heavy metals from water. These include microfiltration (MF), ultrafiltration (UF), nanofiltration (NF), reverse osmosis (RO), membrane-based bioreactors, chlorination, clarification, ion-exchange, activated sludge processes, and disinfection using UV light or ozone (Figure 2.1a). Many of these processes achieve removal efficiencies above 90% for several heavy metal ions. However, certain species, most notably As(III), remain challenging to eliminate effectively, as illustrated in Figure 2.1b (Bolisetty et al. 2019). The overall treatment of polluted water involves three stages: primary, secondary, and tertiary treatment, each designed to progressively remove suspended solids, organic matter, and dissolved contaminants, as shown in Figure 2.2. Primary treatment focuses on removing large particles and emulsions through coagulation and flocculation, chemical precipitation, MF, centrifugation, and sedimentation. Suspended particles, sludge, and oil emulsions are removed in this step, often using MF. Some heavy metals may co-precipitate and accumulate within the resulting sludge.

Secondary treatments employ both aerobic and anaerobic biological processes, wherein microorganisms convert organic pollutants into CO_2 and other less harmful by-products, facilitated by continuous aeration. Recent developments have demonstrated that microbial systems can also contribute to metal sequestration; for example, genetically engineered bacterial biofilm sponges have shown potential for mercury capture (Tay et al. 2017).

Tertiary treatment is the most advanced stage of purification, which incorporates processes such as oxidation, distillation, electrochemical precipitation, photocatalysis,

DOI: 10.1201/9781003599098-2

crystallization, ion-exchange, adsorption, and membrane filtration. Oxidation aims to transform persistent organic molecules into less toxic molecules, and complete conversion into CO_2 and water is referred to as mineralization. Common oxidants used for this purpose include chlorine, chlorine dioxide, permanganate, oxygen, and ozone. In distillation, water and volatile compounds are separated from non-volatile residues. Photocatalysis involves the generation of electron-hole pairs and reactive

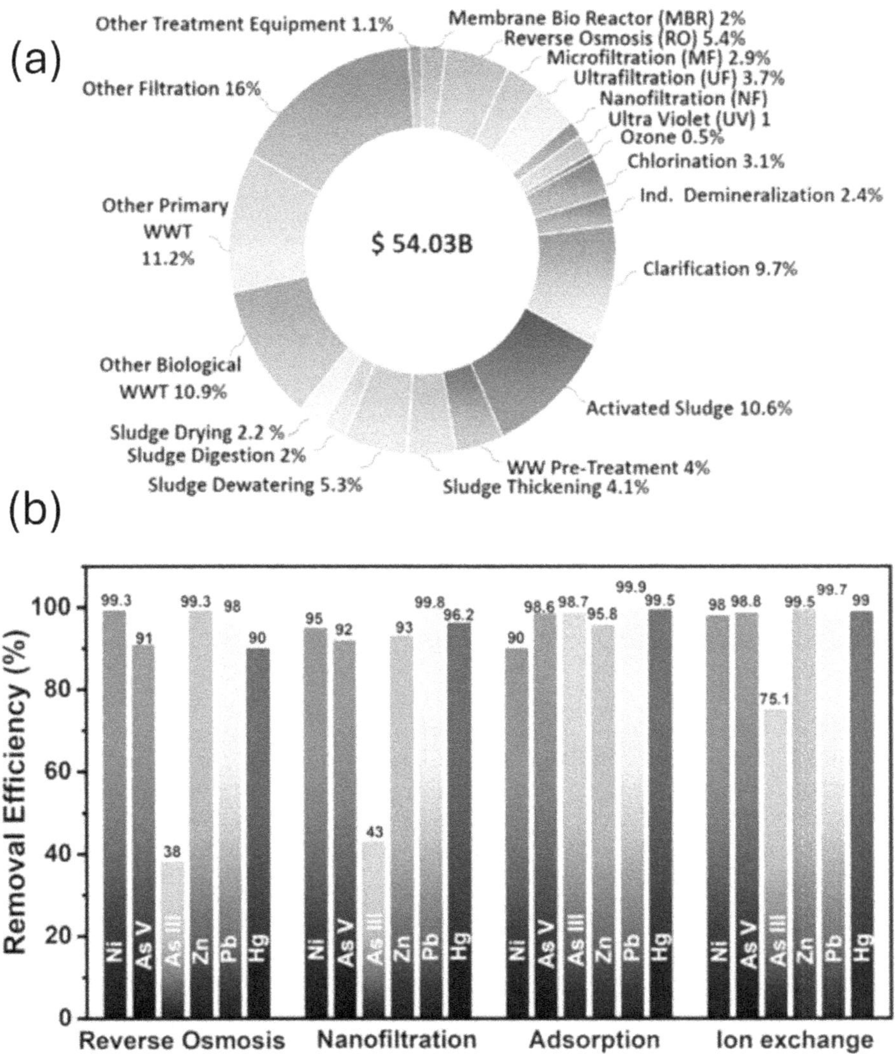

FIGURE 2.1 Overview of current water purification technologies: (a) Global economic turnover by sector; (b) Comparative removal efficiency (reported to one decimal place) of selected purification technologies for a representative set of heavy metal ions. Adapted from Bolisetty et al. (2019; Copyright Royal Society of Chemistry, 2019).

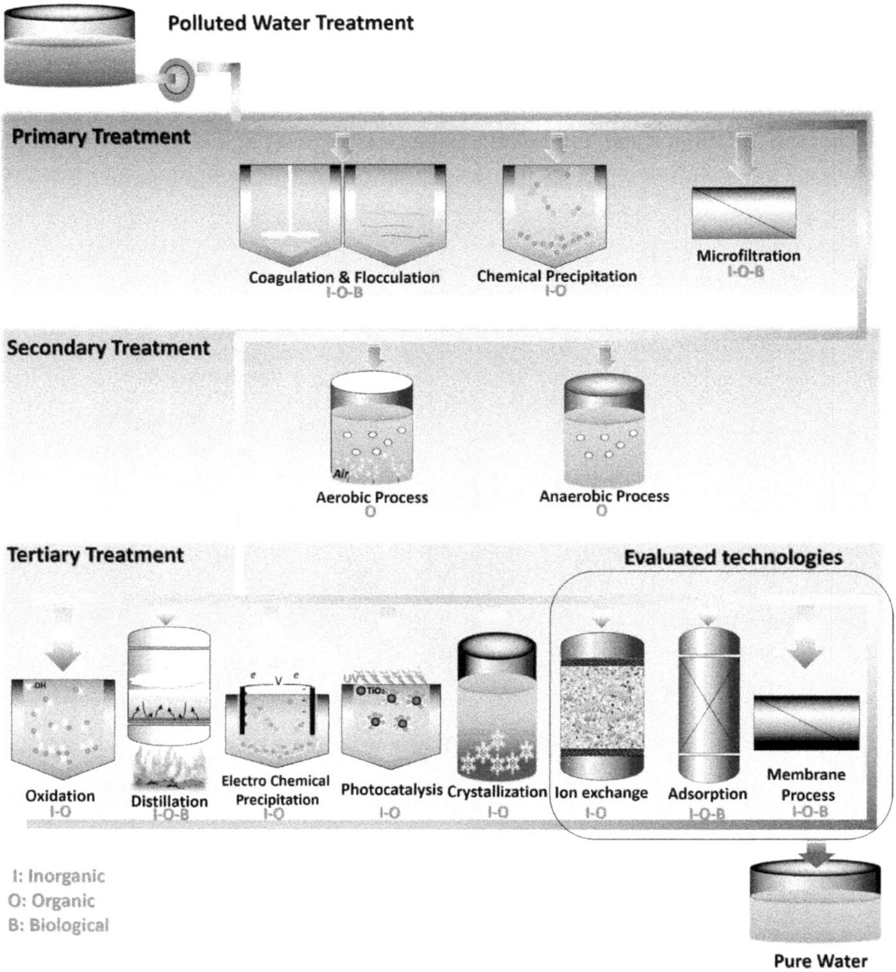

FIGURE 2.2 Schematic flowchart of a typical water purification process, illustrating the integration of primary, secondary, and tertiary treatments for the removal of organic, inorganic, and biological pollutants. Major technologies are highlighted along with their corresponding applications to different pollutant classes. Adapted from Bolisetty et al. (2019; Copyright Royal Society of Chemistry, 2019).

hydroxyl radicals in semiconductor materials suspended in water, enabling the degradation of organic contaminants. Electrochemical precipitation removes metal ions via electrodeposition onto the cathode under an applied voltage. In crystallization, the concentration of some pollutants is increased to promote crystallization. On the other hand, in ion-exchange, weakly bound non-toxic ions on resin surfaces are exchanged by heavy metal ions present in solution. Finally, in the adsorption process, both heavy metal ions and organic pollutants are adsorbed onto the surface of solid adsorbents through electrostatic interactions, van der Waals forces, and other surface-driven mechanisms.

This comprehensive array of treatment methods illustrates both the complexity and the critical importance of effective technologies for heavy metal removal. Their selection and optimization depend on multiple factors, including water composition, target contaminants, treatment goals, and operational constraints, which together shape sustainable wastewater treatment strategies.

2.2 METAL ION CAPTURE PROCESSES

2.2.1 Adsorption-Based Capture

Conventional adsorbents, such as activated carbons, clays, and zeolites, exhibit limited binding affinity toward dissolved metal ions due to weak interactions between surface sites and the target contaminants (Peng et al. 2018). To enhance performance, adsorbents are often functionalized with specific chemical groups that promote stronger coordination or electrostatic binding, thereby improving metal uptake capacity. Metal-organic frameworks (MOFs), in particular, benefit significantly from such functionalization strategies.

2.2.1.1 Mechanism of Cation Adsorption

A widely used approach for removing heavy metal ions from wastewater involves inducing the formation of insoluble particles through precipitation, as illustrated in stage I of Figure 2.3a. The addition of an appropriate precipitating agent causes metal ions to form micro- and macro-particles. In the next step (Figure 2.3b), a surfactant, such as cetyltrimethylammonium bromide (CTAB) or TERGITOL NP-40, is introduced. These surfactants adsorb onto the metal-containing particles through electrostatic attraction or specific chemical interactions. In some cases, metal ions may bind directly to functional groups on surfactant molecules.

Once the surfactant-coated particles are formed, they are removed from solution using the dissolved air flotation (DAF) process. DAF operates by generating air microbubbles that attach to the suspended particles, decreasing their effective density and promoting their separation as a floating sludge, as shown in the floatation column on Figure 2.3c. This clarification method is widely employed in both membrane-based and conventional granular filtration water-treatment systems (Edzwald 2010). DAF is particularly effective at removing a range of particulates, including activated carbon, metal hydroxide flocs generated during coagulation, organic matter aggregates, and particles formed by the oxidation of Fe and Mn ions. A related method, precipitation-flotation, follows a similar principle, where contaminants are first converted into insoluble precipitates, which subsequently attach to microbubbles and are separated from the aqueous phase.

2.2.1.2 Carbon-Based Adsorbents

Carbon-based materials remain the most widely studied adsorbents for heavy-metal removal because of their high specific surface area, adjustable porosity, chemical stability, and ease of surface modification. Traditional activated carbon (AC), made from coal, wood, or agricultural waste, offers abundant microporosity and surface oxygen groups that interact with metal ions through complexation, ion-exchange, and

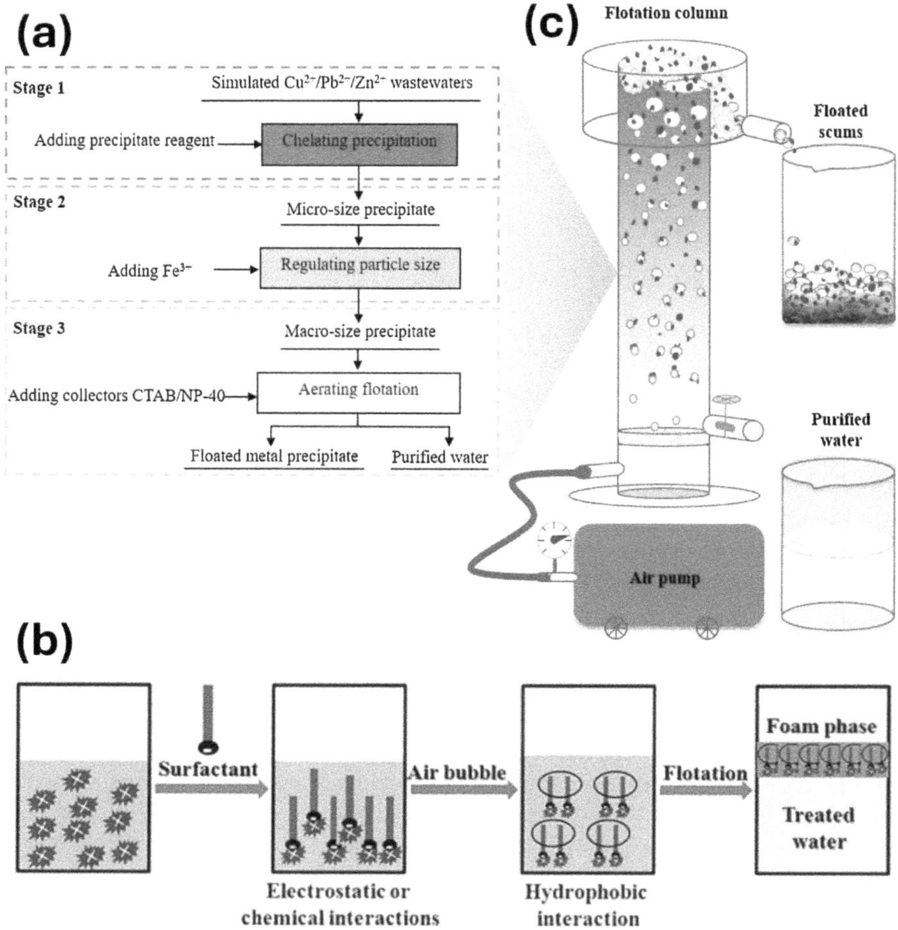

FIGURE 2.3 (a, c) Schematic diagram of the metal-ion removal process from wastewater simulants using the enhanced precipitation-flotation method, along with the corresponding experimental setup. Adapted from Wu et al. (2019; Copyright Royal Society of Chemistry, 2019). (b) Schematic illustration of mechanism of ion flotation process. Adapted from Gahrouei et al. (2024; Copyright Elsevier, 2024).

electrostatic attraction. However, raw AC often has limited selectivity for specific metal ions, which has led to extensive research on chemical functionalization (e.g., amination, thiolation, and oxidation) and the creation of composites to add specific binding sites and improve adsorption speed. Besides AC, newer carbon materials like biochar, carbon nanotubes (CNTs), graphene, and carbon aerogels offer additional benefits. For instance, biochar is a low-cost, feedstock-derived adsorbent with tunable surface chemistry. CNTs and graphene-based materials have very high surface areas and can be tailored with functional groups to boost affinity and selectivity.

Modification strategies are commonly used to improve the capacity and selectivity of carbon-based adsorbents, including grafting with chelating groups such as

$-NH_2$, $-SH$, and $-COOH$, impregnation with metal oxides, and the formation of composites with polymers or inorganic phases to combine high affinity with mechanical stability and ease of handling. An important practical consideration in developing these adsorbents is regeneration. Many carbon adsorbents can be chemically or thermally regenerated, but regeneration efficiency and structural integrity vary widely and must be assessed for each material-metal pair. Details on the modification processes of carbon-based adsorbents are provided in Chapter 5. For a comprehensive discussion of carbon-based adsorbents, their fabrication, and modification methods, readers can consult the reviews on biomass-derived carbons by (Sabzehmeidani et al. 2021).

2.2.1.3 Chitosan-Based Adsorbents

Chitosan, a partially deacetylated derivative of chitin, is a biopolymer that has attracted significant interest for heavy-metal adsorption because of its high density of reactive amino ($-NH_2$) and hydroxyl ($-OH$) groups that can chelate and exchange metal cations. Unmodified chitosan exhibits strong affinity for cations such as Cu^{2+}, Pb^{2+}, Cd^{2+}, and arsenic species but suffers from low mechanical strength and solubility in acidic media. To address these limitations, researchers often use chemical cross-linking agents such as glutaraldehyde and epichlorohydrin, grafting to add functional chelators (e.g., thiol, carboxyl, and ethylenediaminetetraacetic acid (EDTA) groups), and forming composites with carbonaceous materials or magnetic nanoparticles. These modifications enhance stability, increase the number of available binding sites, and often improve selectivity and capacity under realistic wastewater conditions. Chitosan-based beads, membranes, and nanocomposites have been reported to have high uptake capacities and reasonable reusability after desorption cycles. However, the performance of chitosan-based adsorbents strongly depends on pH, ionic strength, and the nature of competing ions. Therefore, careful optimization is necessary for their application in complex effluents. While the use of chitosan in preparing composite adsorbents and their application in heavy metal removal is discussed in Chapter 5, for a detailed overview of the synthesis, functionalization, and use of chitosan in removing heavy metal ions, as well as its biodegradation behaviors under treatment conditions, the authors can refer to the review article by (Omer et al. 2022).

2.2.1.4 Mineral and Magnetic Adsorbents

Mineral adsorbents like clays, zeolites, layered double hydroxides, and iron-based minerals are traditional options for removing heavy metals because they are inexpensive, naturally available, and operate through ion-exchange or surface-complexation mechanisms. Clays and zeolites eliminate cations via ion-exchange at the structural sites and through surface adsorption. Their selectivity can be tailored by altering the cationic form, pillaring, or surfactant loading to produce organo-clays for specific contaminant removal. LDHs offer anion-exchange capacity and can be designed to capture oxyanions of metalloids, such as arsenate. On the other hand, magnetic adsorbents, usually iron oxides like Fe_3O_4 and γ-Fe_2O_3, whether used alone or supported on carbon, polymer, or biopolymer matrices, have become a practical solution for combining high adsorption capacity with easy separation. By functionalizing magnetic cores with chelating ligands, polymers, or biopolymers (including chitosan), composite adsorbents can effectively bind metal ions and be removed from water using an external magnetic field, simplifying solid-liquid separation and eliminating the need for secondary filtration. Magnetic biochars and magnetic activated carbons further integrate the adsorption benefits of carbon materials with rapid magnetic recovery. Important practical considerations for mineral and magnetic adsorbents include colloidal stability, potential iron leaching, the ecotoxicity of nanoparticles, and the engineering of large-scale magnetic separation. While laboratory studies demonstrate high capacities and multiple reuse cycles, implementing these in full-scale water-treatment plants still requires optimizing separation equipment and conducting life-cycle assessments (Liosis et al. 2021).

2.2.1.5 Metal-Organic Framework Adsorbents

Functionalized MOFs are a powerful class of adsorbents for heavy metal removal. One promising approach involves embedding EDTA into a sturdy MOF framework to create strong chelating sites for metal ions (Peng et al. 2018). These EDTA-functionalized frameworks demonstrate exceptional performance, achieving high removal efficiency even in multi-ion environments where competitive binding usually decreases effectiveness. Their high selectivity, adjustable chemistry, and structural stability make MOFs an emerging platform for advanced heavy metal capture.

The mineral, magnetic, and carbon-based platforms described earlier often inspire hybrid strategies. For example, magnetic carbon composites, chitosan-clay beads, or MOFs grown on carbon supports combine high capacity, selectivity, mechanical stability, and easy separation. This paves the way for MOF-based adsorbents, where targeted ligand chemistry (e.g., EDTA functionalization) can deliver exceptionally high single-ion removal (>99%) while maintaining structural flexibility for multi-ion systems, as thoroughly discussed by Sabzehmeidani et al. (2021). In Figure 2.4, the compositions of real-world samples (Rhone River water and Mediterranean Sea

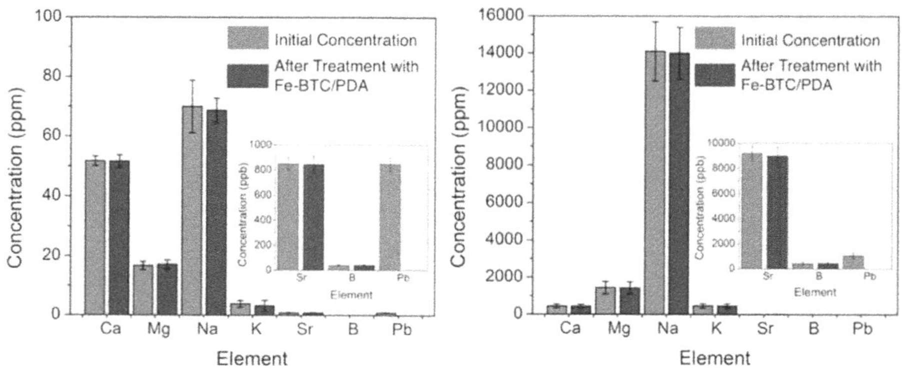

FIGURE 2.4 Compositions of real-world samples. Rhone River water (left) and Mediterranean Sea water (right) were spiked with ~1 ppm of Pb^{2+} and then subsequently treated with Fe-BTC/PDA-19. Ion concentrations before and after treatment are shown in gray and blue, respectively. Adapted from Sun et al. (2018; Copyright American Chemical Society, 2018).

water), spiked with ~1 ppm of Pb^{2+} and subsequently treated with Fe-BTC/PDA-19 (BTC = 1,3,5-benzenetricaboxylate, PDA = polydopamine), are presented. Fe-BTC, also known as MIL-100, is an MOF-polymer composite, used as an efficient adsorbent for removing Pb^{2+} and Hg^{2+} ions from water. Use of Fe-BTC is reported to reduce heavy metal concentrations by up to ~1 ppb in treated water.

2.2.2 Membrane-Based Removal

2.2.2.1 Removal by Microfiltration, Ultrafiltration, and Nanofiltration

Pressure-driven membrane processes, including MF, UF, NF, and RO, are widely employed for the separation and purification of contaminated water, as illustrated in Figure 2.5. These processes utilize semipermeable membranes that selectively allow water and specific solutes to pass through while retaining suspended particles, colloids, and dissolved ions. In all cases, an external pressure applied on the feed side drives water through the membrane, producing two distinct output streams: *permeate*, consisting of purified water, and *retentate*, containing the concentrated contaminants that require subsequent treatment or proper disposal (Liao et al. 2018). While the fundamental operating principle is the same, the separation mechanism differs depending on membrane pore size and surface properties.

As discussed in Chapter 3, MF is mainly effective for removing suspended solids and particulate matter. In the context of wastewater containing heavy metals, MF

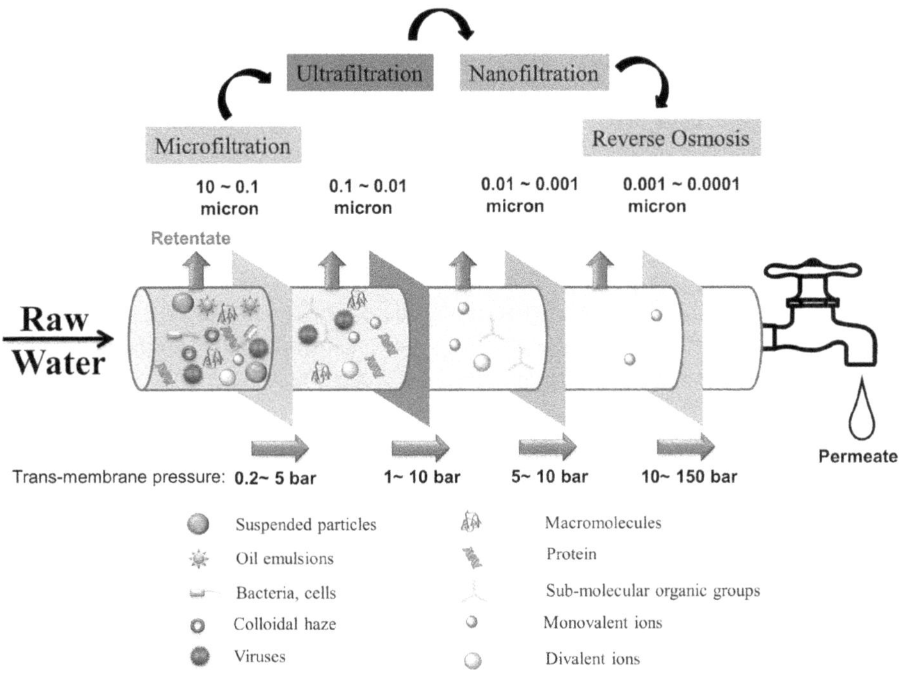

FIGURE 2.5 Schematic diagram of a membrane-based water treatment system. Adapted from Liao et al. (2018; Copyright Elsevier, 2018).

can efficiently remove particles or aggregates that have previously adsorbed heavy-metal ions, acting as a polishing step or a pretreatment stage ("Puricon" 2025). UF provides a tighter pore structure that can retain macromolecules, metal-organic complexes, and colloidal species, helping to reduce downstream membrane fouling and improve overall process performance. Conversely, NF falls between UF and RO. NF membranes display both size-exclusion and charge-based selectivity, allowing the removal of divalent metal ions, including Pb^{2+}, Cd^{2+}, Cu^{2+}, and Ni^{2+}. Meanwhile, RO membranes have the highest selectivity, rejecting even monovalent ions and achieving nearly complete desalination. Despite their high efficiency, NF systems often encounter challenges from membrane fouling caused by organic matter, colloids, or metal precipitates, which decrease permeate flux and raise operational and maintenance costs (Cui et al. 2014). Overall, MF, UF, and NF create a complementary filtration sequence in advanced wastewater treatment, where MF/UF serve as pretreatment stages, and NF/RO enable the final removal of dissolved ionic contaminants to meet regulatory standards.

2.2.2.2 Removal by Forward Osmosis and Reverse Osmosis

Osmosis is the spontaneous flow of a pure solvent into a solution across a semipermeable membrane that permits solvent transport while restricting solute movement. The pressure required to counteract this solvent flow is known as osmotic pressure (Atkins et al. 2018). Building on this natural phenomenon, forward osmosis (FO) uses the osmotic pressure difference between a dilute feed solution and a concentrated draw solution to pull water across the membrane. As FO relies on osmotic gradients rather than hydraulic pressure, it requires lower energy input and exhibits a reduced tendency to membrane fouling (Tian et al. 2017). As a result, FO has emerged as a promising approach for removing heavy metal ions from contaminated water sources, including electroplating wastewater (Cui et al. 2014; ISTEE 2025).

Despite these advantages, FO alone cannot produce drinking-grade water because the draw solution must be reconcentrated or regenerated. Therefore, FO is often integrated with RO or other pressure-driven processes to complete the purification cycle. In RO, external hydraulic pressure exceeding the osmotic pressure forces water from the concentrated feed through a dense, semipermeable membrane, resulting in high ion rejection and producing high-purity permeate. RO cartridges are widely used in drinking water production, agricultural irrigation, municipal wastewater reuse, and industrial water treatment ("Puricon" 2025). Effective RO performance requires protecting the membrane from fouling. Therefore, RO systems typically employ multi-stage pretreatment. In such systems, suspended solids such as dirt, rust, and particulates are first removed via prefiltration. Subsequently, activated carbon filtration eliminates chlorine and organic chemicals that could degrade or oxidize the RO membrane ("FloWater" 2025). This step also helps to remove chlorinated pesticides, volatile organic compounds, and asbestos ("APEC Water" 2025). Once protected, the RO membrane can operate efficiently, especially when treating feed waters of stable composition and low fouling potential. Due to their high desalination capability, RO systems are particularly suitable when stringent purity standards must be met (Mega 2025). However, to ensure long-term membrane integrity, RO membranes require protection from larger particles, making MF pretreatment essential (WaterGroupPure 2025).

By combining FO and RO, modern treatment systems can lower energy use, extend membrane lifespan, and achieve high removal efficiencies for a broad spectrum of heavy metals. FO can serve as a low-fouling pre-concentration step, while RO functions as the final polishing barrier, integrating both technologies into an efficient and complementary treatment process.

2.2.2.3 Electrodialysis

Electrodialysis (ED) is among the most cost-effective methods for removing aqueous ions using electrically driven ion-exchange membranes. ED relies on an electric field gradient to separate minerals from the feed water solution (Mega 2025). Under an electric field, cations and anions migrate selectively through cation-exchange membranes (CEM) and anion-exchange membranes (AEM), respectively, thus separating the ionic content of the feed. As a result, ED produces two distinct output streams: a *diluate*, containing water with reduced ionic concentration, and a *concentrate* or brine stream, in which the separated ions accumulate (Mega 2025). ED operates at low hydraulic pressure (approximately 4 bar), requires relatively simple pre-treatment, exhibits good tolerance to chlorine, and generally has a favorable lifecycle cost (WaterGroupPure 2025).

ED is particularly advantageous for treating feedwater with high fouling potential or unstable quality, such as conditions that challenge pressure-driven systems. Because ion transport occurs under an electric field rather than through membrane pores, fouling and scaling are less severe, making ED well-suited for applications requiring high recovery or highly concentrated brines (up to 200 g/L) (Mega 2025). The introduction of ED reversal (EDR), in which the electrode polarity is periodically reversed, has further enhanced the technology. Periodic reversal drives ionic migration in alternating directions, helping to dislodge scale-forming deposits, reducing the need for acid or anti-scaling pretreatments, and improving tolerance toward difficult feedwaters (WaterGroupPure 2025). These advantages have made ED and EDR increasingly competitive with RO, especially in industrial metal recovery. For instance, laboratory-scale ED systems have been used to treat nickel-rich electroplating wastewater, allowing for both nickel extraction and recovery of its salts (Shrestha et al. 2021).

The configuration of an electrodialytic cell with a bipolar membrane (BPM) is illustrated in Figure 2.6. The cell features alternating cation-exchange membranes (CEMs) and anion-exchange membranes (AEMs), with a bipolar membrane (BPM) positioned centrally, and the anode and cathode located at opposite ends. The BPM consists of an anion-exchange layer (AEL) and a cation-exchange layer (CEL). Unlike traditional ion-exchange membranes, the BPM blocks the transport of both cations and anions. Instead, it promotes the dissociation of water into H^+ and OH^- ions at the interface between the AEL and the CEL. These ions move to balance the charge of species crossing the adjacent membranes (e.g., neutralizing sulfate ions (SO_4^{2-}) on the anode side and sodium ions (Na^+) on the cathode side) (Cerrillo-Gonzalez et al. 2024). The integration of BPMs enables ED systems not only to desalinate but also to generate acids and bases, or to adjust pH, broadening their applicability for selective metal recovery and industrial wastewater treatment.

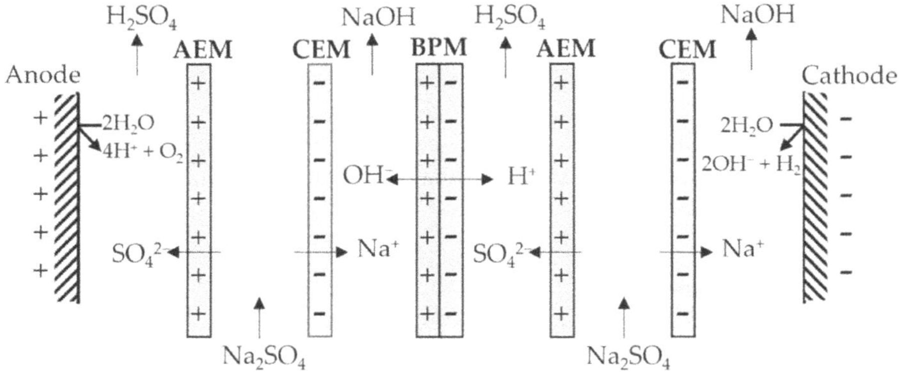

FIGURE 2.6 Schematic presentation of a three-compartment electrodialytic cell equipped with a bipolar membrane. AEM, anion-exchange membranes; CEM, cation-exchange membranes. Adapted from Cerrillo-Gonzalez et al. (2024; Copyright MDPI, 2024).

2.2.3 Chemical and Electrochemical Removal

2.2.3.1 Precipitation- and Coalescence-Based Methods

Chemical precipitation is one of the most widely used industrial techniques for the primary removal of heavy metals from wastewater, owing to its operational simplicity and low cost. In this process, the pH of the contaminated solution is adjusted, typically to alkaline conditions, followed by the addition of precipitating agents such as lime, metal hydroxides, or sulfide salts. These reagents react with dissolved heavy metal ions to form insoluble metal hydroxides or metal sulfides, which subsequently aggregate into particles that can be removed by sedimentation or filtration. The main advantage of chemical precipitation lies in its ease of implementation and low operational cost. However, the method also presents several limitations. Most notably, the treated water often does not achieve heavy metal concentrations low enough to meet discharge standards, necessitating additional post-treatment steps. Moreover, the process generates large volumes of sludge that must be dewatered, handled, and disposed of safely, which not only increases treatment cost but may also introduce risks of secondary pollution (Bolisetty et al. 2019).

Coagulation and flocculation are often used either as complementary steps to precipitation or as separate treatment processes. Coagulation destabilizes suspended and colloidal particles by adding coagulants, usually ferric or aluminum salts, which help particles stick together. Flocculation then promotes the formation of larger flocs that settle more easily. These processes improve sludge settling and help remove particulate pollutants, including metal-bearing solids. However, the heavy reliance on chemicals and the additional sludge produced remain major disadvantages.

2.2.3.2 Electrochemical Methods

Electrochemical technologies, including electrocoagulation, electrochemical reduction, electroflotation, and electrooxidation, offer alternative or complementary approaches for heavy metal removal (Shofia et al. 2025). These methods rely on applying an electric current to promote redox reactions or generate coagulants in situ, enabling efficient metal capture without the extensive use of chemical reagents.

Electrochemical reduction, also known as electrodeposition or electroplating, involves transferring electrons from the cathode to dissolved metal ions, converting them into their metallic form. Often, a reducing agent is added to the solution to improve the reduction process. The reduced metal is deposited on the cathode surface and can be manually removed later. This method is particularly useful for recovering valuable or toxic metals from industrial wastewaters while minimizing the production of chemical sludge.

2.2.3.3 Electrocoagulation

Electrocoagulation combines the principles of coagulation and electrochemistry. In this process, sacrificial metal electrodes—commonly aluminum, iron, zinc, or galvanized iron—are used to generate metal ions in situ when an electric current is applied (Oladimeji et al. 2024). These ions hydrolyze to form metal hydroxides, which act as coagulants that bind and remove dissolved and suspended contaminants, including heavy metal ions (Shofia et al. 2025).

A schematic representation of a typical electrocoagulation system is shown in Figure 2.7. The setup consists of an anode-cathode pair connected to a power supply,

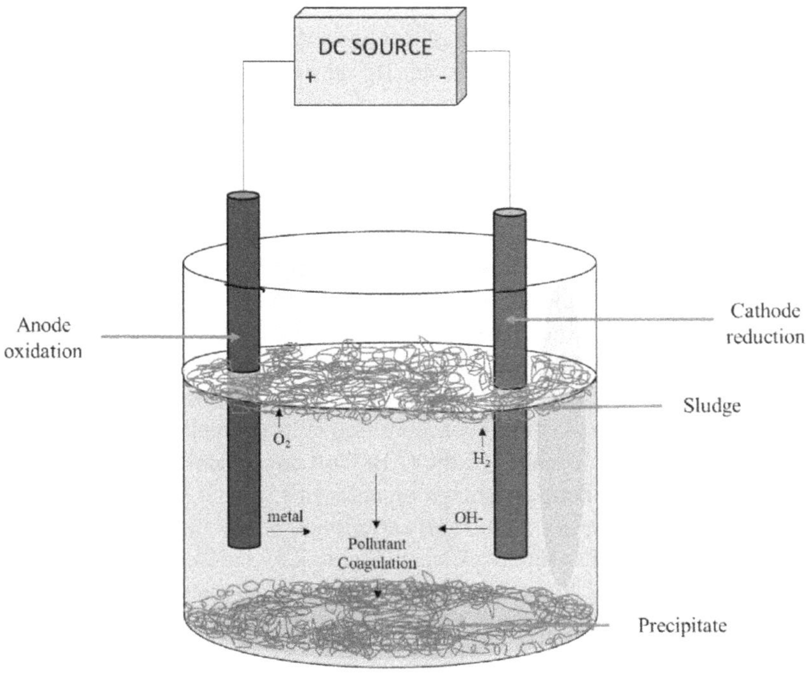

FIGURE 2.7 Schematic representation of electrochemical coagulation process. Adapted from Oladimeji et al. (2024; Copyright Cell Press, 2024).

enabling the generation of coagulant species directly within wastewater. This eliminates the need for additional chemicals and often results in lower sludge production than the conventional coagulation process. As can be noticed, electrocoagulation offers a practical and environmentally attractive approach, particularly for complex wastewater containing mixtures of heavy metals and organic pollutants.

2.3 CASE STUDIES AND SCALEUP OPPORTUNITIES

Case Study 1. *Heavy Metals Pollution of a Tropical Impoundment in Singrauli Industrial Region and Its Management Through Wetland Plants* (Rai 2012).

The Singrauli region, located in Northern India (Figure 2.8), is one of the country's industrial hubs. It hosts numerous large-scale activities, including 11 open-cast mining sites, multiple coal-based thermal power plants, and extensive coal extraction operations. These industries collectively release substantial quantities of heavy metals into the surrounding environment. A significant portion of these pollutants eventually enters Govind Ballabh Pant Sagar (G. B. Pant Sagar), a vast reservoir situated near the Rihan Dam. G.B. Pant Sagar is one of the largest water reservoirs in Asia, with a submergence area of around 46,600 hectares. The reservoir plays a vital role in regional socioeconomic activities by providing water for drinking, irrigation, and fish farming, and by supporting a 300-MW hydropower plant. However, the reservoir faces serious contamination issues. Discharges of coal fly ash and ash slurry from thermal power plants and coal mines are the main sources of heavy metal pollution in the area.

In the referenced study, water samples were collected and filtered, and the concentrations of Cu, Cr, Fe, Mn, Ni, Pb, Zn, Hg, and Cd were measured to assess the level of contamination. Despite the severity of pollution, the region lacks the financial resources needed for traditional wastewater treatment methods such as chemical precipitation, membrane filtration, or electrochemical metal removal systems. As a result, local communities have turned to wetland plants as a low-cost, accessible alternative for adsorbing and sequestering heavy metal ions from contaminated water bodies. This phytoremediation approach, although not as rapid or comprehensive as conventional treatments, offers a practical and economically feasible way to lower pollutant levels. The concentrations of heavy metals in industrial effluents before discharge in G.B. Pant Sagar are presented graphically in Figure 2.9, which highlights the scale of the contamination problem facing the region. The Anpara effluent exhibits the highest overall concentration of heavy metals. Heavy metal concentrations measured at five sites of the G.B. Pant Sagar (Belwadah, Dongia Nala, Ash Pond, and Rihand dam) are shown by color in Figure 2.8 and summarized in Table 2.1. The highest heavy metal concentration is at the Belwadah site, attributable to its proximity to the Anpara effluent (Figure 2.8). On the other hand, elevated concentrations of Cu, Fe, Ni, and Zn are observed at locations situated near the Bina effluent. The concentration of Hg is notably higher at Dongia Nala, likely due to its proximity to the Kanoria effluent.

Case Study 2. *Heavy Metal Removal Using Zeolite-based Materials in the Tito Scalo Industrial Area (Basilicata Region, Southern Italy) (Mancinelli et al. 2020).*

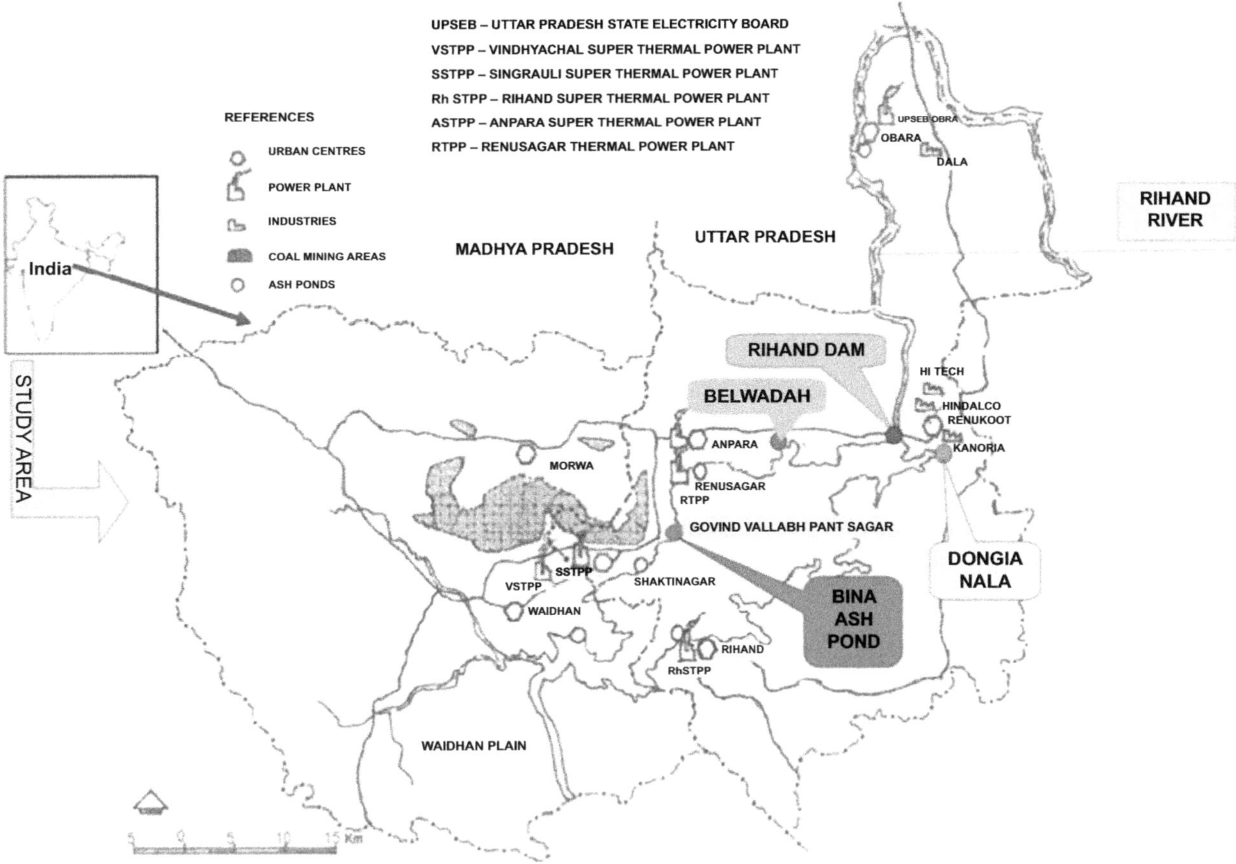

FIGURE 2.8 Study area and geographical locations of the sampling sites referred to in Case Study 1. Adapted from Rai (2012; Copyright Springer Nature 2012).

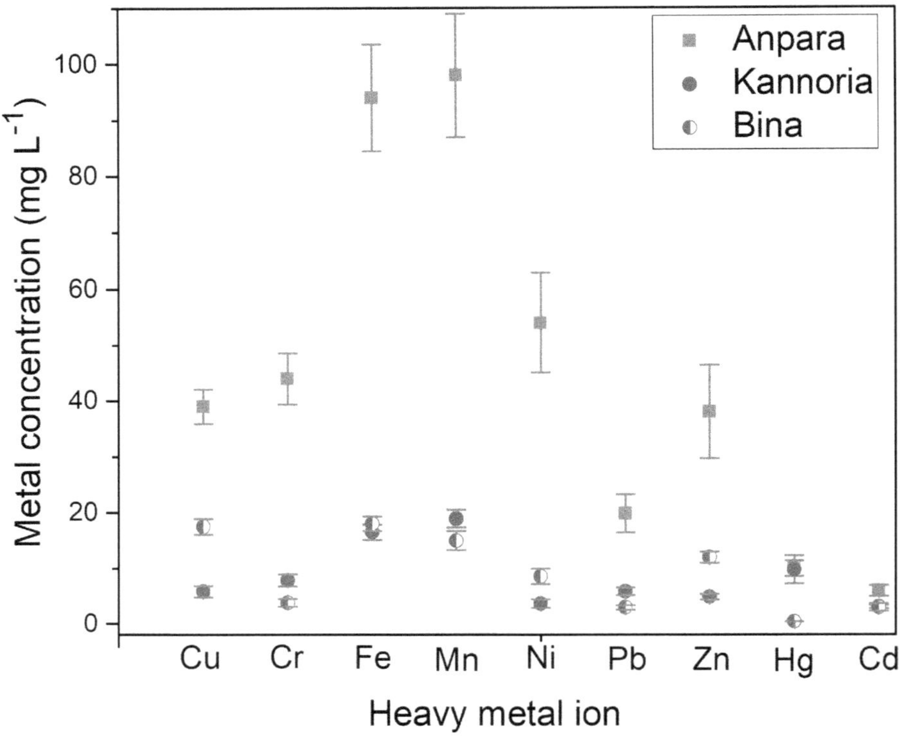

FIGURE 2.9 Metal concentrations (mg L^{-1}) in three industrial effluents (Anpara, Kannoria, Bina) in India before being discharged into a water reservoir. Data taken from Rai (2012).

This case study examined the effectiveness of a 13X zeolite ($Na_{90.04}[Al_{92.64}Si_{99.78}O_{384}]\cdot 200.03H_2O$) in removing Fe and Mn ions from groundwater in the Tito Scalo Industrial Area, in the Basilicata Region of southern Italy. Groundwater samples were collected from five different monitoring sites (PM1, PM10, PM17, PM21, and RW8), each representing distinct hydrological conditions within the industrial zone.

The physicochemical characterization of the 13X zeolite revealed a BET surface area of 791 $m^2\ g^{-1}$, a total pore volume of 0.301 $cm^3\ g^{-1}$, and a micropore volume of 0.267 $cm^3\ g^{-1}$. These properties reflect the material's highly porous, microporous nature. The structural arrangement of the zeolite, composed of interconnected micropores, is illustrated in Figure 2.10, highlighting the network through which metal ions can diffuse and be adsorbed.

The study measured both the initial concentrations of Fe and Mn in the groundwater samples and the corresponding adsorption capacities of the zeolite (Table 2.2). The results demonstrate that Mn removal was substantially high in three of the five sampling sites, indicating a strong affinity between Mn ions and the microporous 13X framework. For Fe, the performance was even more remarkable. Removal efficiencies reached 100% at four of the five sites, highlighting the excellent capacity of zeolite for Fe sequestration under the tested conditions. The findings confirm that

TABLE 2.1
Metal Concentrations (mg L⁻¹) in G.B. Pant Sagar Water

Month	Cu	Cr	Fe	Mn	Ni	Pb	Zn	Hg	Cd
Belwadah									
March	22 ± 2.1	29 ± 2.7	33 ± 2.4	18 ± 1.4	26 ± 1.9	13 ± 0.9	13 ± 0.1	1.8 ± 0.1	4.1 ± 0.6
June	27 ± 1.4	34 ± 4.1	41 ± 2.9	22 ± 1.9	36 ± 3.1	17 ± 1.7	21 ± 1.1	2.0 ± 0.3	4.5 ± 0.9
Sept.	16 ± 1.4	19 ± 1.6	22 ± 2.2	11 ± 1.1	16 ± 1.2	8 ± 0.79	11 ± 0.9	1.1 ± 0.3	2.9 ± 0.4
Dec.	24 ± 1.8	32 ± 1.8	37 ± 2.7	19 ± 1.0	29 ± 1.8	15 ± 1.2	14 ± 1.1	1.7 ± 0.3	3.7 ± 0.8
Dongia Nala									
March	0.9 ± 0.1	0.7 ± 0.2	4.6 ± 0.06	2.9± 0.7	0.9 ± 0.1	0.7 ± 0.1	3.0± 0.1	3.9 ± 0.3	0.7 ± 0.9
June	1.2 ± 0.2	0.8 ± 0.2	5.1 ± 0.4	3.1 ± 0.9	1.3 ± 0.2	1.1 ± 0.2	3.2 ± 0.3	4.1± 0.4	0.9 ± 0.1
Sept.	0.6 ± 0.1	0.3 ± 0.01	1.2 ± 0.1	1.8 ± 0.2	0.4 ± 0.09	0.3 ± 0.09	1.9 ± 0.9	2.7 ± 0.1	0.3 ± 0.05
Dec.	0.8 ± 0.2	0.5 ± 0.09	3.9 ± 0.9	2.6 ± 0.1	0.6 ± 0.07	0.4 ± 0.1	1.7 ± 0.1	2.9 ± 0.1	0.5 ± 0.3
Ash Pond									
March	5.4 ± 1.3	0.18 ± 0.01	41.3 ± 1.7	0.2 ± 0.01	3.2 ± 0.6	0.3 ± 0.09	4.2 ± 0.90	0.17 ± 0.01	1.0 ± 0.3
June	6.8 ± 1.1	0.2 ± 0.01	38 ± 3.1	0.4 ± 0.09	3.7 ± 0.7	1.1 ± 0.2	5.1 ± 0.8	0.19 ± 0.01	1.2 ± 0.2
Sept.	4.3 ± 1.2	0.09 ± 0.01	27.5 ± 1.4	0.02 ± 0.0	1.9 ± 0.4	0.3 ± 0.09	2.9 ± 0.3	0.01 ± 0.0	0.6 ± 0.01
Dec.	4.9 ± 0.9	0.1 ± 0.04	34 ± 2.7	0.3 ± 0.04	2.1 ± 0.9	0.4 ± 0.1	3.2 ± 0.4	0.1 ± 0.01	0.8 ± 0.1
Rihand dam									
March	0.01 ± 0.01	0.02 ± 0.021	1.1 ± 0.2	1.3 ± 0.4	0.07 ± 0.01	0.06 ± 0.01	1.0 ± 0.2	-	0.01 ± 0.0
June	0.02 ± 0.02	0.03 ± 0.031	1.2 ± 0.4	1.5 ± 0.2	0.09 ± 0.01	0.08 ± 0.01	1.2 ± 0.3	-	0.02 ± 0.0
Sept.	0.01 ± 0.01	-	0.9 ± 0.1	1.1 ± 0.2	0.01 ± 0.0	0.02 ± 0.0	0.8 ± 0.1	-	0.01 ± 0.0
Dec.	0.01 ± 0.01	0.01 ± 0.01	1.1 ± 0.2	1.4 ± 0.3	0.08 ± 0.01	0.09 ± 0.01	0.9 ± 0.1	-	0.01 ± 0.01

Source: Rai (2012).

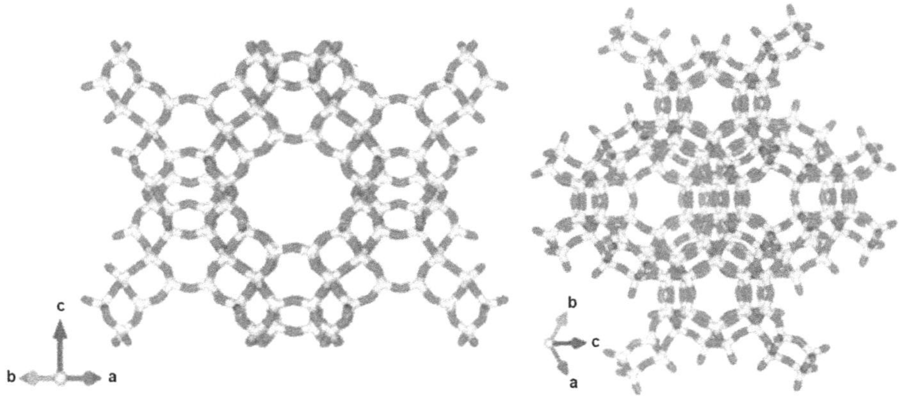

FIGURE 2.10 Structural representation of 13X zeolite. Adapted from Mancinelli et al. (2020; Copyright MDPI, 2020).

TABLE 2.2
Initial Contaminant Concentration (C_0), Contaminant Uptake (q), and Removal Efficiency of 13X Zeolite for Fe and Mn Ions

	Mn ion			Fe ion		
Sample	C_0 (mg L^{-1})	q (μg g^{-1})	Removal efficiency (%)	C_0 (mg L^{-1})	q (μg g^{-1})	Removal efficiency (%)
PM1	962 ± 5.2	694 ± 1.91	73.00	-	-	-
PM10	1104 ± 7.2	1014 ± 3.80	94.35	56.8 ± 0.56	29.4 ± 0.47	53.27
PM17	578.5 ± 0.90	554 ± 1.26	98.40	2.8 ± 0.24	2.8 ± 0.35	100
PM21	414 ± 1.00	412 ± 1.23	99.83	1.5 ± 0.11	1.5 ± 0.31	100
PM21	-	-	-	4.9 ± 0.18	4.8 ± 0.35	100
RW8	1745 ± 4.3	1172 ± 3.97	67.99	6.6 ± 0.58	6.5 ± 0.38	100

13X zeolite is a highly effective, low-cost, and scalable adsorbent for the remediation of Fe- and Mn-contaminated groundwater in industrial regions such as Tito Scalo.

Case Study 3. *Electrochemical Treatment of Industrial Wastewater Effluent from Boat Harbour Lagoon, Canada (Adesida et al. 2022).*

Boat Harbour is a former stabilization lagoon that historically received effluents from a pulp and paper mill located in Picton Landing, Nova Scotia, Canada. For several decades, wastewater from the mill was discharged into this treatment facility, leading to the accumulation of a complex mixture of contaminants (Adesida et al. 2022). The effluent contains persistent organic pollutants, including polychlorinated dibenzo-*p*-dioxins, polychlorinated dibenzofurans, and polycyclic aromatic hydrocarbons (PAHs), as well as inorganic contaminants such as As and Pb.

In this study, an electrochemical filtration system based on a continuous-flow electrochemical reactor was employed to simultaneously remove toxic organic compounds and heavy metal ions from the contaminated wastewater. Table 2.3 presents

TABLE 2.3
Heavy Metal Ion Concentrations before and after Electrochemical Filtration Treatment of Wastewater Effluent from Boat Harbour Lagoon

Heavy metal ion	Concentration before the treatment (mg L^{-1})	Concentration after the treatment (mg L^{-1})	WHO drinking water standard (mg L^{-1})
Cr	8.26	0.097	0.05
Ni	8.40	0.85	0.07
As	21.38	0.036	0.01
Cd	7.44	0.0073	0.003
Pb	8.11	0.49	0.01

the concentrations of heavy metal ions before and after treatment using the electrochemical systems. The results show that the concentrations of Cr, As, and Cd in the treated effluent approach the WHO's maximum permissible limits. However, for Ni and Pb, post-treatment concentrations remain above recommended thresholds, indicating that additional or complementary treatment technologies are required to reduce their levels further and achieve full compliance with WHO guidelines.

2.4 CHALLENGES AND PROSPECTS

2.4.1 Drawbacks and Limitations

Rai (2012) compared biosystem-based and chemical treatment approaches for the removal of heavy metals from contaminated water. Biosystems processes, including those mediated by phytochemicals and proteins, use naturally occurring compounds to bind, sequester, or transform metal ions. Phytochemicals such as phytochelatins, metallothionein, and various organic acids can significantly influence metal mobility by altering their chemical forms and physicochemical behavior in the environment. In contrast, conventional chemical processes often rely on synthetic chelating agents such as EDTA and diethylenetriaminepentaacetic acid. Although these chelators effectively form stable complexes with metal ions, their use can generate secondary pollutants, posing risks to aquatic ecosystems. Moreover, precipitation-based methods, one of the most common chemical treatments, tend to produce large volumes of metal-laden sludge. Managing and disposing of this sludge increases operational costs, raises additional environmental concerns, and thus limits the overall sustainability of chemical remediation strategies.

2.4.2 Prospects

Despite these limitations, advancements in material design, biotechnology, and integrated treatment technologies offer promising pathways for more efficient and sustainable heavy metal removal. Emerging biosorption systems that harness engineered microorganisms, plant-based materials, and bio-derived ligands offer selective and eco-friendly alternatives with reduced secondary pollution. Similarly, developments in nanomaterials, including functionalized AC, zeolites, MOFs, and high-entropy oxides, enable higher adsorption capacities, faster kinetics, and tunable selectivity. Hybrid technologies that couple electrochemical, adsorption, and membrane-based processes are especially promising, as they can target multiple classes of contaminants simultaneously, while reducing energy consumption.

Further prospects also include process intensification, sensor-integrated monitoring systems, and circular-economy strategies that recover and valorize captured metals, transforming waste streams into valuable economic resources.

2.5 CONCLUSION

Heavy metal removal remains a critical challenge due to the persistent, toxic, and non-biodegradable nature of metal ions in aquatic environments. While biological and chemical methods each offer distinct advantages, their limitations (secondary pollution and high disposal costs) highlight the need for more sustainable alternatives. Ongoing innovation in functional materials, biotechnology, and integrated treatment technologies provides a promising outlook for achieving efficient, cost-effective, and environmentally responsible remediation solutions.

REFERENCES

Adesida, Adekolapo A., Andrew Carrier, Michelle Adams, Tony R. Walker, Ken Oakes, Collins Nganou, Muhammad Fahad Ehsan, and Xu Zhang. 2022. "Simultaneous Degradation of Persistent Organic Pollutants and Heavy Metal Removal via an Electrochemical Filtration System: A Case Study on a Pulp Mill Wastewater Effluent." *Case Studies in Chemical and Environmental Engineering* 6: 100258. https://doi.org/10.1016/j.cscee.2022.100258.

"APEC Water." 2025. https://www.apecwater.com/blogs/water-health/quality-water-filtration-method?srsltid=AfmBOoq4IKU3L7O6_kckO4Vxnvs7TBNCIfs9EFALdo-jL93qOF80uiSP.

Atkins, P., J. De Paula, and J. Keeler. 2018. *Physical Chemistry*. 11th ed. Oxford, UK: Oxford University Press.

Bolisetty, Sreenath, Mohammad Peydayesh, and Raffaele Mezzenga. 2019. "Sustainable Technologies for Water Purification from Heavy Metals: Review and Analysis." *Chemical Society Reviews* 48 (2): 463–487. https://doi.org/10.1039/c8cs00493e.

Cerrillo-Gonzalez, Maria Mar, Juan Manuel Paz-Garcia, Maria Villen-Guzman, and Jose Miguel Rodriguez-Maroto. 2024. "Metal Recovery from Wastewater Using Electrodialysis Separation." *Metals* 14: 38. https://doi.org/ 10.3390/met14010038

Cui, Yue, Qingchun Ge, Xiang Yang Liu, and Tai Shung Chung. 2014. "Novel Forward Osmosis Process to Effectively Remove Heavy Metal Ions." *Journal of Membrane Science* 467: 188–194. https://doi.org/10.1016/j.memsci.2014.05.034.

Edzwald, James K. 2010. "Dissolved Air Flotation and Me." *Water Research* 44: 2077–2106. https://doi.org/10.1016/j.watres.2009.12.040.

"FloWater." 2025. https://drinkflowater.com/the-future-of-water-filtration-emerging-technologies-to-watch/.

Gahrouei, Amirreza Erfani, Armita Rezapour, Majid Pirooz, and Sina Pourebrahimi. 2024. "From Classic to Cutting-Edge Solutions: A Comprehensive Review of Materials and Methods for Heavy Metal Removal from Water Environments." *Desalination and Water Treatment* 319 (April): 100446. https://doi.org/10.1016/j.dwt.2024.100446.

ISTEE. 2025. "A New Solution for Electroplating Wastewater Treatment: High-Performance Forward Osmosis Membrane Developed from Hydrophilic Materials." https://istee.vn/en/a-new-solution-for-electroplating-wastewater-treatment-high-performance-forward-osmosis-membrane-developed-from-hydrophilic-materials/.

Liao, Yuan, Chun Heng Loh, Miao Tian, Rong Wang, and Anthony G. Fane. 2018. "Progress in Electrospun Polymeric Nanofibrous Membranes for Water Treatment: Fabrication, Modification and Applications." *Progress in Polymer Science* 77: 69–94. https://doi.org/10.1016/j.progpolymsci.2017.10.003.

Liosis, Christos, Athina Papadopoulou, Evangelos Karvelas, Theodoros E Karakasidis, and Ioannis E Sarris. 2021. "Heavy Metal Adsorption Using Magnetic Nanoparticles for Water Purification: A Critical Review." *Materials* 14: 7500. https://doi.org/10.3390/ma14247500.

Mancinelli, Maura, Antonella Arfè, Annalisa Martucci, Luisa Pasti, Tatiana Chenet, Elena Sarti, Giulia Vergine, and Claudia Belviso. 2020. "Evaluation for the Removal Efficiency of Vocs and Heavy Metals by Zeolites-Based Materials in the Wastewater: A Case Study in the Tito Scalo Industrial Area." *Processes* 8: 1519. https://doi.org/10.3390/pr8111519.

Mega. 2025. "Electrodialysis." https://www.mega.cz/electrodialysis-in-water-treatment/.

Oladimeji, T. E., M. Oyedemi, M. E. Emetere, O. Agboola, and J. B. Adeoye. 2024. "Review on the Impact of Heavy Metals from Industrial Wastewater Effluent and Removal Technologies." *Heliyon* 10 (23): e40370. https://doi.org/10.1016/j.heliyon.2024.e40370.

Omer, Ahmed M., Rana Dey, Abdelazeem S. Eltaweil, Eman M. Abd El-monaem, and Zyta M. Ziora. 2022. "Insights into Recent Advances of Chitosan-Based Adsorbents for Sustainable Removal of Heavy Metals and Anions." *Arabian Journal of Chemistry* 15: 103543. https://doi.org/10.1016/j.arabjc.2021.103543.

Peng, Yaguang, Hongliang Huang, Yuxi Zhang, Chufan Kang, Shuangming Chen, Li Song, Dahuan Liu, and Chongli Zhong. 2018. "A Versatile MOF-Based Trap for Heavy Metal Ion Capture and Dispersion." *Nature Communications* 9: 187. https://doi.org/10.1038/s41467-017-02600-2.

"Puricon." 2025. https://www.puricom.com/en/new/top-5-water-filtration-technologies-for-cleaner-safer-drinking-water.

Rai, Prabhat Kumar. 2012. "An Eco-Sustainable Green Approach for Heavy Metals Management: Two Case Studies of Developing Industrial Region." *Environmental Monitoring and Assessment* 184: 421–448. https://doi.org/10.1007/s10661-011-1978-x.

Raouf, M. S., Abdel, and A. R. M. Abdul Raheim 2016. "Removal of Heavy Metals from Industrial Waste Water by Biomass-Based Materials: A Review." *Journal of Pollution Effects & Control* 05: 1–13. https://doi.org/10.4172/2375-4397.1000180.

Sabzehmeidani, Mohammad Mehdi, Sahar Mahnaee, Mehrorang Ghaedi, Hadi Heidari, and Vellaisamy A L Roy. 2021. "Carbon Based Materials: A Review of Adsorbents for Inorganic and Organic Compounds." *Advanced Materials* 2: 598–627. https://doi.org/10.1039/d0ma00087f.

Shofia, Saghya Infant, A. S. Vickram, A. Saravanan, V. C. Deivayanai, and P. R. Yaashikaa. 2025. "Sustainable Separation Technologies for Heavy Metal Removal from Wastewater: An Upgraded Review of Physicochemical Methods and Its Advancements." *Sustainable Chemistry for the Environment* 10: 100264. https://doi.org/10.1016/j.scenv.2025.100264.

Shrestha, Rakesh, Sagar Ban, Sijan Devkota, Sudip Sharma, Rajendra Joshi, Arjun Prasad Tiwari, Hak Yong Kim, and Mahesh Kumar Joshi. 2021. "Technological Trends in Heavy Metals Removal from Industrial Wastewater: A Review." *Journal of Environmental Chemical Engineering* 9: 105688. https://doi.org/10.1016/j.jece.2021.105688.

Sun, Daniel T., Li Peng, Washington S. Reeder, Seyed Mohamad Moosavi, Davide Tiana, David K. Britt, Emad Oveisi, and Wendy L. Queen. 2018. "Rapid, Selective Heavy Metal Removal from Water by a Metal-Organic Framework/Polydopamine Composite." *ACS Central Science* 4 (3): 349–356. https://doi.org/10.1021/acscentsci.7b00605.

Tay, Pei Kun R., Peter Q. Nguyen, and Neel S. Joshi. 2017. "A Synthetic Circuit for Mercury Bioremediation Using Self-Assembling Functional Amyloids." *ACS Synthetic Biology* 6: 1841–1850. https://doi.org/10.1021/acssynbio.7b00137.

Tian, Miao, Yi Ning Wang, Rong Wang, and Anthony G. Fane. 2017. "Synthesis and Characterization of Thin Film Nanocomposite Forward Osmosis Membranes Supported by Silica Nanoparticle Incorporated Nanofibrous Substrate." *Desalination* 401: 142–150. https://doi.org/10.1016/j.desal.2016.04.003.

WaterGroupPure. 2025. "Electro Dialysis." https://purewatergroup.com/electrodialysis/.

Wu, Hongyang, Wenjuan Wang, Yanfang Huang, Guihong Han, Shuzhen Yang, Shengpeng Su, Hafiza Sana, Weijun Peng, Yijun Cao, and Jiongtian Liu. 2019. "Comprehensive Evaluation on a Prospective Precipitation-Flotation Process for Metal-Ions Removal from Wastewater Simulants." *Journal of Hazardous Materials* 371: 592–602. https://doi.org/10.1016/j.jhazmat.2019.03.048.

3 Membrane-Based Separation of Heavy Metal Ions

Jaya Sikder, Susmita Pandit,
Joyeeta Bose, and Sudip Chakraborty

3.1 INTRODUCTION

The rising contamination of aquatic environments by toxic metals, driven by rapid industrial growth, mineral extraction, and urban expansion, has significant implications for human health, the environment, sustainability, and survival. Elements like cadmium (Cd), copper (Cu), nickel (Ni), lead (Pb), zinc (Zn), mercury (Hg), arsenic (As), and chromium (Cr), along with other potentially toxic elements such as cobalt (Co), boron (B), titanium (Ti), strontium (Sr), tin (Sn), vanadium (V), and molybdenum (Mo) etc., are non-biodegradable and accumulate in living tissues. These elements have different toxic effects, and it is crucial to remove them from water (Saravanan et al., 2024). Cadmium (Cd), arsenic (As), chromium (Cr), mercury (Hg), and lead (Pb) are among the most dangerous heavy metals. Even at trace amounts, they can cause neurotoxic effects, kidney diseases, and cancer. On the other hand, metals like Cu and Zn are micronutrients that cause harmful effects when present in high amounts (Oladimeji et al., 2024). These metals are discharged into water bodies through various effluents such as electroplating, battery manufacturing units, textile dyeing effluents, mining, and other effluents from chemical manufacturing. Given the shortcomings of conventional treatments, which include the risk of secondary pollution from chemical precipitation and the high energy demands of adsorption, there is a pressing demand for better solutions (Akpor & Muchie, 2010). Membrane-based processes have therefore gained significant interest for heavy metal ion removal. This technology is recognized for its high efficiency, scalability, and environmentally benign operation, positioning it as a high-performing and sustainable alternative (Azmi et al., 2025).

The author provides an in-depth understanding of the membrane-based separation process in this chapter. In this regard, the fundamental principle governing the process is described, along with numerous applications and the importance of the process in achieving the vision of sustainable water and supporting the circular economy. The scope of membrane technology is vast and encompasses several membrane-based technologies, including ultrafiltration (UF), forward osmosis (FO), nanofiltration (NF), membrane distillation (MD), reverse osmosis (RO), and electrodialysis (ED),

DOI: 10.1201/9781003599098-3

which are used for different water problems. Through semipermeable barriers, membrane selectivity governs the separation in these processes, which allows water molecules to pass but prevents heavy metal ions. Membranes may reject heavy metal ions through size-based exclusion, electrostatic forces, or specific chemical interactions. Among all these processes, ED stands out as the most effective in recovering valuable metals from waste streams. There is growing interest in extracting valuable metals from end-of-life lithium-ion batteries. A three-stage ED system with different anion-exchange membranes (cationic supports doped with sulfonic acid groups) and EDTA in the concentrate chamber obtained 99.3% Ni^{2+}, 87.3% Co^{2+}, and 99% Li^{+} separation efficiencies. Interestingly, the same ED technology, in addition to producing pure water, can also focus on recovering valuable metals as sustainable processing demands. MD utilizes thermal gradients to produce high-purity permeate, achieving over 99% removal of As^{3+} and As^{5+} and greater than 96% for Ca^{2+}, Mg^{2+}, Fe^{2+}, and Fe^{3+} through its microporous hydrophobic membrane (Chan et al., 2022). FO using Na-ZSM-5 zeolite membranes achieves 99% rejection of heavy metals such as Cr(VI), Se(VI), Se(IV), As(V), and As(III), which can be removed via molecular sieving and electrostatic repulsion, with excellent antifouling properties against organic and inorganic (Sakai et al., 2025). While the chapter focuses on highly toxic metals like Pb, Cd, Hg, As, Cr, Cu, Zn, due to their prevalence in industrial effluents and regulatory priority, produced water can be treated using hydrophilic NF and RO membranes, achieving compliance with about 96% of drinking water quality criteria. However, they are less effective for boron, molybdenum, and ammonia, with NF as a pre-treatment to enhance RO performance and reduce fouling (Alzahrani et al., 2013).

Developing new membrane materials has greatly improved the operational performance of these processes. Notably, graphene oxide (GO) modified with onion extract (GO/OE) and quercetin (GO/Q) forms lamellar membranes with high rejection efficiencies for heavy metal ions (e.g., ~95.2% for Pb^{2+}, ~88.0% for Cd^{2+} with GO/OE) and >70% for small ions like NaCl, alongside excellent water permeance (~460 L m^{-2} h^{-1} bar^{-1} for GO/OE), enabling effective treatment of industrial wastewater and desalination of river water for potable use (Janwery et al., 2023). The elimination of heavy metal ions from wastewater, using technologies like graphene oxide-modified membranes, advances multiple United Nations Sustainable Development Goals, including SDG 6 (Clean Water and Sanitation) by improving water access, SDG 3 (Good Health and Well-being) by enhancing public health, SDG 12 (Responsible Consumption and Production) by minimizing environmental harm, and further supports SDGs 1, 2, 7, 8, 9, 11, 13, and 14 through sustainable resource management and pollution reduction (Obaideen et al., 2022). Ensuring access to clean water, these technologies mitigate health risks associated with heavy metal toxicity, protecting communities from chronic diseases. Additionally, these systems help advance circular economy objectives through heavy metal recovery and reuse, reducing resource depletion and environmental pollution (Qasem et al., 2021). For instance, ED can concentrate heavy metal ions for subsequent recovery in mining and electronics industries to promote closed-loop systems. Additionally, producing eco-friendly, biodegradable, or recyclable membrane materials such as cellulose polymers can improve the sustainability of membrane processes by lowering the ecological footprint of water treatment processes (Naeem et al., 2024).

This chapter provides a comprehensive overview of membrane technologies used to remove heavy metal ions from wastewater, emphasizing sustainability and application. First, this section outlines the fundamental principles, benefits, and limitations of key membrane processes used for separating toxic metal ions, including NF, FO, ED, MD, UF, and RO (Zulu et al., 2025) (Cerrillo-Gonzalez et al., 2024). The role of membrane technologies in advancing the Sustainable Development Goals (SDGs) and circular economy through water reuse, resource recovery, and environmentally friendly treatment is then discussed. Several global case studies of membrane-based technologies in mining, semiconductors, food and beverage, and water treatment are also presented concerning their real applications and scalability. Finally, the challenges associated with the technologies, such as membrane fouling, energy costs, and policy constraints, are discussed, and future research directions and policy interventions toward more efficient, affordable, and broader application of these technologies are proposed.

3.2 MEMBRANE-BASED TECHNOLOGIES FOR THE REMOVAL OF HEAVY METAL IONS

Because of their potential for highly efficient, selective, and scalable separation processes, membrane-based technologies have received considerable attention for the treatment of water polluted with heavy metals. Semi-permeable membranes selectively retain the heavy metal ions through size exclusion, charge repulsion, and concentration gradient-driven transport processes. A comparative schematic showing the structure and mechanism of ion rejection for six typical technologies, such as NF, FO, MD, ED, UF, and RO, which are widely applied, is shown in Figure 3.1, and their typical applications in the removal of heavy metals. In this chapter, an attempt has been made to explain these technologies in detail, including their process mechanisms, advantages, and limitations. Emphasis will be on removing toxic heavy metals, including Cr^{6+}, Pb^{2+}, Cd^{2+}, and Hg^{2+}, from wastewater streams.

3.2.1 Reverse Osmosis

3.2.1.1 Mechanisms of Metal Ion Removal

RO employs high-pressure-driven membranes with pore sizes classically less than 0.10 nm, enabling the removal of dissolved heavy metals, ions, salts, and other impurities from water (Qasem et al., 2021). The primary mechanisms of heavy metal ion removal in RO include size exclusion and the solution-diffusion mechanism. Size exclusion prevents larger hydrated metal ions (e.g., Pb^{2+}, Cd^{2+}, Hg^{2+}, and Cr^{+2}) from passing through the dense polyamide-based membrane matrix. While the negatively charged surface of RO membranes may influence ion transport, positively charged metal ions are not repelled. However, they are primarily retained due to their low diffusivity and inability to dissolve within the membrane phase. Selective water transport across the membrane occurs through the solution-diffusion mechanism, leaving heavy metal ions in the concentrated brine (Foorginezhad et al., 2025). These fundamental mechanisms are schematically illustrated in Figure 3.2, the pathways for

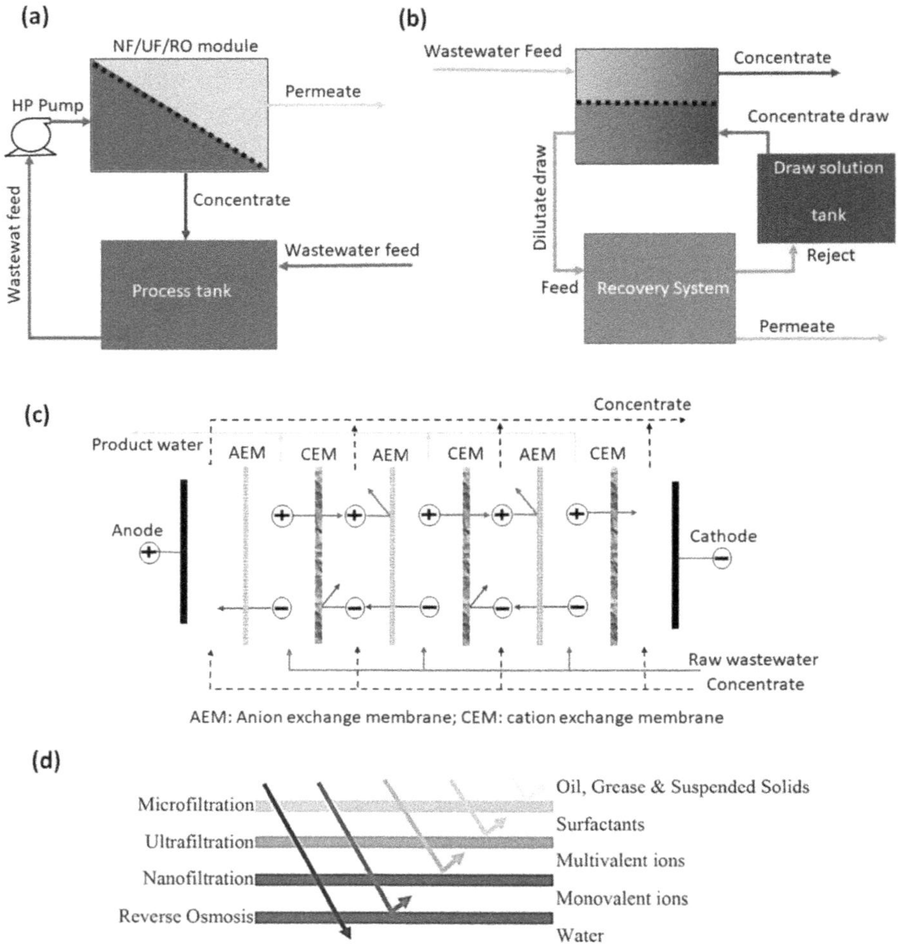

FIGURE 3.1 Membrane structures, ion rejection mechanisms, and typical heavy metal removal applications are depicted in this comparative schematic of membrane-based technologies (RO, NF, UF, ED, MD, and FO). Qasem et al. (2021).

contaminant removal in membrane filtration: (a) size exclusion, (b) hydrophobic interactions, (c) electrostatic forces, and (d) adsorption. Additionally, Figure 3.3 provides a detailed representation of size exclusion, charge repulsion, and solution-diffusion mechanisms in the RO process, emphasizing the relationship between membrane structure and ion rejection efficiency for heavy metal removal.

3.2.1.2 Advantages and Disadvantages

RO is one of the most popular water purification systems, and its main advantage is that it is quite effective in removing heavy metal ions. One of the advantages is its remarkable capacity for rejection, which helps in the almost complete elimination of these dangerous metals. The result of this process is a high-purity permeate, which

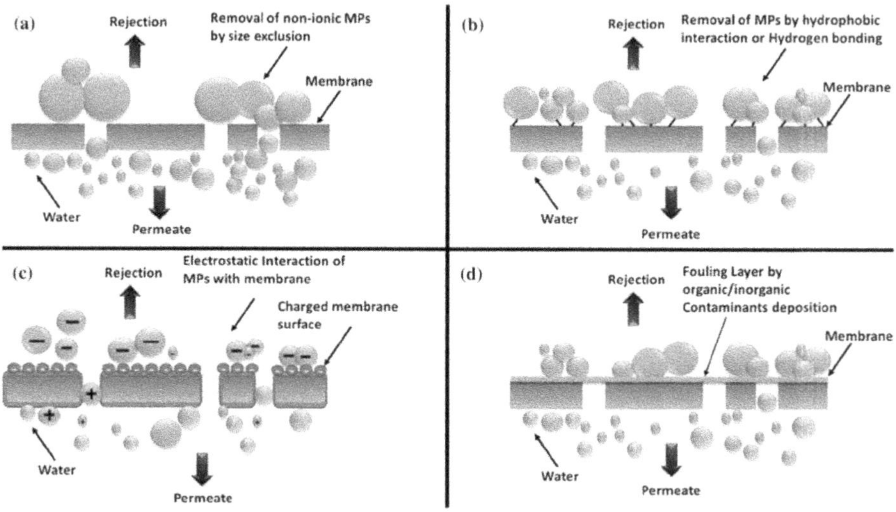

FIGURE 3.2 Pathways of contaminant removal in membrane filtration: (a) size exclusion, (b) hydrophobicity, (c) electrostatic interaction, and (d) adsorption. Della Rocca et al. (2021).

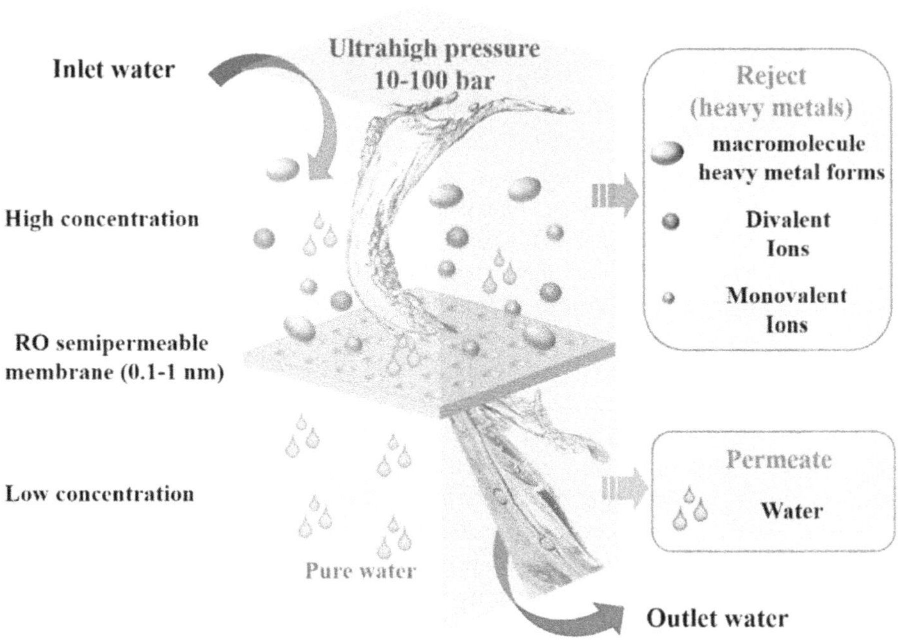

FIGURE 3.3 Diagram demonstrating size exclusion, charge repulsion, and solution-diffusion mechanisms with an emphasis on membrane structure and ion rejection in the RO process for heavy metal ion removal.

Xiang et al. (2022a).

satisfies the most rigorous requirements of drinkable water and recycling to industry. Beyond heavy metals, the technology is equally proficient at extracting dissolved salts and organic pollutants, demonstrating its broad utility across different water treatment scenarios. As a well-established process, RO continues to evolve through advancements in membrane science. For instance, modern thin-film nanocomposite (TFN) membranes now integrate specialized interlayers, such as those made with TNT-Fe_3O_4. These innovations significantly boost water flow rates, achieving permeance levels between 1.63 and 2.82 liters per square meter per hour per bar, all while consistently rejecting over 98% of contaminants like cadmium (Cd^{2+}), lead (Pb^{2+}), and assorted trace organic compounds (Soo et al., 2020).

Additionally, prolonged high-pressure application leads to membrane fatigue or mechanical deformation. Several solutions to the energy problem have been attempted, one of which involves the addition of energy recovery devices (ERDs) like ERD turbo pumps of Energy Recovery Inc. (ERI) to lower the energy input by recovering pressure energy from the brine stream (Wang et al., 2022; Liang et al., 2020). Although these solutions have been achieved, the problem of discharging concentrated brine that may contain toxic heavy metals still exists. Failing to dispose of brine properly can cause secondary pollution. A comparative summary of RO membrane performance for heavy metal ion removal, including rejection rates for common metals, operating pressures, permeate flux, and key operational limitations, is presented in Table 3.1.

3.2.2 Nanofiltration

3.2.2.1 Mechanisms of Metal Ion Removal

NF operates at 5–20 bars and 0.5–2 nm pores membranes are used for removal of selectively multivalent heavy metal ions. Smaller monovalent ions like Na^+ & Cl^- are passed through it. Size exclusion, Donnan exclusion, and dielectric exclusion are the primary mechanisms responsible for heavy metal removal in NF membrane.

Size exclusion occurs when larger hydrated metal ions are physically retained due to their molecular dimensions. Donnan exclusion operates through electrostatic repulsion between a negatively charged NF membrane and cationic heavy metals; this effect has been shown to enhance rejection for ions like Pb^{2+}, Cu^{2+}, and Ni^{2+} (Ayub & Othman, 2022). Dielectric exclusion further reduces the transport of highly charged ions by creating an energy barrier associated with their entry into low-dielectric nanopores. These mechanisms are illustrated in Figure 3.4, which schematically presents: (a) the overall NF process, (b) trace metal ion elimination via pore size exclusion, (c) elimination via Donnan exclusion, and (d) elimination via dielectric exclusion. Figure 3.4 depicts the combined effects of pore size, surface charge, and ion valence on NF membrane selectivity for heavy metal ion removal (Roy et al., 2017). Nanomaterials such as graphene oxide (GO) have recently improved the performance of NF membranes by enhancing surface charge, selectivity, and antifouling resistance. For instance, NF membranes modified with ethylenediamine for cross-linking, followed by post-treatment with hyperbranched polyethyleneimine, exhibited high water permeability (5.01 L/m²/h/bar) and elimination efficiencies greater than 98% for cations Pb^{2+}, Cd^{2+}, Ni^{2+}, and Zn^{2+} (Zhang et al., 2015).

TABLE 3.1
Comparative Performance of Reverse Osmosis for Heavy Metal Ion Removal, Including Rejection Rates for Common Metals, Performances, and Operational Challenges

Heavy metal ion	RO membrane type	Rejection rate (%)	Operating pressure	Permeate flux/ permeability	Operational limitations	Reference
Lead (Pb^{2+})	ESPA1–4040 (TFC)	98.55	11 bar (1.1 MPa)	-	Potential scaling and fouling at higher concentrations	Omran & Ebrahim (2025)
Cadmium (Cd^{2+})	ESPA1–4040 (TFC)	97.97	11 bar (1.1 MPa)	-	Sensitive to pH; slightly reduced at low pH	Omran & Ebrahim (2025)
Cobalt (Co^{2+})	ESPA1–4040 (TFC)	97.31	11 bar (1.1 MPa)	-	Competes with other cations; efficiency is affected by ionic strength	Omran & Ebrahim (2025)
Chromium (Cr^{3+})	ESPA1–4040 (TFC)	97.11	11 bar (1.1 MPa)	-	Low-pH conditions reduce efficiency	Omran & Ebrahim (2025)
Arsenic (As^{3+})	TFC-PA	20–55% (lower)	4–6 bar (0.4–0.6 MPa)	-	Neutral species (As^{3+}); poorly rejected due to a lack of charge	Kapepula & Luis (2024)
Arsenic (As^{5+})	TFC-PA	91–99	4–6 bar (0.4–0.6 MPa)	-	Effective under neutral to slightly alkaline pH	Kapepula & Luis (2024)
Nickel (Ni^{2+})	TFC	99.5	5.03 bar (0.5 MPa)	Moderate (exact value not stated)	May need chelating agents like EDTA to improve efficiency	Kapepula & Luis (2024)
Zinc (Zn^{2+})	TFM	98.9	Not specified	Moderate	Enhanced with EDTA; not pH-sensitive	Kapepula & Luis (2024)
Copper (Cu^{2+})	TFC/Polyamide	99.5	5.03 bar (0.5 MPa)	-	Influenced by ionic size and EDTA presence	Kapepula & Luis (2024)
Antimony (Sb^{3+})	RO X-20TM	98.8	50 bar (5 MPa)	98–132 L/m²·h	Moderate performance under real water; stable under broad pH	Lumami Kapepula et al., (2022)

(Continued)

TABLE 3.1 (*Continued*)
Comparative Performance of Reverse Osmosis for Heavy Metal Ion Removal, Including Rejection Rates for Common Metals, Performances, and Operational Challenges

Heavy metal ion	RO membrane type	Rejection rate (%)	Operating pressure	Permeate flux/ permeability	Operational limitations	Reference
Chromium (Cr^{3+})	RO X-20TM (TFC)	99	50 bar (5 MPa)	98–132 L/m²·h	High removal in synthetic water; sensitive to fouling	Lumami Kapepula et al., (2022)
Cadmium (Cd^{2+})	RO X-20TM (TFC)	99	50 bar (5 MPa)	98–132 L/m²·h	Sensitive to pre-treatment and competing ions	Lumami Kapepula et al., (2022)
Nickel (Ni^{2+})	RO X-20TM (TFC)	99	50 bar (5 MPa)	98–132 L/m²·h	Consistently high rejection; potential fouling	Lumami Kapepula et al., (2022)
Lead (Pb^{2+})	RO X-20TM (TFC)	99	50 bar (5 MPa)	98–132 L/m²·h	Effective at high concentrations	Lumami Kapepula et al., (2022)
Calcium (Ca^{2+})	Polyamide	97	6–22 bar (0.6–2.2 MPa)	29.3 L/m²·h	Hardness contributes to scaling	Kapepula & Luis (2024)
Magnesium (Mg^{2+})	Polyamide	Not specified	6–22 bar (0.6–2.2 MPa)	29.3 L/m²·h	Often fully rejected; needs pretreatment	Kapepula & Luis (2024)

Note: EDTA, ethylenediaminetetraacetic acid; RO, reverse osmosis.

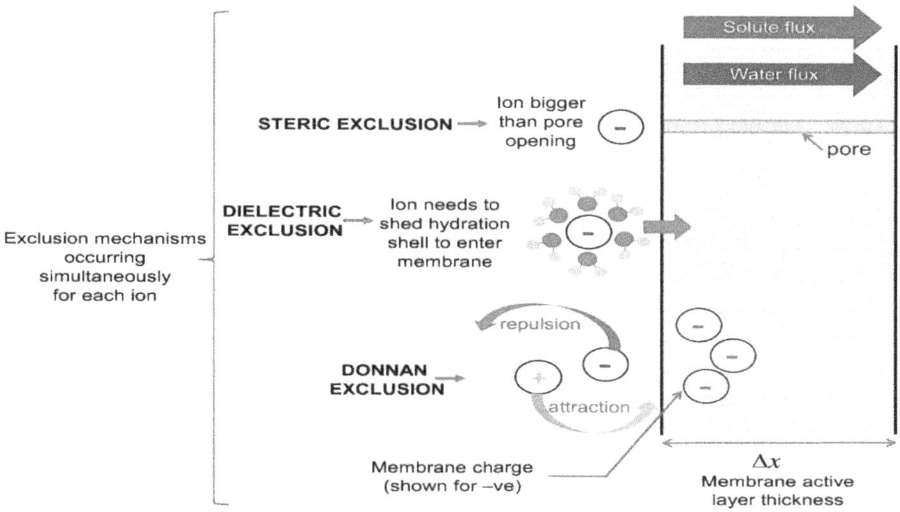

FIGURE 3.4 Schematic of the NF membrane process including size-based filtration, electrostatic (Donnan) exclusion, and dielectric interactions in heavy metal ion removal, highlighting the relationship between pore size and membrane charge. Roy et al. (2017).

3.2.2.2 Advantages and Disadvantages

NF offers a balanced combination of performance and operational efficiency, making it a promising option for industrial and municipal wastewater treatment involving heavy metals. This moderate-pressure operation enables easier integration into existing infrastructure and lowers equipment wear. Furthermore, NF exhibits selective permeability, allowing it to retain divalent and trivalent metal ions like Pb^{2+}, Cu^{2+}, and Ni^{2+} while permitting beneficial monovalent salts to pass, thus preserving essential minerals in treated water, an advantage in agricultural and domestic reuse scenarios (Soo et al., 2020).

Regarding practical maintenance, NF membranes are generally less prone to severe fouling than RO membranes due to their slightly larger pore sizes and lower compaction requirements. This can lead to longer operational lifespans and reduced cleaning frequency, especially when appropriate pretreatment steps are followed (Xu et al., 2006). NF is also considered more cost-effective for partial decontamination, making it suitable for pretreatment stages before more intensive purification steps.

However, NF systems do face some technical and economic limitations. Their lower rejection rates for monovalent ions, particularly for neutral or weakly charged species such as As^{3+}, limit their standalone applicability for removing all types of heavy metals. In addition, membrane costs can be high, especially for advanced NF configurations incorporating nanomaterials such as graphene oxide, which may challenge their use in decentralized or small-scale operations (Roy et al., 2017). NF performance is also susceptible to feedwater pH, requiring pH control and potential chemical dosing to maintain rejection efficiency and membrane integrity Moreover, while fouling is generally lower than in RO, biofouling and scaling risks remain, particularly in long-term industrial applications without rigorous pretreatment (Tiwary et al., 2024).

3.2.3 Ultrafiltration

3.2.3.1 Mechanisms of Metal Ion Removal

UF employs membranes with pore sizes ranging from 2 to 100 nm, which can retain macromolecules, colloids, and suspended solids but typically allow smaller ions, including heavy metals, to pass through. UF is frequently used with methods such as polymer-enhanced ultrafiltration (PEUF) and micellar-enhanced ultrafiltration (MEUF), or complexation-ultrafiltration, to improve the removal of heavy metals. MEUF forms micelles that selectively bind divalent metal ions, for example, Mn^{2+}, Ni^{2+}, Cd^{2+}, Zn^{2+} (Yaqub & Lee, 2019). This enhancement enables rejection efficiencies exceeding 99%, with metal removal performance primarily influenced by the surfactant-to-metal ratio (S:M) and pH. In MEUF, cationic heavy metal ions such as Cu^{2+}, Ni^{2+}, and Co^{2+} interact with micelles formed by anionic surfactants, such as sodium dodecyl sulfate (SDS), forming metal micelle complexes large enough to be retained by UF membranes. The dominant mechanisms involve size exclusion of these complexes and, in some cases, electrostatic interactions. UF systems employing MEUF can achieve metal rejections of 91–99%, with performance depending on factors such as pH, S/M ratio, and membrane characteristics. These mechanisms are visually represented in Figure 3.5 of the

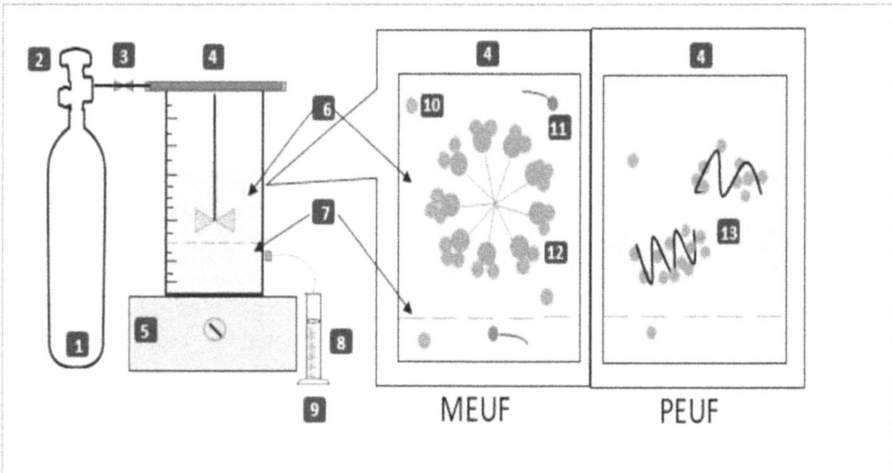

FIGURE 3.5 Schematic of the UF process incorporating MEUF/PEUF, depicting micelle- and polymer–metal complex formation and their retention by the UF membrane. The diagram illustrates size exclusion and complexation mechanisms. Components: 1—nitrogen gas, 2—regulator, 3—pressure control valve, 4—ultrafiltration cell, 5—magnetic stirrer, 6—feed solution, 7—ultrafiltration membrane, 8—permeate, 9—volumetric cylinder, 10—sulfate ions, 11—surfactant monomer, 12—surfactant micelle bound to sulfate ions, 13—polymer bound to sulfate ions. Lin et al. (2017).

study, which illustrates the MEUF process, including micelle–metal complex formation and their retention by the UF membrane over size exclusion and micelle interactions (Lin et al., 2017).

3.2.3.2 Advantages and Disadvantages

UF offers several advantages for wastewater treatment, particularly when dealing with high volumes of effluent containing suspended solids, colloids, and macromolecules. Operating at relatively low pressures (typically 1–3 bar), (Barakat & Schmidt, 2010). UF membranes typically have pore sizes of 5–20 nm, allowing for high permeate flux, making the process suitable for large-scale applications where throughput is a key concern. Moreover, when combined with complexation strategies, such as MEUF or PEUF, UF demonstrates remarkable efficiency in eliminating dissolved trace element ions, even at trace concentrations. These hybrid techniques facilitate the formation of larger surfactant-metal or polymer-metal complexes, which are readily retained by the membrane (Sandu et al., 2022).

Despite these strengths, UF also presents limitations. On its own, UF is generally ineffective at removing small, free metal ions due to its pore size, necessitating the use of chemical additives to form larger complexes (Sandu et al., 2022). Though polymers or surfactants will improve removal efficiency, they will increase operational costs and may pose environmental issues if not properly recovered or degraded (Barakat & Schmidt, 2010). The other critical problem is membrane fouling, especially with industrial wastewater that contains high concentrations of organic matter or micellar structures. The fouling decreases the performance and permeability of the membranes with time thus necessitating frequent cleaning, back flushing or pre-treatment measures to maintain the long-term functioning (Shi et al., 2014). Nevertheless, with proper system design and maintenance, UF combined with complexation technologies remains a viable and scalable option for selective heavy metal removal.

3.2.4 Electrodialysis

3.2.4.1 Mechanisms of Metal Ion Removal

Ions can migrate selectively through ion-exchange membranes under an applied electric potential thanks to the electrically powered membrane process known as ED. The system is constructed by alternating cation-exchange membranes (CEMs) and anion-exchange membranes (AEMs) between two electrodes. Ion accumulation in concentrate channels and in purified water in the diluted stream results from negatively charged anions migrating toward the anode through AEMs and from positively charged metal ions, such as Pb^{2+}, Ni^{2+}, and Cu^{2+}, migrating toward the cathode through CEMs (Arana Juve et al., 2022).

The removal efficiency in ED depends on factors such as membrane selectivity, ion valence, electrophoretic mobility, and the applied voltage. ED is particularly effective for divalent and trivalent metal ions, with reported removal efficiencies of up to >85%, >99%, and 95%, for Cu^{2+}, Pb^{2+}, Ni^{2+}, as demonstrated in both laboratory and industrial-scale studies (Arana Juve et al., 2022). Advanced configurations, such as Selective Electrodialysis (SED), incorporate monovalent-selective membranes or

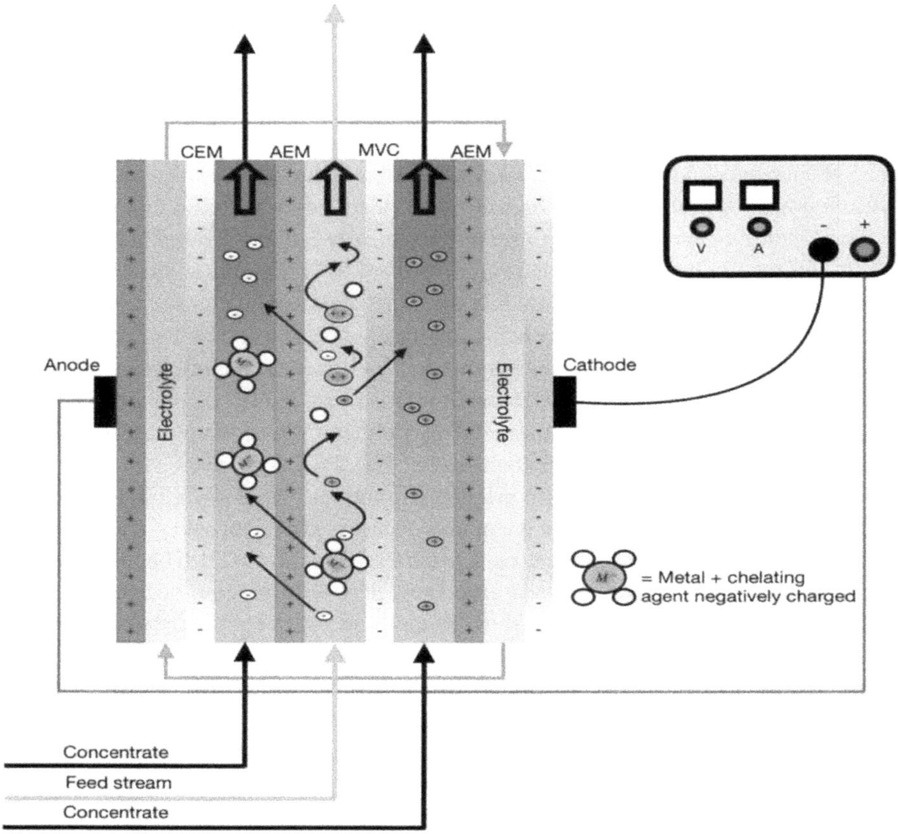

FIGURE 3.6 Diagram of a three-compartment electrodialysis (ED) system with integrated monovalent-selective ED (SED) units for the separation of a divalent metal cation, a monovalent metal cation, and an SED cell assisted by a chelating agent for improved metal recovery selectivity (Arana Juve et al., 2022).

use chelating agents to enhance ion separation performance. This approach allows for the simultaneous recovery of specific monovalent and multivalent ions from complex wastewater streams. Figure 3.6 illustrates such a system, a three-compartment ED unit integrated with monovalent-SED cells and a chelating-assisted SED cell, designed to separate a divalent metal cation, a monovalent metal cation, and to improve recovery selectivity through chelation (Arana Juve et al., 2022).

3.2.4.2 Advantages and Disadvantages

ED presents many advantages for removing heavy metals. It allows for selective separation and concentration of ionic species, particularly divalent and trivalent heavy metals like Cu^{2+}, Zn^{2+}, and Ni^{2+}, making it effective for both purification and metal recovery (Cerrillo-Gonzalez et al., 2024). The process operates with minimal chemical input, reducing secondary waste compared to precipitation-based methods

However, ED also faces limitations. The energy demand can be high, particularly when treating high-salinity or high-conductivity effluents. Membrane fouling and scaling from metal hydroxide precipitation or organics can reduce performance and necessitate pre-treatment and frequent cleaning. Moreover, ED is ineffective for removing non-ionic or weakly ionized metal complexes, limiting its applicability in specific wastewater matrices.

3.2.5 Membrane Distillation

3.2.5.1 Mechanisms of Metal Ion Removal

Using hydrophobic microporous membranes, MD is a temperature-driven membrane technique that excludes non-volatile solutes, including heavy metal ions, and only permits water vapor to flow (F. E. Ahmed et al., 2021). Membranes made of PTFE or PVDF, typically with pores ranging from 0.2 to 1.0 μm, are commonly used. For nearly all salts, rejection rates usually surpass 99%. Several variables affect process efficiency, including feed temperature, membrane characteristics, and MD configuration (e.g., DCMD and AGMD). Recent reviews indicate that MD has shown encouraging outcomes in treating complex wastewater streams from industries such as oil and gas, mining, and textiles (R. V. Patel et al., 2025). As an example, from the oil and gas sector, a pre-pilot hybrid system combining FO and MD successfully treated produced water, resulting in notable reductions in volume and heavy metal concentrations and enabling energy recovery through waste heat (Linares et al., 2016).

3.2.5.2 Advantages and Disadvantages

MD offers notable advantages in heavy metal ion removal. Its high rejection efficiency for non-volatile solutes makes it well-suited for producing ultrapure water from high-salinity or toxic wastewater sources. Unlike pressure-driven processes like RO, NF operates near-atmospheric pressure, reducing mechanical stress and enabling simpler design and integration with low-grade or waste heat sources. Additionally, MD systems are adaptable to hybrid configurations that enhance performance and energy efficiency, as demonstrated in FO–MD applications for industrial water reuse (Li et al., 2020). However, MD faces challenges, particularly high thermal energy consumption and susceptibility to membrane fouling and wetting, which can impair performance and increase maintenance requirements. These limitations hinder scalability, although ongoing advancements in membrane surface engineering and hybrid system designs actively address these barriers.

3.2.6 Forward Osmosis

3.2.6.1 Mechanisms of Metal Ion Removal

In FO, water moves from a feed solution containing heavy metal ions through a semi-permeable membrane. This is driven by the osmotic pressure difference created between the feed and a concentrated draw solution (e.g., NaCl or $MgCl_2$). The primary mechanism of heavy metal ion removal is size exclusion, where the dense FO membrane (like RO membranes) retains metal ions like Cd^{2+}, Pb^{2+}, Cu^{2+}, and

Zn^{2+}. In contrast, enabling the transfer of water molecules into the draw solution (You et al., 2017; Chekli et al., 2012). Charge repulsion also contributes, as negatively charged membranes repel cationic heavy metal ions. FO achieves 94% rejection rates for divalent heavy metals, with efficiency dependent on draw solution properties, membrane fouling, and reverse solute flux (You et al., 2017). Recent developments in thin-film composite (TFC) and biomimetic membranes (e.g., aquaporin-based) have enhanced FO performance by improving water flux and ion selectivity.

3.2.6.2 Advantages and Disadvantages

FO provides many benefits compared to traditional pressure-driven membranes. It requires minimal external pressure, relying primarily on osmotic gradients, which significantly reduces energy costs compared to RO (Mohammadifakhr et al., 2020). The absence of hydraulic pressure also minimizes membrane fouling, thereby enhancing operational stability and extending membrane lifespan (Mohammadifakhr et al., 2020). Additionally, FO demonstrates versatility in treating complex wastewater with high fouling potential, such as industrial effluents, and can be seamlessly incorporated into hybrid systems, such as FO-RO setups.

However, FO also presents certain disadvantages. One major limitation is draw solution recovery; FO generates a diluted draw solution that necessitates an additional separation process, such as RO, to obtain clean water, which increases both cost and process complexity (Johnson et al., 2018). A further issue is reverse solute flux, where draw solutes can transfer into the feed stream, potentially contaminating the water and reducing process efficiency (Johnson et al., 2018). Furthermore, FO generally exhibits lower water flux than RO or NF, necessitating a larger membrane area to achieve equivalent treatment capacity.

3.3 INTEGRATION OF MEMBRANE-BASED PROCESSES WITH SUSTAINABLE DEVELOPMENT GOALS AND CIRCULAR ECONOMY CONCEPTS

Global sustainability goals are significantly advanced by membrane-based heavy metal ion removal technologies, encompassing technologies such as FO, ED, UF, NF, MD, and RO. By guaranteeing access to clean water, lowering environmental pollution, and encouraging resource recovery, these technologies support the circular economy principles and the SDGs of the United Nations. This section explores how these processes contribute to specific SDG targets, deliver circular economy benefits, and enhance environmental and economic sustainability, supported by references from SCI and Scopus-indexed journals.

3.3.1 Contribution to SDG Targets

Membrane-based technologies directly support several SDGs, particularly SDG 6 (Clean Water and Sanitation), SDG 3 (Good Health and Well-being), SDG 12 (Responsible Consumption and Production), and SDG 14 (Life Below Water). For instance, RO and NF produce high-quality permeate for potable use, while UF and

FO treat industrial effluents, reducing pollution in water bodies. Additionally, these processes contribute to SDG 12 by promoting sustainable industrial practices through efficient wastewater treatment and resource recovery, such as concentrating heavy metals for reuse in the metal industry (Hatayama, 2022). By preventing heavy metal discharge into aquatic ecosystems, technologies like MD and ED support SDG 6, protecting marine life from bioaccumulation (Carmona & Abejón, 2023).

3.3.2 Circular Economy Benefits

Membrane-based technologies advance circular economy principles by enabling resource recovery, waste minimization, and closed-loop systems. Unlike conventional methods like chemical precipitation, which generate sludge, technologies such as ED and FO concentrate heavy metal ions into recoverable streams, allowing metals, including Zn, Pb, As, Ni, Cu, Cr, and Cd, to be reused in industries (Azmi et al., 2025). For example, ED can produce concentrated metal solutions for direct industrial reuse, reducing raw material extraction and aligning with circular economy goals (Cerrillo-Gonzalez et al., 2024). RO and NF, when integrated with recovery systems, facilitate water reuse in industrial processes, minimizing freshwater demand. Additionally, advancements in membrane materials, such as biodegradable or recyclable polymers, reduce waste from membrane disposal. With its low fouling propensity, FO extends membrane lifespan, further supporting resource efficiency.

3.3.3 Environmental and Economic Sustainability

Membrane-based technologies enhance environmental sustainability by reducing the ecological footprint of wastewater treatment. Using eco-friendly membrane materials, such as nanoparticles or nanomaterial-enhanced membranes, reduces environmental impacts from production and disposal (Shukla et al., 2024). Additionally, these technologies protect ecosystems by preventing heavy metal discharge, preserving biodiversity, and reducing remediation costs economically. Membrane-based processes offer long-term cost savings through water and metal recovery, particularly in water-scarce regions or high-value industries. For instance, NF and RO reduce operational costs by enabling water reuse in textile industries (M. Ahmed et al., 2022). However, challenges like high initial capital costs and membrane fouling require ongoing research into cost-effective materials and antifouling strategies. Hybrid systems, such as FO-MD, optimize energy and cost efficiency, enhancing economic viability (Suleman et al., 2025). These technologies balance environmental protection with economic benefits, supporting sustainable industrial growth.

3.4 PROSPECTS OF MEMBRANE-DRIVEN METHODS FOR HEAVY METAL ION SEPARATION

Membrane-oriented technologies, such as FO, ED, UF, NF, MD, and RO, are poised to transform heavy metal ion separation across diverse industries. The alignment, scalability, and efficiency with sustainable goals make managing global water contamination challenges acute. The section explains the prospects of these technologies, concentrating on industrial goals, regulatory framework, and industry-specific case studies.

3.4.1 Case Studies

Membrane-based technologies, such as trans membrane hydraulic pressure, including MF, UF, osmotic pressure, RO, FO, and NF, are good in addressing heavy metal contaminations across different industries. These are customized, specific wastewater characteristics, regulatory requirements, and sustainability goals, making eliminating heavy metals like cadmium, copper, and chromium straightforward. The case studies from different countries have been highlighted, representing some industries like metallurgical and mining, where membrane treatment is followed. The two presented case studies show the working of the RO membrane in industrial wastewater with different contaminants in different settings. Slovenia's metal finishing plant practices a UF/RO system to treat electroplating wastewater, which achieved 91.3–99.8% removal of heavy metals like nickel, chromium, and total elimination of chemical oxygen demand and suspended particles. The UF step reduced fouling, while RO produced high-quality permeate for reuse, though frequent chemical cleaning was required (Petrinic et al., 2015). In contrast, an Australian gold-antimony mine used RO membranes to treat mining wastewater, achieving 85% turbidity, 96% TDS, and 95% antimony removal. However, inconsistent pre-treatment initially led to membrane fouling and occasional noncompliance with discharge limits until a clarifier was added to stabilize feed quality. Both cases highlight RO's adaptability but emphasize the importance of proper pre-treatment and maintenance to ensure efficiency and regulatory compliance (Samaei et al., 2020).

Despite the increasing use of membrane-oriented technologies like FO, ED, UF, NF, MD, and RO for removing heavy metal ions, several challenges impede their large-scale application. These include technological limitations, high economic costs, inconsistent regulatory frameworks, and the need for further research and development.

Technologically, membrane fouling, scaling, wetting, and limited selectivity in complex water matrices reduce system efficiency and increase operational demands. RO and NF membranes are particularly susceptible to fouling from organic compounds and metal hydroxides, and FO experiences reverse solute flux that contaminates treated water (Xiang et al., 2022b). Despite its potential for lower energy consumption during water extraction, FO faces significant economic challenges due to the high cost of draw solution regeneration, which remains a technical and financial bottleneck in full-scale deployment. Weak enforcement and limited policy incentives in some regions reduce the urgency for industries to invest in sustainable membrane technologies. Lengthy permitting processes and a lack of incentives for zero-liquid discharge (ZLD) practices further inhibit progress (Tong & Elimelech, 2016). Tackling these issues demands rigorous research and development initiatives focused on antifouling, selective, and durable membranes enhanced with nanomaterials like graphene oxide, metal-organic frameworks (MOFs), or molecularly imprinted polymers (Qu et al., 2021). International harmonization of discharge regulations and dedicated support through tax incentives or subsidies are recommended to accelerate adoption. Programs like Horizon Europe and national funding agencies can be pivotal in scaling up promising technologies through pilot studies and industry-academia collaborations (European Commission, 2024). A coordinated strategy involving material innovation, economic optimization, regulatory alignment, and sustainability assessments will be crucial for unlocking the complete capabilities of membrane-oriented technologies to achieve safe, effective, and sustainable heavy metal elimination.

3.5 KEY CHALLENGES AND PROSPECTIVE APPROACHES

Despite the expanding utilization of membrane-driven approaches like FO, MD, ED, UF, NF, and RO for heavy metal ion removal, several challenges impede their large-scale application. These include technological limitations, high economic costs, inconsistent regulatory frameworks, and the need for further research and development.

Technologically, membrane fouling, scaling, wetting, and limited selectivity in complex water matrices reduce system efficiency and increase operational demands. RO and NF membranes are particularly vulnerable to fouling caused by organic substances and metal hydroxides, while MD suffers from membrane wetting, and FO experiences reverse solute flux that contaminates treated water (You et al., 2017; Son et al., 2025). Moreover, membranes often lack the specificity needed to selectively remove heavy metal ions, particularly in the presence of competing ions, for example Na^+ or Ca^{2+}. Durability is also an issue, as harsh wastewater conditions degrade membrane materials over time.

Economically, the high capital cost of membrane modules, particularly TFC membranes employed in RO systems and ion-exchange membranes for ED, combined with energy-intensive operations and frequent maintenance requirements, limits its widespread adoption (S. K. Patel et al., 2024; Lejarazu-Larrañaga et al., 2022). FO, despite its potential for lower energy consumption during water extraction, faces significant economic challenges due to the high cost of draw solution regeneration, which remains a technical and financial bottleneck in full-scale deployment (Reddy et al., 2024). Additionally, ED becomes less economically viable in high-salinity streams due to increased energy demand and membrane replacement cycles.

From a policy standpoint, disparate global discharge limits, such as the EU's strict thresholds (e.g., Cd 0.001 or 0.005 mg/L) compared to more lenient standards in other regions, complicate technology deployment and compliance (Vos & Janssen, 2008). Weak enforcement and limited policy incentives in some regions reduce the urgency for industries to invest in sustainable membrane technologies. Lengthy permitting processes and a lack of incentives for metal recovery or ZLD practices further inhibit progress.

Overcoming these challenges demands extensive research and development efforts focused on antifouling, selective, and enduring membranes enhanced with nanomaterials like graphene oxide, MOFs, or molecularly imprinted polymers (Tiwary et al., 2024). FO-RO, UF-NF hybrid system, which is economical and energy-saving, can work better if integrated with heat recovery waste. AI and IoT technologies are used for intelligent monitoring. Real-time optimization and predictive maintenance are also monitored through it. Nevertheless, these are under development, but monitoring will be easier once they come into the spotlight. Furthermore, sustainability-focused innovations such as biodegradable membrane materials and green manufacturing processes are needed to minimize environmental impacts. International harmonization of discharge regulations and dedicated support through tax incentives or subsidies are recommended to accelerate adoption. Coordinated strategies encompassing material innovation, economic optimization, regulatory compliance, and sustainability assessments will be vital to fully harness membrane technologies for the harmless, effective, and sustainable removal of heavy metals.

3.6 CONCLUSION

Membrane-based technologies have demonstrated significant potential in addressing heavy metal contamination across diverse industrial sectors. Their high rejection efficiency, compatibility with hybrid systems, and potential for metal recovery position them as vital tools for advancing sustainable water management. A multi-faceted approach is essential to align their application with broader sustainable development goals. Industries should adopt cleaner production practices at the source level by replacing toxic metal-based inputs with environmentally benign alternatives. During processing, in-situ techniques like ED and FO should be leveraged for early-stage concentration and recovery of heavy metals, minimizing their entry into wastewater streams. Advanced treatment systems incorporating RO, NF, or MD, optimized through hybrid configurations and energy recovery, can support ZLD objectives for unavoidable discharges while enabling water and resource reuse.

To maintain the sustainability of these technologies over time, future efforts must focus on developing low-cost, fouling-resistant, and biodegradable membrane materials and integrating AI-driven monitoring systems for real-time process optimization. Harmonizing environmental regulations and offering policy-driven incentives can further support widespread adoption. Collaborative initiatives between academia, industry, and policymakers will drive innovation, fund pilot-scale demonstrations, and scale deployment. Membrane-based systems can be central in achieving a circular, resource-efficient industrial framework by integrating source control, in-process metal recovery, and sustainable treatment technologies. This holistic pathway mitigates heavy metal pollution's ecological and health risks and reinforces the chapter's core message, advancing sustainable environmental stewardship through technological innovation and responsible industrial practice.

ACKNOWLEDGMENTS

The authors thank the Department of Science and Technology (DST), Government of India (File No. DST/TMD/EWO/WTI/DM/2021/83) and the Department of Science & Technology and Biotechnology, Government of West Bengal [Memo No.: 49 (Sanc.)/ST/P/S & T/15G-6/2018] for their support, which contributed to the broader research environment. Fellowship support from the Ministry of Education (formerly MHRD), Government of India, is gratefully acknowledged.

REFERENCES

Ahmed, F. E., Khalil, A. & Hilal, N. (2021). Emerging desalination technologies: Current status, challenges and future trends. *Desalination*, *517*(June), 115183. https://doi.org/10.1016/j.desal.2021.115183

Ahmed, M., Mavukkandy, M. O., Giwa, A., Elektorowicz, M., Katsou, E., Khelifi, O., Naddeo, V. & Hasan, S. W. (2022). Recent developments in hazardous pollutants removal from wastewater and water reuse within a circular economy. *NPJ Clean Water*, *5*(1), 1–25. https://doi.org/10.1038/s41545-022-00154-5

Akpor, O. B. & Muchie, M. (2010). Remediation of heavy metals in drinking water and wastewater treatment systems: Processes and applications. *International Journal of Physical Sciences*, *5*(12), 1807–1817.

Alzahrani, S., Mohammad, A. W., Hilal, N., Abdullah, P. & Jaafar, O. (2013). Comparative study of NF and RO membranes in the treatment of produced water-Part I: Assessing water quality. *Desalination*, *315*, 18–26. https://doi.org/10.1016/j.desal.2012.12.004

Arana Juve, J. M., Christensen, F. M. S., Wang, Y. & Wei, Z. (2022). Electrodialysis for metal removal and recovery: A review. *Chemical Engineering Journal*, *435*(P2), 134857. https://doi.org/10.1016/j.cej.2022.134857

Ayub, M. & Othman, M. H. D. (2022). Graphene oxide-based nanofiltration membranes for separation of heavy metals. In *Emerging Techniques for Treatment of Toxic Metals from Wastewater* (First Edit). Elsevier Ltd. https://doi.org/10.1016/B978-0-12-822880-7.00013-3

Azmi, L. S., Jabit, N. 'Ain, Ismail, S., Ku Ishak, K. E. H. & Abdullah, T. K. (2025). Membrane filtration technologies for sustainable industrial wastewater treatment: A review of heavy metal removal. *Desalination and Water Treatment*, *323*(July). https://doi.org/10.1016/j.dwt.2025.101321

Barakat, M. A. & Schmidt, E. (2010). Polymer-enhanced ultrafiltration process for heavy metals removal from industrial wastewater. *Desalination*, *256*(1–3), 90–93. https://doi.org/10.1016/j.desal.2010.02.008

Carmona, B. & Abejón, R. (2023). Innovative membrane technologies for the treatment of wastewater polluted with heavy metals: Perspective of the potential of electrodialysis, membrane distillation, and forward osmosis from a bibliometric analysis. *Membranes*, *13*(4). https://doi.org/10.3390/membranes13040385

Cerrillo-Gonzalez, M. del M., Villen-Guzman, M., Rodriguez-Maroto, J. M. & Paz-Garcia, J. M. (2024). Metal recovery from wastewater using electrodialysis separation. *Metals*, *14*(1). https://doi.org/10.3390/met14010038

Chan, K. H., Malik, M. & Azimi, G. (2022). Separation of lithium, nickel, manganese, and cobalt from waste lithium-ion batteries using electrodialysis. *Resources, Conservation and Recycling*, *178*(August 2021). https://doi.org/10.1016/j.resconrec.2021.106076

Chekli, L., Phuntsho, S., Shon, H. K., Vigneswaran, S., Kandasamy, J. & Chanan, A. (2012). A review of draw solutes in forward osmosis process and their use in modern applications. *Desalination and Water Treatment*, *43*(1–3), 167–184. https://doi.org/10.1080/19443994.2012.672168

Della Rocca, D. G., Peralta, R. M., Peralta, R. A., Rodríguez-Castellón, E. & de Fatima Peralta Muniz Moreira, R. (2021). Adding value to aluminosilicate solid wastes to produce adsorbents, catalysts and filtration membranes for water and wastewater treatment. *Journal of Materials Science*, *56*(2), 1039–1063. https://doi.org/10.1007/s10853-020-05276-0

European Commission. (2024). *Evaluation Study of the European Framework Programmes for Research and Innovation for Innovative Europe* (Vol. 2). https://doi.org/10.2777/499132

Foorginezhad, S., Zerafat, M. M., Ismail, A. F. & Goh, P. S. (2025). Emerging membrane technologies for sustainable water treatment: A review on recent advances. *Environmental Science: Advances*, *4*(4), 530–570. https://doi.org/10.1039/d4va00378k

Hatayama, H. (2022). The metals industry and the Sustainable Development Goals: The relationship explored based on SDG reporting. *Resources, Conservation and Recycling*, *178*(November 2021), 106081. https://doi.org/10.1016/j.resconrec.2021.106081

Janwery, D., Memon, F. H., Memon, A. A., Iqbal, M., Memon, F. N., Ali, W., Choi, K. H. & Thebo, K. H. (2023). Lamellar graphene oxide-based composite membranes for efficient separation of heavy metal ions and desalination of water. *ACS Omega*, *8*(8), 7648–7656. https://doi.org/10.1021/acsomega.2c07243

Johnson, D. J., Suwaileh, W. A., Mohammed, A. W. & Hilal, N. (2018). Osmotic's potential: An overview of draw solutes for forward osmosis. *Desalination*, *434*(August 2017), 100–120. https://doi.org/10.1016/j.desal.2017.09.017

Kapepula, V. L. & Luis, P. (2024). Removal of heavy metals from wastewater using reverse osmosis. *Frontiers in Chemical Engineering*, *6*(February). https://doi.org/10.3389/fceng.2024.1334816

Lejarazu-Larrañaga, A., Landaburu-Aguirre, J., Senán-Salinas, J., Ortiz, J. M. & Molina, S. (2022). Thin film composite polyamide reverse osmosis membrane technology towards a circular economy. *Membranes*, *12*(9). https://doi.org/10.3390/membranes12090864

Li, M., Li, K., Wang, L. & Zhang, X. (2020). Feasibility of concentrating textile wastewater using a hybrid forward osmosis-membrane distillation (FO-MD) process: Performance and economic evaluation. *Water Research*, *172*, 115488. https://doi.org/10.1016/j.watres.2020.115488

Liang, K., Ma, Q., Lu, H., Fang, H., Yang, P. & Fan, J. (2020). The research and application progress of the isobaric erd technique for swro desalination plant. *Desalination and Water Treatment*, *202*, 14–26. https://doi.org/10.5004/dwt.2020.26100

Lin, W., Jing, L., Zhu, Z., Cai, Q. & Zhang, B. (2017). Removal of heavy metals from mining wastewater by micellar-enhanced ultrafiltration (MEUF): Experimental investigation and Monte Carlo-based artificial neural network modeling. *Water, Air, and Soil Pollution*, *228*(6). https://doi.org/10.1007/s11270-017-3386-5

Linares, R. V., Li, Z., Yangali-Quintanilla, V., Li, Q., Vrouwenvelder, J. S., Amy, G. L. & Ghaffour, N. (2016). Hybrid SBR–FO system for wastewater treatment and reuse: Operation, fouling and cleaning. *Desalination*, *393*, 31–38. https://doi.org/10.1016/j.desal.2016.03.015

Lumami Kapepula, V., García Alvarez, M., Sang Sefidi, V., Buleng Njoyim Tamungang, E., Ndikumana, T., Musibono, D. D., Van Der Bruggen, B. & Luis, P. (2022). Evaluation of commercial reverse osmosis and nanofiltration membranes for the removal of heavy metals from surface water in the Democratic Republic of Congo. *Clean Technologies*, *4*(4), 1300–1316. https://doi.org/10.3390/cleantechnol4040080

Mohammadifakhr, M., Grooth, J. de, Roesink, H. D. W. & Kemperman, A. J. B. (2020). Forward osmosis: A critical review. *Processes*, *8*(4), 1–29. https://doi.org/10.3390/PR8040404

Naeem, A., Saeed, B., AlMohamadi, H., Lee, M., Gilani, M. A., Nawaz, R., Khan, A. L. & Yasin, M. (2024). Sustainable and green membranes for chemical separations: A review. *Separation and Purification Technology*, *336*(October 2023). https://doi.org/10.1016/j.seppur.2024.126271

Obaideen, K., Shehata, N., Sayed, E. T., Abdelkareem, M. A., Mahmoud, M. S. & Olabi, A. G. (2022). The role of wastewater treatment in achieving sustainable development goals (SDGs) and sustainability guideline. *Energy Nexus*, *7*(July). https://doi.org/10.1016/j.nexus.2022.100112

Oladimeji, T. E., Oyedemi, M., Emetere, M. E., Agboola, O., Adeoye, J. B. & Odunlami, O. A. (2024). Review on the impact of heavy metals from industrial wastewater effluent and removal technologies. *Heliyon*, *10*(23), e40370. https://doi.org/10.1016/j.heliyon.2024.e40370

Omran, R. R. & Ebrahim, S. E. (2025). Comparison between the performance efficiencies of reverse osmosis and nanofiltration membrane systems in removing heavy metal ions from industrial wastewater. *Results in Nonlinear Analysis*, *8*(1), 60–72. https://doi.org/10.31838/rna/2025.08.01.004

Patel, R. V., Yadav, A. & Shahi, V. K. (2025). Advances in membrane distillation for wastewater treatment: Innovations, challenges, and sustainable opportunities. *Science of the Total Environment*, *969*(September 2024). https://doi.org/10.1016/j.scitotenv.2025.178749

Patel, S. K., Lee, B., Westerhoff, P. & Elimelech, M. (2024). The potential of electrodialysis as a cost-effective alternative to reverse osmosis for brackish water desalination. *Water Research*, *250*(October 2023), 121009. https://doi.org/10.1016/j.watres.2023.121009

Petrinic, I., Korenak, J., Povodnik, D. & Hélix-Nielsen, C. (2015). Reuse in the metal finishing in a feasibility study of ultrafiltration/reverse osmosis (UF/RO)-based wastewater treatment and reusdustry. *Journal of Cleaner Production*, *101*, 292–300. https://doi.org/10.1016/j.jclepro.2015.04.022

Qasem, N. A. A., Mohammed, R. H. & Lawal, D. U. (2021). Removal of heavy metal ions from wastewater: A comprehensive and critical review. *NPJ Clean Water*, *4*(1). https://doi.org/10.1038/s41545-021-00127-0

Qu, H. J., Huang, L. J., Han, Z. Y., Wang, Y. X., Zhang, Z. J., Wang, Y., Chang, Q. R., Wei, N., Kipper, M. J. & Tang, J. G. (2021). A review of graphene-oxide/metal–organic framework composites materials: Characteristics, preparation and applications. *Journal of Porous Materials*, *28*(6), 1837–1865. https://doi.org/10.1007/s10934-021-01125-w

Reddy, S., Dey, K., Dsilva Winfred Rufuss, D., Arulvel, S. & Akinaga, T. (2024). Forward osmosis desalination: A critical review focussing on recent advancements in draw solution recovery techniques for enhanced efficiency and regeneration. *Journal of Environmental Chemical Engineering*, *12*(5), 113968. https://doi.org/10.1016/j.jece.2024.113968

Roy, Y., Warsinger, D. M. & Lienhard, J. H. (2017). Effect of temperature on ion transport in nanofiltration membranes: Diffusion, convection and electromigration. *Desalination*, *420*(July), 241–257. https://doi.org/10.1016/j.desal.2017.07.020

Sakai, M., Negishi, E. & Matsukata, M. (2025). Rejection of heavy metal ions in water by zeolite forward osmosis membrane. *Separation and Purification Technology*, *357*(July 2024), 0–7. https://doi.org/10.1016/j.seppur.2024.130163

Samaei, S. M., Gato-Trinidad, S. & Altaee, A. (2020). Performance evaluation of reverse osmosis process in the post-treatment of mining wastewaters: Case study of Costerfield mining operations, Victoria, Australia. *Journal of Water Process Engineering*, *34*(January), 101116. https://doi.org/10.1016/j.jwpe.2019.101116

Sandu, T., Sârbu, A., Căprărescu, S., Stoica, E. B., Iordache, T. V. & Chiriac, A. L. (2022). Polymer membranes as innovative means of quality restoring for wastewater bearing heavy metals. *Membranes*, *12*(12), 1–27. https://doi.org/10.3390/membranes12121179

Saravanan, P., Saravanan, V., Rajeshkannan, R., Arnica, G., Rajasimman, M., Baskar, G. & Pugazhendhi, A. (2024). Comprehensive review on toxic heavy metals in the aquatic system: Sources, identification, treatment strategies, and health risk assessment. *Environmental Research*, *258*(November 2023), 119440. https://doi.org/10.1016/j.envres.2024.119440

Shi, X., Tal, G., Hankins, N. P. & Gitis, V. (2014). Fouling and cleaning of ultrafiltration membranes: A review. *Journal of Water Process Engineering*, *1*, 121–138. https://doi.org/10.1016/j.jwpe.2014.04.003

Shukla, B. K., Sharma, P. K., Yadav, H., Singh, S., Tyagi, K., Yadav, Y., Rajpoot, N. K., Rawat, S. & Verma, S. (2024). Advanced membrane technologies for water treatment: Utilization of nanomaterials and nanoparticles in membranes fabrication. *Journal of Nanoparticle Research*, 26(9). Springer Netherlands. https://doi.org/10.1007/s11051-024-06117-w

Son, H. S., Alpatova, A., Nawaz, M. S., Soukane, S., Medina, S. C., Gudideni, V., Al-Qahtani, A., Anjum, D. H. & Ghaffour, N. (2025). Pre-pilot forward osmosis—Membrane distillation hybrid system for sustainable produced water treatment and reducing volume of hazardous waste in oil and gas industry. *Journal of Water Process Engineering*, *69*(November 2024), 106628. https://doi.org/10.1016/j.jwpe.2024.106628

Soo, K. W., Wong, K. C., Goh, P. S., Ismail, A. F. & Othman, N. (2020). Efficient heavy metal removal by thin film nanocomposite forward osmosis membrane modified with geometrically different bimetallic oxide. *Journal of Water Process Engineering*, *38*(August), 101591. https://doi.org/10.1016/j.jwpe.2020.101591

Suleman, M., Al-Rudainy, B. & Lipnizki, F. (2025). Overcoming the limitations of forward osmosis and membrane distillation in sustainable hybrid processes managing the water–energy nexus. *Membranes*, *15*(6), 1–45. https://doi.org/10.3390/membranes15060162

Tiwary, S. K., Singh, M., Chavan, S. V. & Karim, A. (2024). Graphene oxide-based membranes for water desalination and purification. *NPJ 2D Materials and Applications*, *8*(1), 1–19. https://doi.org/10.1038/s41699-024-00462-z

Tong, T. & Elimelech, M. (2016). The global rise of zero liquid discharge for wastewater management: Drivers, technologies, and future directions. *Environmental Science and Technology*, *50*(13), 6846–6855. https://doi.org/10.1021/acs.est.6b01000

Vos, H. & Janssen, M. (2008). *EU-Wide Control Measures to Reduce Pollution from WFD Relevant Substances: Copper and Zinc in the Netherlands*. RIVM.

Wang, C., Meng, P., Wang, S., Song, D., Xiao, Y., Zhang, Y., Ma, Q., Liu, S., Wang, K. & Zhang, Y. (2022). Comparison of two types of energy recovery devices: Pressure exchanger and turbine in an island desalination project case. *Desalination*, *533*(March), 115752. https://doi.org/10.1016/j.desal.2022.115752

Xiang, H., Min, X., Tang, C. J., Sillanpää, M. & Zhao, F. (2022a). Recent advances in membrane filtration for heavy metal removal from wastewater: A mini review. *Journal of Water Process Engineering*, *49*(August). https://doi.org/10.1016/j.jwpe.2022.103023

Xiang, H., Min, X., Tang, C. J., Sillanpää, M. & Zhao, F. (2022b). Recent advances in membrane filtration for heavy metal removal from wastewater: A mini review. *Journal of Water Process Engineering*, *49*(March). https://doi.org/10.1016/j.jwpe.2022.103023

Xu, P., Drewes, J. E., Kim, T. U., Bellona, C. & Amy, G. (2006). Effect of membrane fouling on transport of organic contaminants in NF/RO membrane applications. *Journal of Membrane Science*, *279*(1–2), 165–175. https://doi.org/10.1016/j.memsci.2005.12.001

Yaqub, M. & Lee, S. H. (2019). Heavy metals removal from aqueous solution through micellar enhanced ultrafiltration: A review. *Environmental Engineering Research*, *24*(3), 363–375. https://doi.org/10.4491/EER.2018.249

You, S., Lu, J., Tang, C. Y. & Wang, X. (2017). Rejection of heavy metals in acidic wastewater by a novel thin-film inorganic forward osmosis membrane. *Chemical Engineering Journal*, *320*, 532–538. https://doi.org/10.1016/j.cej.2017.03.064

Zhang, Y., Zhang, S. & Chung, T. S. (2015). Nanometric graphene oxide framework membranes with enhanced heavy metal removal via nanofiltration. *Environmental Science and Technology*, *49*(16), 10235–10242. https://doi.org/10.1021/acs.est.5b02086

Zulu, E., Ramasamy, S., Khoabane Sikhwivhilu, K. & Syampungani, S. (2025). A comprehensive review of recent advances in membrane innovations for efficient heavy metal removal from mine effluents. *Scientific African*, *27*(September 2024), e02510. https://doi.org/10.1016/j.sciaf.2024.e02510

4 Heavy Metal Removal Using Inorganic Nanostructures

Jose Luis Ortiz-Quiñonez and Umapada Pal

4.1 INTRODUCTION

Heavy metal contamination in water bodies poses a serious threat to the environment and human health due to the toxic, persistent, and non-biodegradable characteristics of metals such as cadmium (Cd), lead (Pb), mercury (Hg), arsenic (As), and chromium (Cr). Besides natural sources such as volcanic eruptions and rock weathering, these pollutants originate from industrial processes such as mining, electroplating, battery manufacturing and disposal, and textile production. Conventional treatment methods, such as chemical precipitation (see Chapter 2), ion-exchange, and membrane filtration (see Chapter 3), while widely used, have several limitations, including low efficiency at low concentrations, high operational costs, and the generation of secondary waste.

Recently, inorganic nanostructures have emerged as promising materials for the efficient removal of heavy metals from water. Thanks to their high surface area-to-volume ratio (specific surface area), tunable surface chemistry, and unique physicochemical properties, nanomaterials such as metal oxides (e.g., TiO_2, Fe_3O_4, or ZnO), metal sulfides (e.g., FeS, Na_2S, SnS_2, or MoS_2), layered double hydroxides (LDHs), nanoclays, and metal-organic frameworks (MOFs) exhibit superior adsorption and catalytic abilities. These materials can selectively interact with heavy metal ions through electrostatic attraction, surface complexation, ion-exchange, and redox reactions. Using these specially designed nanostructures made from advanced materials not only enhances removal efficiency but also enables material regeneration and reuse, making them an attractive option for sustainable water treatment.

In this chapter, we examine the ability of several inorganic nanomaterials to capture metal ions, highlighting their key features and the mechanisms underlying this process. Finally, we compare their efficiency and discuss their strengths and weaknesses.

4.2 ION CAPTURE OVER CLAYS

The use of clays in water purification to remove heavy metals is advantageous because of their moderate surface area, structural diversity, modifiable surface chemistry, and high adsorption capacity (Yang et al. 2025). Moreover, clays are low-toxic

DOI: 10.1201/9781003599098-4

materials that are abundant in soil and relatively inexpensive. Furthermore, they can be functionalized with surfactants or ions such as orthophosphate to enhance their adsorption capacity (Adebowale et al. 2006). Clays absorb ions through electrostatic interactions, ion-exchange, and surface complexation (Yang et al. 2025). Clays can also be integrated into ceramic water filters (Solomon et al. 2023). Nanocomposites based on clays, such as clay-chitosan, are highly effective for removing heavy metals (Biswas et al. 2021).

4.2.1 Classifications and Structure of Clays

Clays are hydrous aluminum silicates characterized by a structure consisting of a two-dimensional tetrahedral sheet of silicates (SiO_4) linked to a two-dimensional octahedral sheet of aluminates (Al_2O_3) (Jiménez-Castañeda and Medina 2017). The octahedral and tetrahedral layers are joined by sharing O^{2-} ions. In some clays, aluminum ions are partially replaced by Fe^{3+} and Mg^{2+} ions. Figure 4.1 illustrates a schematic of how tetrahedral and octahedral layers are arranged in clays, along with the structures of different clay types. Clay minerals are classified as layered silicates, chain silicates, sesquioxides, and other minerals (Figure 4.2). The layered silicates are divided into 1:1, 2:1, and 2:1:1 clays. Examples of 1:1 layered silicates include kaolinite and halloysite, which have a structure consisting of one tetrahedral and one octahedral layer (Figure 4.1a–e). The 2:1 clays consist of one octahedral layer sandwiched between two tetrahedral layers (Figure 4.1f). These include silicates such as montmorillonite, hectorite, and vermiculite, as well as muscovite, mica, and illite. The 2:1:1 clays, such as chlorite, have a layered structure composed of tetrahedral and octahedral sheets, as shown in Figure 4.1g. The clay palygorskite features two bands of silica tetrahedra linked by Al(III) and Mg(II) ions in octahedral coordination, forming nanoscale channels approximately 0.37 nm × 0.64 nm and 0.56 nm × 1.1 nm in size, as shown in Figure 4.3a.

When an Al^{3+} ion in the clay replaces a Si^{4+} ion, it creates a negative charge in the clay. Similarly, when Mg^{2+} ions substitute Al^{3+} ions in the tetrahedral sites, a permanent negative charge is formed. To balance this negative charge, clays incorporate ions such as Na^+, K^+, or Ca^{2+} into octahedral and tetrahedral sites. Additionally, water molecules are also situated between the layers. Figure 4.3a shows the octahedral and tetrahedral sites in clays, along with the interlayer ions and water. Table 4.1 shows the point of zero charge for various clays. The point of zero charge is the pH value above which the clay exhibits a negative charge. Clays can exchange interlayer ions with other ions in aqueous solutions. Figure 4.3b illustrates the exchange of Na^+ ions for M^{2+} ions in the clay montmorillonite. Another notable feature of several clays is their high cation exchange capacity (CEC), as shown in Table 4.1. Specifically, vermiculite and montmorillonite have superior CEC. Consequently, clays can adsorb heavy metals from water through exchangeable cations (e.g., Na^+, K^+, or Ca^{2+}) and electrostatically bound counter-ions in the diffusion layer. After heavy metal adsorption, the concentration of exchangeable ions in the solution increases (Kumrić et al. 2013).

Clays are effective materials for removing heavy metals from water. For example, kaolinite has been used to remove Pb^{2+} and Cd^{2+} ions from water, with adsorption capacities of 15.04 and 9.23 mg g^{-1}, respectively (Adebowale et al. 2006). Table 4.2

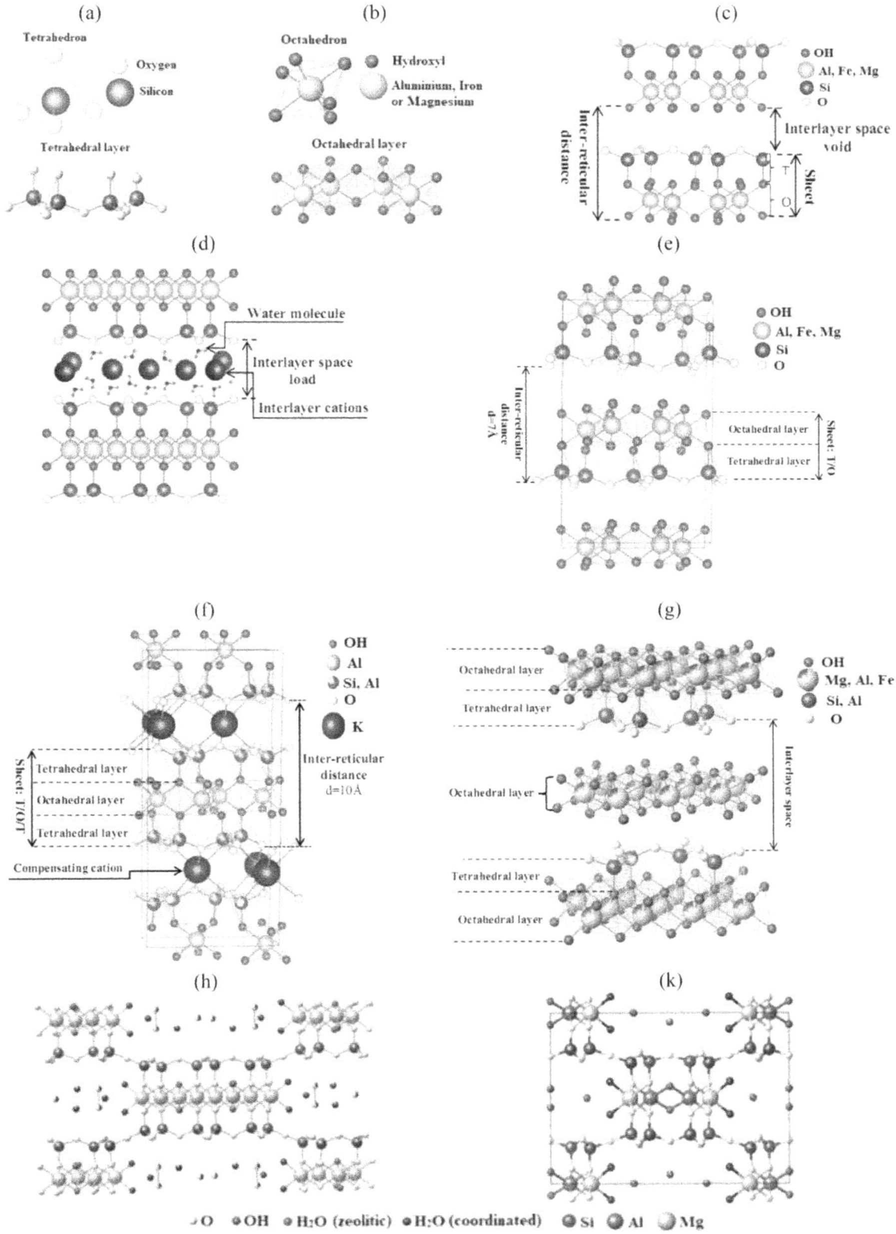

FIGURE 4.1 Schematic illustration of structural arrangements of different layered silicates: (a) Tetrahedral layer; (b) Octahedral layer; (c) Uncharged layer; (d) Charged layer; (e) Tetrahedral/Octahedral (T/O) clay minerals (e.g., kaolinite); (f) T/O/T clay minerals (e.g., muscovite); (g) T/O/T with an additional octahedral layer (T/O/T-O) clay minerals (e.g., chlorite). Reproduced from El-habacha et al. (2023; Copyright Elsevier, 2023).

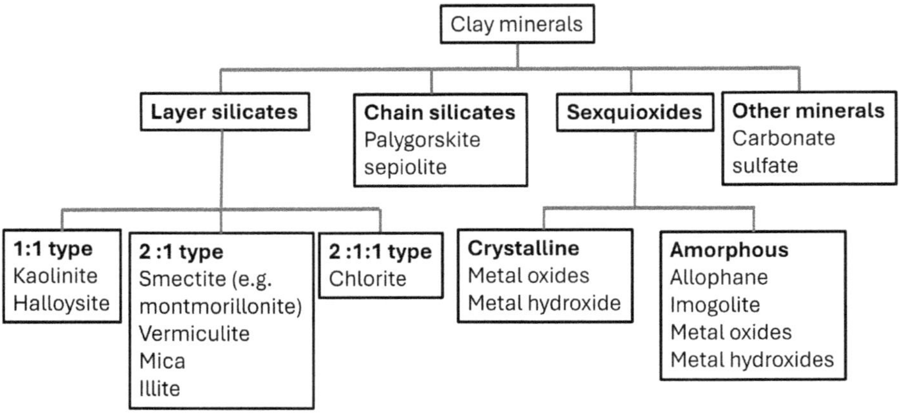

FIGURE 4.2 Classification scheme of clay minerals. Adapted from Kumari and Mohan (2021).

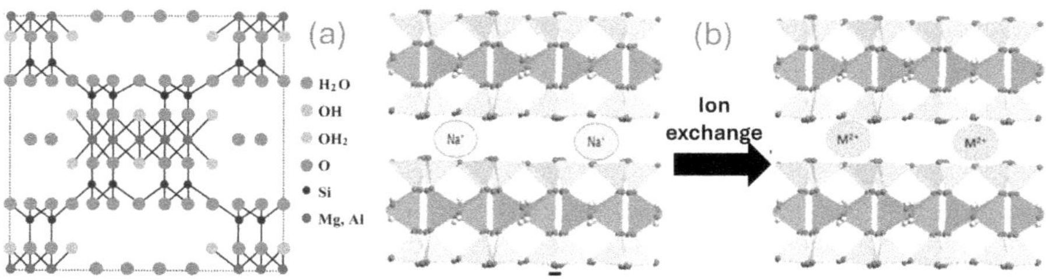

FIGURE 4.3 (a) Schematic representation of the crystal structure of the clay mineral palygorskite. Adapted from Li et al. (2015; Copyrights Royal Society of Chemistry, 2015). (b) Ion-exchange process in montmorillonite: replacement of Na^+ ions by M^{2+} ions. Adapted from Chu et al. (2019; Copyright ACS, 2019).

lists selected studies on the removal of heavy metals using clay-based adsorbents. The adsorption capacity of clays improves when they are made into nanocomposites with polymers such as chitosan, polyaniline, polypyrrole, cellulose, PVA, PMEA, polyacrylic acid, polyacrylamide, and lignin (Yadav et al. 2019; Chen et al. 2014; Das et al. 2024; Moreno-Sader et al. 2019; Zhou et al. 2017). Clays can also be functionalized with molecules like *N*-β-aminoethyl-γ-aminopropyltrimethoxysilane (KH-792), tetraethoxysilane (TEOS), and 3-mercaptopropyltrimethoxysilane (MPTMS) (Tchinda et al. 2009; Luo et al. 2011).

The sorption characteristics, such as active surface area, porosity, pore volume, and CEC of nanocomposite materials, can be enhanced by functionalizing one or more components (Yadav et al. 2019). Moreover, clays can be functionalized with amines, phosphates, or other molecules to enhance their adsorption capacity. For instance, grafting 3-(aminopropyl)triethoxysilane or trimethoxy[3-(methylamino)

TABLE 4.1
Summary of the Structural and Key Properties of Various Clay Minerals

Clay mineral type	Chemical formula	Zero charge	CEC (mg/100 g)	Specific Surface Area (m^2/g)	Reference
Kaolinite	$Al_4 [Si_4O_{10}] (OH)_8$	2.7–3.2	1–5	8–15	Yang et al. (2025)
		–	3–15	10–20	Adebowale, et al. (2006)
Halloysite	$Al_4[Si_4O_{10}](OH)_8 \cdot nH_2O$ (n =0 or 2)	2.5–3.5	13	45–75	Yang et al. (2025)
Montmorillonite	$(Na, Ca)_{0.33}(Al, Mg)_2Si_4O_{10}(OH)_2 \cdot nH_2O$	4.5–5.9	80–120	600–800	Yang et al. (2025)
				62	Macht et al. (2011)
Illite	$(K,H_3O)(Al,Mg,Fe)_2 (Si,Al)_4O_{10}[(OH)_2(H_2O)]$			41	Macht et al. (2011)
Vermiculite	$Mg_5(H_2O)_4[Si_4O_{10}]_2(OH)_2$	7–9	100–150	5–400	Yang et al. (2025)
Muscovite	$KAl_2(AlSi_3O_{10})(OH)$	~3.5	10–40	10–100	Yang et al. (2025)
Sepiolite	$Mg_8(H_2O)_4[Si_4O_{10}]_3 (OH)_4 \cdot 8H_2O$	7–9	20–30	170–370	Yang et al. (2025)
Palygorskite	$Si_8(Mg,Al,Fe)_5O_{20} (OH)_2(H_2O)_4 \cdot 4H_2O$	6–8	30–40	100–216	Yang et al. (2025); Wang et al. (2019a)
Chlorite	$Y_3[Z_4O_4](OH)_6$ $Y(Mg^{2+},Fe^{2+},Al^{3+}, Fe^{3+})Z(Si,Al)$	3–5	10–40	14	Yang et al. (2025)

propyl]silane onto halloysite enhances its Hg(II) adsorption capacity through interactions between the amine group and Hg(II) ions (Das et al. 2018).

In the adsorption of heavy metals by clays, competition often occurs between heavy metal ions and other common ions present in the water for available adsorption sites. For example, during the adsorption of Tl^+ onto illite, K^+ and NH_4^+ ions compete for the same exchange sites on the clay surface (Wick et al. 2018), whereas Na^+ and Ca^{2+} do not show such competition. One reason for this competition is the similarity in ionic radii between K^+ and Tl^+ ions. The pH of the solution is another key factor affecting the adsorption of heavy metals. For instance, Chen et al. reported that the removal efficiency of Cr(VI) by sepiolite functionalized with polypyrrole decreases as the solution pH increases (Chen et al. 2014). Additionally, the adsorption capacity of clays can be significantly enhanced through mechanochemical treatment. For example, Kumrić et al. (2013) demonstrated that ball-milling a mixed-layer montmorillonite/kaolinite clay increased its Pb(II) adsorption capacity from 27.2 to 42.6 $mg \cdot g^{-1}$.

Figure 4.4a shows the arrangement of tetrahedral and octahedral sheets and interlayer ions in the clays kaolinite, smectite, vermiculite, illite, and chlorite. In the case of kaolinite, there are no interlayer ions between the sheets. Figure 4.4b presents a

TABLE 4.2
Adsorption Capacities of Various Clay Minerals for Removing Heavy Metals from Wastewater

Absorbent	Heavy metal	pH	Adsorbent dose	Concentration of the heavy metal	Adsorption isotherm model and performance		Reference
					Langmuir	Freundlich	
Fe_3O_4@HNT	As(V)	4	$1\ g{\cdot}L^{-1}$	$5\ mg{\cdot}L^{-1}$	$q_m = 12.73 \pm 1.16$	$K_F = 3.35 \pm 0.18$	Deb et al. (2022)
Palygorskite/alginate	Cu(II)	4.7	$1\ g{\cdot}L^{-1}$	$100\ mg{\cdot}L^{-1}$	$q_m = 119\ mg{\cdot}g^{-1}$	$K_F = 50.20\ mg{\cdot}g^{-1}$	Wang et al. (2018b)
	Cd(II)	6.4	$1\ g{\cdot}L^{-1}$	$100\ mg{\cdot}L^{-1}$	$q_m = 160\ mg{\cdot}g^{-1}$	$K_F = 33.44\ mg{\cdot}g^{-1}$	Wang et al. (2018b)
Sepiolite/Fe	Ni(II)	5.6	$1\ g{\cdot}L^{-1}$	$100\ mg{\cdot}L^{-1}$	$q_m = 0.317\ mmol{\cdot}g^{-1}$	$K_F = 0.273\ mmol^{1-1/n} \cdot dm^{3/n} \cdot g^{-1}$	Lazarević et al. (2010)
Sepiolite/polypyrrole	Cr(V)	2	$1\ g{\cdot}L^{-1}$	$200\ mg{\cdot}L^{-1}$	$q_m = 302\ mg{\cdot}g^{-1}$	$K_F = 148.28\ mg{\cdot}g^{-1}$	Chen et al. (2014)
Montmorillonite/Kaolinite	Pb(II)	5.5	$2\ g{\cdot}L^{-1}$	$12.5\ mg{\cdot}L^{-1}$	$q_m = 42.6 \pm 1.1\ mg{\cdot}g^{-1}$	$K_F = 13.6 \pm 0.6$	Kumrić et al. (2013)
	Cd(II)	5.5	$2\ g{\cdot}L^{-1}$	$12.5\ mg{\cdot}L^{-1}$	$q_m = 14.1 \pm 0.6\ mg{\cdot}g^{-1}$	$K_F = 6.1 \pm 0.2$	Kumrić et al. (2013)
	Cu(II)	5.5	$2\ g{\cdot}L^{-1}$	$12.5\ mg{\cdot}L^{-1}$	$q_m = 34.6 \pm 0.8\ mg{\cdot}g^{-1}$	$K_F = 12.1 \pm 0.7$	Kumrić et al. (2013)
	Zn(II)	5.5	$2\ g{\cdot}L^{-1}$	$12.5\ mg{\cdot}L^{-1}$	$q_m = 16.0 \pm 0.8\ mg{\cdot}g^{-1}$	$K_F = 6.5 \pm 0.3$	Kumrić et al. (2013)
Montmorillonite/Illite	Pb(II)	4	$1.3\ g{\cdot}L^{-1}$	$100\ mg{\cdot}L^{-1}$	$q_m = 54.12\ mg{\cdot}g^{-1}$	$K_F = 9.11$	Oubagaranadin and Murthy (2009)

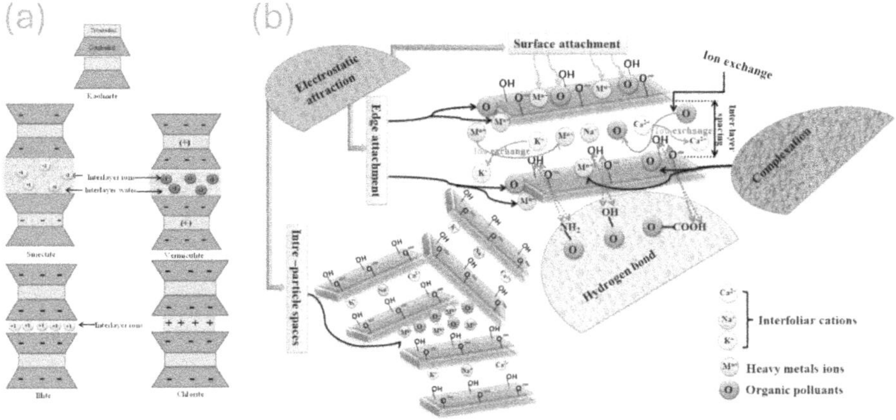

FIGURE 4.4 (a) Overview of different types of clay minerals. Reproduced from Kumari and Mohan (2021; Copyright IntechOpen, 2021). (b) Schematic illustration of the adsorption mechanisms of dyes and heavy metals at different sites on the clay surface. Reproduced from El-habacha et al. (2023; Copyright Elsevier, 2023).

schematic illustration of the adsorption mechanisms of dyes and heavy metals at different sites on the clay surface. The clays contain adsorbed ions through electrostatic interaction between the sheets and at the edges and surface of the sheets. The adsorbed ions can be exchanged with heavy metals present in aqueous solutions. The presence of amine, hydroxyl, and other functional groups in modified clays enhances the adsorption of heavy metals.

4.2.2 Ion Capture over Zeolites

4.2.2.1 Classifications and Structure of Zeolites

Among ion-capturing adsorbents, naturally occurring clays, particularly natural zeolites, are among the most widely used due to their abundance, diverse structural forms, chemical compositions, and effective ion-capturing capabilities (Tran et al. 2015). Natural zeolites are porous, hydrated aluminosilicates with pore networks that range from macropores to mesopores. These materials display ion-exchange, adsorption, and dehydration properties. They form through the alteration of volcanic glass under different geochemical conditions and temperatures. Structurally, natural zeolites consist of a tetrahedral framework in which silicon or aluminum atoms are coordinated to oxygen atoms. This crystalline arrangement creates large internal channels and cavities that host water (H_2O) molecules and exchangeable cations such as K^+, Na^+, Ca^{2+}, and Mg^{2+}, both of which (H_2O and cations) have considerable freedom of movement, permitting ion-exchange and reversible dehydration. Other ions can readily replace cations in zeolites via ion-exchange processes. The general chemical formula of natural zeolites is given as $(M^+_x, M^{2+}_y)[Al_{(x+2y)}\ Si_{n-(x+2y)}O_{2n}]$ mH_2O, where M^+ and M^{2+} represent monovalent and divalent cations, respectively, x and y are their mole fractions, and m is the number of water molecules.

Natural zeolites occur in various alkaline environments, including alkaline lakes, soils, and land surfaces, as well as in deep-sea sediments, hydrological systems, and low-grade metamorphic or diagenetic rocks. More than 80 types of natural zeolites have been found and reported in the literature (Bish and Carey 2001). Common types of natural zeolites include analcime, chabazite, clinoptilolite, erionite, ferrierite, faujasite, mordenite, laumontite, and phillipsite, many of which are typically found in volcanic rock formations.

Zeolites can be classified in several ways. For example, based on pore size, they can be categorized as (i) small-pore zeolites (containing 8-membered rings and about 3–4 Å pore size), (ii) medium-pore zeolites (containing 10-membered rings and roughly 5–6 Å pore size), and (iii) large-pore zeolites (containing 12-membered rings and approximately 7–8 Å pore size). Due to their small pore size, the small-pore zeolites, such as zeolite A (LTA), chabazite (CHA), and erionite (ERI), are often used for separating small molecules like H_2O, CO_2, and NH_3. On the contrary, medium-pore zeolites like ZSM-5 (MFI) and ferrierite (FER) can host aromatic hydrocarbons and are used for shape-selective catalysis, hydrocarbon isomerization, and cracking. Additionally, large-pore zeolites such as faujasite (FAU, also known as Zeolite X and Zeolite Y) and mordenite (MOR) have high ion-exchange capacities. They are widely employed in catalysis, heavy-metal ion adsorption, hydrocarbon cracking, and other applications.

Another way to classify zeolites is by their origin, which includes natural and synthetic types. Natural zeolites, such as clinoptilolite, chabazite, and mordenite, are abundant in nature, low-cost, and have a high ion-exchange capacity. They are widely used for water purification, gas adsorption, and soil conditioning. However, the most common way to classify zeolites is by their framework structure, as proposed by the International Zeolite Association (IZA) and known as IZA nomenclature. In this system, zeolites are categorized as: (i) LTA, which includes Zeolite A, NaA, and CaA, with a cubic structure and small pore size, widely used in detergents and drying processes; (ii) FAU, which includes Zeolite X and Zeolite Y, featuring a large-pore supercage structure, high cation exchange capacity, and common in petroleum refining; (iii) MFI, or ZSM-5, characterized by medium-sized pores, high silica content, and high hydrothermal stability. and acid resistance, used in shape-selective catalysis in petrochemistry; iv) MOR (Mordenite), with large pores and 1D channels, offering high adsorption capacity for large molecules; (v) CHA (Chabazite), with small pore cage-like structures, capable of capturing CO_2 molecules and used in gas separation; and (vi) BEA (Zeolite Beta), which has large pore 3D structures, frequently utilized for alkylation, fine chemical production, and hydro-isomerization. Some common porous zeolite structures with multiple-ring units are shown in Figure 4.5.

Defining the exact structure of natural zeolite is often challenging due to its gradational properties and compositional variability. Additionally, the structures of several natural zeolites remain undetermined. Table 4.3 provides a comprehensive overview of naturally occurring zeolites, including their chemical formulas and structural characteristics. For more detailed structural analyses, readers may refer to W.L. Bragg's Atomic Structure of Minerals (Cornell University Press, 1937), the Mineralogical Society of America Special Paper No. 1 (1963) by J.V. Smith, and

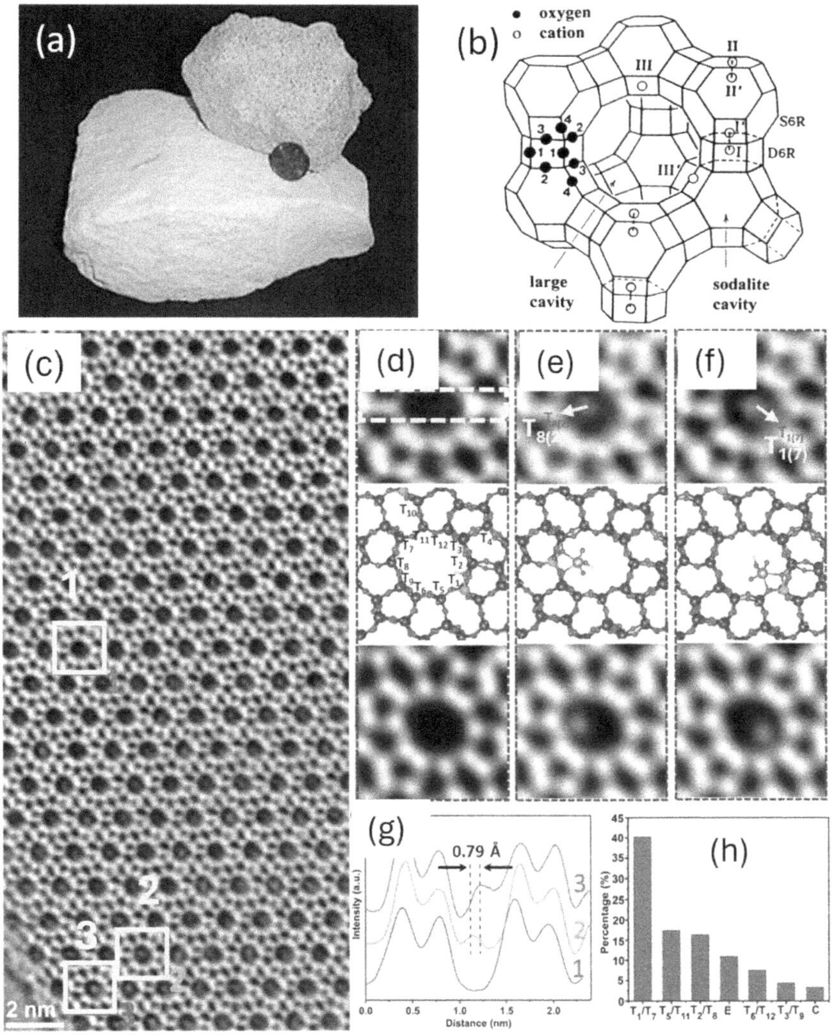

FIGURE 4.5 (a) Photograph of a zeolite-containing volcanic rock. (b) Schematic framework structure of zeolite X. Zhu and Seff (1999; Copyright ACS, 1999). Oxygen atoms are located near the center of each line segment, with nonequivalent atoms labeled 1–4. Silicon and aluminum atoms alternate at tetrahedral sites, except for 4% Si substitution at Al positions. Extra-framework cation sites are denoted by Roman numerals. (c) A representative integrated differential phase contrast (iDPC)-STEM image of Mo/ZSM-5, showing off-center contrast in several 10-MR channels. (d–f) Zoomed-in areas denoted by 1, 2, and 3 of (c), illustrating an empty channel (d) and channels containing an MoO_3H cluster bound at T_8 and T_1 (f) sites. Each panel displays the raw image (top), calculated structural model (middle), and simulated electrostatic potential (bottom). Atom colors: Si (blue), O (red), Al (green), Mo (pink), H (white). (g) Intensity line profiles of the images in (d), (e), and (f), across the red dashed rectangles. (h) Statistical distribution of Al occupancy at different T sites from 100 analyzed channels. The labels "E" and "C" represent empty and on-center contrast channels, respectively. Reproduced with permission Liu et al. (2020a; Copyright Wiley, 2020).

TABLE 4.3
Chemical Formula, Lattice Structure, and Lattice Parameters of Some Zeolites

Zeolite (code/ common name)	Chemical formula (reference unit)*	Structure/Space group	Lattice parameters	Crystal structure example	Reference
LTA (Zeolite A, Na-A).	$[Al_{12}Si_{12}O_{48}].Na_{12}.(H_2O)_{27}$	Cubic, *Fm-3c* (cubic cage framework)	a = b=c= 24.61 Å, α = β = γ = 90°		IZA-SC (2017); Bolotina et al. (2023)
FAU (Faujasite: ZeoliteX/Y).	$[Al_{95}Si_{97}O_{384}]$.(cations/water)	Cubic, *Fd-3m* (12-ring pores)	a = b=c » 24.60–24.80 Å, α = β = γ = 90°		IZA-SC (2017)
MFI (ZSM-5/Sillicate-1)	$Na_x(H_2O)_{16}[Al_xSi_{96-x}O_{192}]$ (varying x)	Orthorhombic, *Pnma* (10-ring intersecting channels)	a = 20.07Å; b= 19.92 Å, c = 13.42 Å, α = β = γ = 90°		IZA-SC (2017)
BEA (Zeolite Beta)	$Na_7[Al_7Si_{57}O_{128}]$	Tetragonal *P4₁22* (12-ring channels)	a = 12.661 Å; c = 26.406 Å, α = β = γ = 90°		IZA-SC (2017)
MOR (Mordenite)	$Na_8(H_2O)_{24}[Al_8Si_{40}O_{96}]$	Orthorhombic *Cmcm* (12-ring main channels + 8-ring side channels)	a = 18.10 Å; b= 20.50 Å, c = 7.50 Å, α = β = γ = 90°		IZA-SC (2017)

IZA-SC = Structure Commission of the International Zeolite Association, Copyright 2017–2025.

*The formulas shown in the second column are typical reference compositions (cell unit with compensation cations and hydration waters). In practice, the composition (Si/Al, cation type and number, degree of hydration) varies and slightly modifies the cell parameters.

proceedings from the Third General Meeting of the International Mineralogical Association.

4.2.2.2 Ion Capture over Zeolites and Their Composites

Natural zeolites are promising materials for removing heavy metals from different water sources. Understanding the structural characteristics, chemical composition, and metal ion removal performance of these materials is essential for designing effective water treatment systems. Natural zeolites have been extensively studied and employed for adsorptive removal of a range of heavy metal ions, including arsenic (As(III), As(V)), cadmium (Cd(II)), chromium (Cr(III), Cr(VI)), lead (Pb(II)), mercury(Hg(II)), and nickel (Ni(II)).

Frequently, these natural zeolites are chemically modified by treating them with acid, base, salts, cationic surfactants, and metallic species or reagents to improve their adsorption efficiency. Among different types, clinoptilolite is the most commonly used natural zeolite for removing heavy metals. Interestingly, zeolites of geological origin can vary significantly in their metal-ion adsorption properties and capacities (Table 4.4), making the origin of the material an important factor to consider.

Several factors affect the adsorption performance of natural zeolites, including both the properties of the metal ions and the zeolites themselves, as well as external adsorption conditions such as pH, temperature, agitation speed, and initial metal concentration (Misaelides 2011). For example, the adsorption capacity of clinoptilolite for Pb^{2+}, Cd^{2+}, Zn^{2+}, and Fe^{3+} cations increases with solution pH (Vishnu et al. 2021). Therefore, optimizing all these parameters is essential to maximize removal efficiency. As mentioned earlier, the effectiveness of natural zeolites in removing metal ions from water depends not only on the characteristics of the adsorbent and operating conditions but also on the specific type of metal ion involved. For example, zeolites may show a higher affinity for divalent cations like Pb(II) and Cd(II) compared to oxyanions such as As(V), due to differences in charge type (negative or positive), ionic radius, charge density, and hydration energy. Additionally, the competition among coexisting ions in the water matrix can influence the zeolite's selectivity and overall adsorption capacity. Therefore, a thorough understanding of how zeolites interact with specific metal ions is vital for developing efficient and selective water treatment methods.

The removal of heavy metals by zeolites occurs through three possible mechanisms: (i) Ion-exchange, (ii) Surface adsorption, and (iii) Intraparticle diffusion. Ion-exchange is the most frequently occurring mechanism in zeolites. In the zeolite framework, the substitution of aluminum for silicon creates a net negative charge. This charge is balanced by loosely bound exchangeable cations (Na^+, K^+, Ca^{2+}, Mg^{2+}) located in the zeolite pores or on surfaces (Obaid et al. 2018). When zeolites come into contact with metal ion-containing solutions, these cations are exchanged for heavy metal ions such as Pb^{2+}, Cu^{2+}, Cd^{2+}, Zn^{2+}, and so forth (Figure 4.6a). On the other hand, both physisorption and chemisorption occur in surface adsorption. Physisorption involves weak van der Waals forces on the external surface or within the pores of zeolite, which are non-specific and reversible (Figure 4.6c). Chemisorption involves the formation of coordinate or covalent bonds between the metal ions and surface functional groups (e.g., hydroxyl groups) on the zeolite or modified zeolite. The

TABLE 4.4
Performance of Metal Ion Adsorption of Some Natural and Modified Zeolites

	Adsorbent	Heavy metals	Adsorption isotherm model and performance		Reference
			q_m, Langmuir model	K_F, Freundlich model	
Natural zeolites	Zeolite 4A	Pb(II) Cd(II) Zn(II) Cu(II)	Pb(II): 170 ± 30 mg·g^{-1} Cd(II): 23 ± 0.75 mg·g^{-1} Zn(II): 1.15 ± 0.075 mg·g^{-1} Cu(II): 4.30 ± 0.45 mg·g^{-1}	Pb(II): 160.82 ± 9.76 mg·g^{-1} Cd(II): 13.27 ± 0.654 mg·g^{-1} Zn(II): 10.76 ± 1.50 mg·g^{-1} Cu(II): 4.68 ± 0.20 mg·g^{-1}	Vasconcelos da Silva et al. (2023)
	Clinoptilolite	Co(II), Cu(II), Zn(II), Mn(II)	Co(II): 244.13 mmol·kg^{-1} Cu(II): 141.12 mmol·kg^{-1} Zn(II):133.85 mmol·kg^{-1} Mn(II): 76.78 mmol·kg^{-1}	Co(II): 1.905 mg·g^{-1} Cu(II): 1.348 mg·g^{-1} Zn(II): 1.548 mg·g^{-1} Mn(II): 6.456 mg·g^{-1}	Erdem et al. (2004)
	Mordenite	Pb(II), Cd(II)	Pb(II): 375.94 mg·g^{-1} Cd(II): 197.63 mg·g^{-1}	—	Pan et al. (2025)
	Zeolite A-Na	Pb(II), Cd(II) pH=4.0	Pb(II): 555.6 mg·g^{-1} Cd(II): 158.7 mg·g^{-1}	—	Galindo Valbuena et al. (2023)
	SM-5	Pb(II)	Pb(II): 46.34 ± 6.25 mg·g^{-1}	Pb(II): 1.12 ± 0.25 mg·g^{-1}	Y. Zhang et al. (2021)
	Faujasite	Pb(II), Cd(II), Cu(II)	Pb(II): 801 mg·g^{-1} Cd(II): 340 mg·g^{-1} Cu(II): 231 mg·g^{-1}	Pb(II): 27.2 mg·g^{-1} Cd(II): 18.3 mg·g^{-1} Cu(II): 19.0 mg·g^{-1}	Jiménez et al. (2024)
Modified zeolites	Zeolite Y modified with HDTMA	Cr(VI) As(V)	Cr(VI): 37.59 mmol·kg^{-1}) As(V): 17.89 mmol·kg^{-1}	—	Yusof and Malek (2009)
	Mordenite modified with HDTMA	As(V)	As(V): 97.33 mmol·kg^{-1}	As(V): 1.24 mg·g^{-1}	Chutia et al. (2009)
	Clinoptilolite modified with HDTMA.	As(V)	As(V): 45.33 mmol·kg^{-1}	As(V): 1.59 mg·g^{-1}	Chutia et al. (2009)
	Clinoptilolite modified with HDTMA	Cr(VI)	Cr(VI): 5.07 mg·g^{-1}	—	Leyva-Ramos et al. (2008)

Aluminosilicate azide cancrinite	Pb(II), Cd(II), Ni(II)	Pb(II): 38.46 $mg \cdot g^{-1}$ Cd(II): 37.04 $mg \cdot g^{-1}$ Ni(II): 30.30 $mg \cdot g^{-1}$	Pb(II): 3.82 $mg \cdot g^{-1}$ Cd(II): 2.87 $mg \cdot g^{-1}$ Ni(II): 3.19 $mg \cdot g^{-1}$	Borhade et al. (2015)
Zeolite/Activated carbon	Cu(II)	Cu(II): 118 $mg \cdot g^{-1}$	Cu(II): 108 $mg \cdot g^{-1}$	Li et al. (2020)
Zeolite/Fe_3O_4 composite	Pb(II), Cu(II), Zn(II),	Pb(II): 140 $mg \cdot g^{-1}$ Cu(II): 19.7 $mg \cdot g^{-1}$ Zn(II): 20.1 $mg \cdot g^{-1}$	—	Phouthavong et al. (2024)
Zeolite/chitosan	Cs(I)	Cs(I): 9.5 $mg \cdot g^{-1}$	—	Araujo et al. (2023)
ZnO/zeolite	Cr(VI) Cd(II)	Cr(VI): 81.4 $mg \cdot g^{-1}$ Cd(II) 35.9 $mg \cdot g^{-1}$	Cr(VI): 12.5 $mg \cdot g^{-1}$ Cd(II) 5.5 $mg \cdot g^{-1}$	Vipin et al. (2023a)

Note: HDTMA, hexadecyltrimethylammonium bromide.

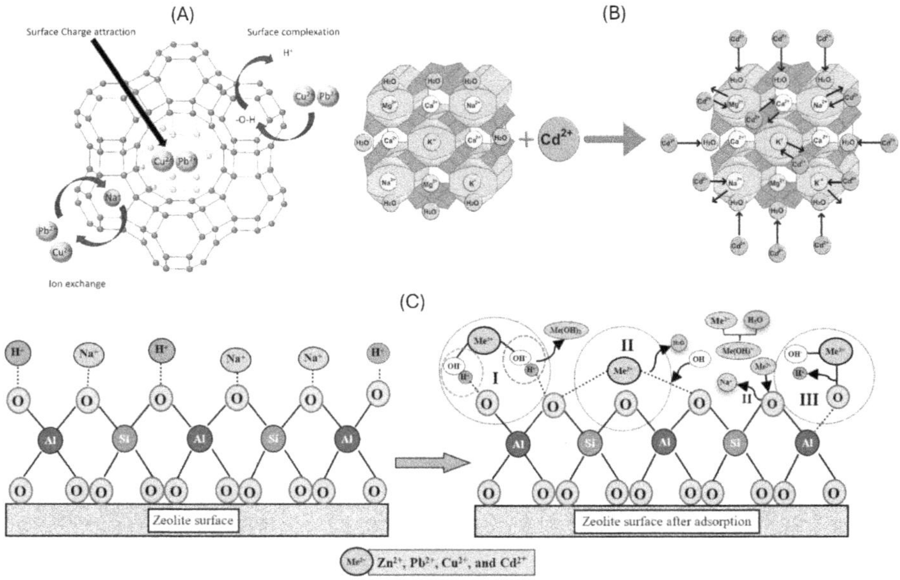

FIGURE 4.6 Metal ion adsorption by natural zeolite through (a) Ion-exchange. Reproduced from Shah et al. (2024; Copyright Springer, 2024). (b) Inter-particle diffusion. Reproduced from Velarde et al. (2024; Copyright Cell Press, 2024). (c) surface adsorption. Reproduced from Ibrahim et al. (2023; Copyright MDPI, 2023).

chemisorption is stronger than physisorption and is usually irreversible. An example of chemisorption at the zeolite surface is the reaction of Si-OH or Al-OH groups with metal ions, forming –O–M bonds.

In the case of inter-particle diffusion (Figure 4.6b), metal ions move from one zeolite particle to another during adsorption or ion-exchange. This movement, driven by diffusion, can limit the overall rate at which metal ions are removed from the solution. In practical applications, zeolites are often used as particles or beads that are packed into a column or bed. When a solution containing metal ions is passed through the zeolite column or bed, the ions first need to diffuse from the bulk solution to the outer surface of the zeolite. Then, they must diffuse from the outer surface into the inner pores, where adsorption or ion-exchange takes place. If inter-particle diffusion is slow, it can become the limiting step for the entire ion removal process. In this situation, the rate of metal ion removal is mainly controlled by diffusion rather than the kinetics of adsorption or ion-exchange itself. This can result in the zeolite particles within the column or bed not being fully saturated with metal ions, even after prolonged contact. The overall adsorption capacity of the zeolite may be lower than expected because metal ions cannot reach the interior parts of the zeolite particles quickly enough.

To address some of the limitations of natural zeolites, researchers have developed zeolite-based composites by integrating them with other functional materials (Vipin et al. 2023b). These composites often demonstrate improved adsorption performance due to synergistic effects. For instance, combining zeolites with iron oxides (Phouthavong et al. 2024), activated carbon (Li et al. 2020; Smržová et al. 2022), SiO_2 (Wang et al. 2019b), or biopolymers (Nakamoto et al. 2017) can enhance both the surface area and the chemical affinity for specific metal ions.

Magnetic nanoparticles have frequently been incorporated into zeolite structures (Yuan et al. 2018). Those "magnetic zeolite composites" offer the advantage of easy separation from treated water using external magnetic fields, making the process more practical for large-scale applications. On the other hand, zeolites have been modified with organic groups, surfactants, or metal oxides to increase their selectivity and adsorption capacity. Such modified zeolites are often designated as "functionalized zeolites." Embedding zeolites in biopolymers or synthetic polymers to prepare "polymer-zeolite composites" is another strategy to improve their mechanical stability and reusability. The performance, stability, selectivity, and adsorption capacity of these modified zeolites are listed in Table 4.4. Zeolite-based membranes are also efficient materials for removing heavy metals from water (Tufail et al. 2025). However, in this chapter, we will not discuss that as Membrane-based Separation of Heavy Metal Ions has been discussed in Chapter 3.

Modifying zeolites with surfactants enhances their ability to adsorb heavy metals from water. For example, clinoptilolite treated with NaCl solution increases its efficiency in removing Ni^{2+} from 73.2% to 96.6% (Biblioteca et al. 2023). Using surfactants like hexadecyltrimethylammonium bromide (HDTMA-Br) or octadecyltrimethylammonium bromide (ODTMA-Br) improves the sorption of chromates from water. When clinoptilolite is modified with HDTMA-Br, it can remove 2.27 $mg{\cdot}g^{-1}$ of Cr(VI) (Warchoł et al. 2006), while surfactant-functionalized ODTMA-Br reaches 2.18 $mg{\cdot}g^{-1}$. Mordenite modified with HDTMA increases Cr(VI) removal from zero to approximately 2.2 $mg{\cdot}g^{-1}$ (Campos et al. 2007). Functionalizing clinoptilolite with HDTMA increases As(V) adsorption capacity by 12.5 times compared to unmodified clinoptilolite (Mendoza-Barrón et al. 2011). Modifications of mordenite and clinoptilolite with HDTMA raise their As(V) adsorption capacity from 17.33 to 97.33 $mmol{\cdot}kg^{-1}$ and from 9.33 to 45.33 $mmol{\cdot}kg^{-1}$, respectively (Chutia et al. 2009). Pohang zeolite and Haruna zeolite modified with hexadecylpyridinium bromide (HDPB) demonstrate more than three times higher Cr(VI) adsorption capacity than the untreated zeolites (Zeng et al. 2010). On the other hand, zeolites exhibit varying affinities for the removal of different heavy metals. For example, zeolite A has a higher adsorption capacity for Pb^{2+} (115.2 $mg{\cdot}g^{-1}$, using 0.2 g of adsorbent suspended in 10 mL of $Pb(NO_3)_2$ 1 $mmol{\cdot}L^{-1}$) than for other monovalent or divalent cations (Meng et al. 2017).

Despite these advances, the regeneration and reuse of both natural and modified zeolites remain active research topics. The ability to restore their adsorption capacity through chemical or thermal regeneration without significant performance loss is crucial for the economic viability and environmental sustainability of zeolite-based water treatment systems.

4.3 ION CAPTURE OVER CARBONACEOUS MATERIALS

Carbon, the sixth most abundant element in the universe, exists in various allotropes, including activated carbon (AC), graphite/graphene, diamond, fullerenes, carbon nanotubes CNTs), etc. Among these, activated carbon, graphene, and CNTs stand out as highly effective adsorbents due to their large surface areas, unique electronic structures, and distinctive chemical and physical properties.

Activated carbon is an amorphous material with a three-dimensional (3D) network of carbon atoms and a highly porous surface rich in diverse functional groups. This structure enables the adsorption of a broad range of ions and molecules. In contrast, graphene consists of one or a few layers of sp^2-hybridized carbon atoms arranged in a two-dimensional (2D) honeycomb lattice. It demonstrates exceptional mechanical strength, thermal conductivity, optical transparency, and electrical conductivity, making it a highly promising material for many applications.

A notable derivative of graphene is graphene oxide (GO), a single-layer form of graphite oxide that is functionalized with multiple oxygen-containing groups, such as carboxyl, phenol, and quinone, at the sheet edges. GO can be partially reduced to form reduced graphene oxide (rGO), which retains many of graphene's desirable properties while offering enhanced chemical functionality. Both GO and rGO, either on their own or as part of composite materials, have been effectively employed for water purification, particularly in removing heavy metal ions from industrial or wastewater.

CNTs, another widely studied carbon-based material for water treatment, are essentially rolled-up graphene sheets that form one-dimensional (1D) sp^2-bonded carbon nanostructures. Their high surface area, unique shape, and chemical tunability make them ideal for ion adsorption.

In the following subsections, we will analyze the performance of these three key carbon allotropes: activated carbon, graphene (and its derivatives), and CNTs, in capturing heavy metal ions from aqueous environments.

4.3.1 Ion Capture over Activated Carbon

Since its first use in the 1940s, activated carbon has been a standard adsorbent in the water industry for treating municipal and industrial wastewater to produce potable water (Cheremisinoff and Ellerbush 1979; Weber and Vanvliet 1980). The use of carbon as an adsorbent for treating liquid industrial waste streams is relatively recent, but it has been successfully used for removing metals from electroplating wastewater (Lyman 1987). Activated carbon is considered a highly effective adsorbent due to its high surface area and porous structure. Activated carbon is characterized by its high surface area (600–2000 m^2/g), predominant microporous structure (pore size < 2 nm), and the presence of surface functional groups such as hydroxyl, carbonyl, and carboxyl that enhance metal binding. Despite previously mentioned advantages, the large-scale use of activated carbon for water treatment is limited by its high production costs, an expensive regeneration process, and limited economic feasibility for large-scale wastewater treatment. The limitations of activated carbon have driven the search for cost-effective alternatives from different carbon-rich sources. These organic precursors are often by-products from agriculture or the food industry, such as bark, tannin-rich materials, sawdust, rice husk, fruit stones, coconut shells, seafood processing wastes, and more. Industrial and natural wastes, including scrap tires, petroleum waste, fertilizer by-products, peat moss, chitosan, algae, seaweed, resins, and fly ash, have also been used to produce activated carbon.

The preparation of activated carbon from these carbon-rich wastes involves a two-step process: carbonization and activation. Carbonization involves pyrolyzing the precursor without oxygen at temperatures between 500°C and 800°C to produce fixed carbon structures. Activation is performed using physical activators such as steam, CO_2, and air at high temperatures, or chemical agents like $ZnCl_2$, H_3PO_4, and KOH. The choice of activator and temperature influences the surface area, pore volume, and adsorption efficiency of the resulting activated carbon. During carbonization, the precursor undergoes pyrolytic decomposition by removing non-carbon elements such as H, N, O, and S (Fitzer et al. 1971). Initially, low-molecular-weight volatile compounds are released, followed by lighter aromatic species, and finally hydrogen gas, leading to the formation of a fixed carbonaceous char (Lewis 1982). However, the pores formed during this process often become clogged with tarry pyrolysis residues. As a result, an activation step is essential to increase the material's internal surface area and develop its porous structure.

A wide range of carbon-rich feedstocks has been utilized to produce activated carbon, typically through the catalytic activation of pyrolyzed char. Commercial-scale production primarily relies on pine wood (~130,000 tons/year), anthracite and bituminous coal (~100,000 tons/year), and coconut shells (~35,000 tons/year), along with lignite and peat (~50,000 tons/year). Alternative sources, such as olive stones and almond shells, contribute approximately 10,000 tons/year. These precursors typically contain 40–90% carbon (w/w) and have bulk densities ranging from 0.40 to 1.45 g/cm³. In Figure 4.7, the steps involved in the synthesis of activated carbon from carbon-rich waste materials are depicted schematically.

In certain cases, chemical activation is initiated prior to the pyrolysis or carbonization stage. This approach involves impregnating the feedstock with a nearly saturated solution of a chemical (often referred to as a catalyst) such as $ZnCl_2$, H_3PO_4, H_2SO_4, KOH, K_2S, and KCNS, followed by drying. The impregnated chemicals alter the subsequent pyrolysis process by suppressing tar formation and reducing volatilization. The material is then subjected to a combined carbonization and activation procedure, typically carried out in a single step involving two temperature stages. Carbonization starts around 400 °C, followed by activation at higher temperatures reaching up to 800 °C, a process commonly used in the production of industrial-grade wood-based activated carbons. The pore-size distribution of the final product is strongly affected by the level of chemical impregnation during the initial stage (Bansal et al. 1988; Li et al. 2019).

As mentioned earlier, activated carbon can have various functional groups on its surface, depending on the used precursor. It can also carry different surface charges, positive or negative, depending on the activation process used. Typically, acid-activated carbon has a positive surface charge, while base-activated carbon has a negative surface charge, although the pH of the reaction mixture can influence this. Therefore, the adsorption of heavy metal ions onto activated carbon involves several mechanisms, including: (i) ion-exchange, where heavy metal ions replace other ions on the carbon surface; (ii) complexation, where heavy metal ions form complexes with functional groups on the activated carbon surface; and (iii) electrostatic interactions, which attract or repel heavy metal ions depending on the pH of the solution.

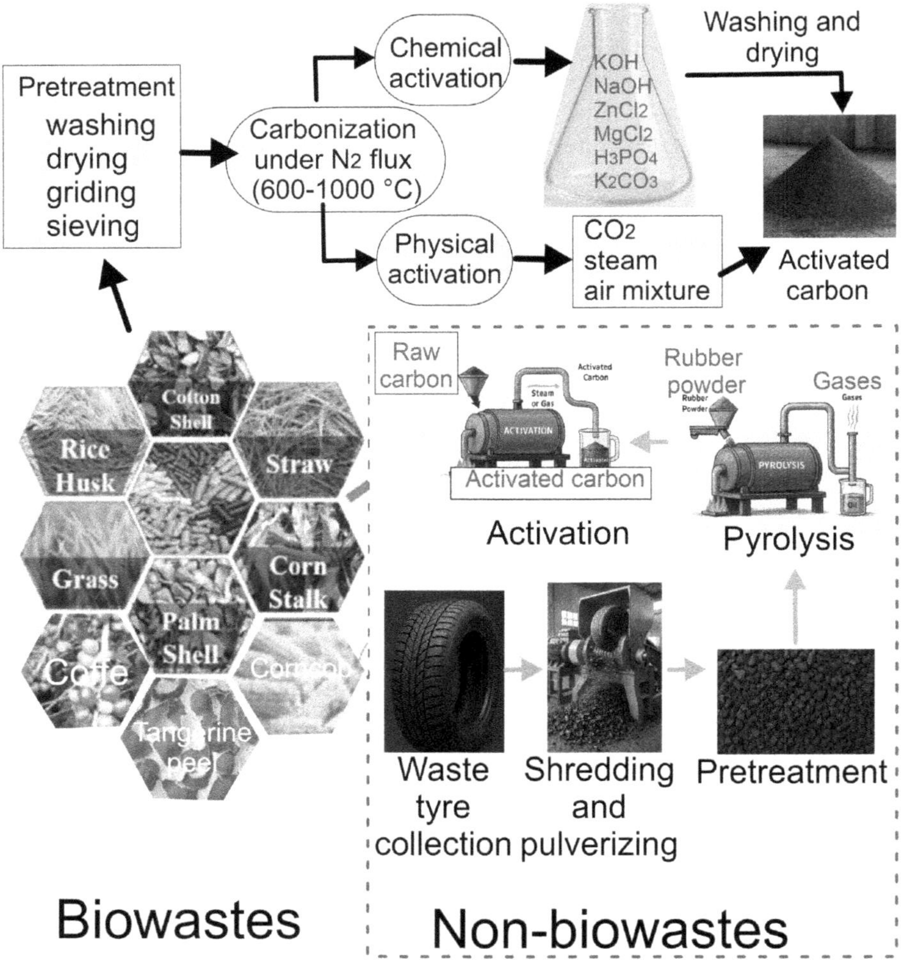

FIGURE 4.7 Scheme for the steps involved in the preparation process of activated carbon using bio- and non-bio wastes. Adapted from Naji and Tye (2022; Copyright, Elsevier 2022); Kaur et al. (2021b; Copyright, Nature 2021); Nurmalita et al. (2022; Copyright, Editora UFLA 2022); Muttil et al. (2023; Copyright, MDPI 2023).

As case studies, we can consider a few examples of activated carbons from various origins, prepared using different surface modification processes. Srinivasan and Balasubramanian (2023) prepared activated carbon from rice husk by steam activation for the removal of Cd(II) ions. The adsorbent could remove 94.4% of Cd(II) ions from water compared to only 25.4% removal by commercial activated carbon Filtrasorb 400. Orhan and Büyükgüngör (1993) prepared activated carbon from pine bark and nut shell, which delivered performance similar to that of commercial activated carbon for removing Cd(II), Cu(II), Hg(II), and Pb(II) metal ions. Guo et al. (2008) prepared activated carbon rich in carboxylic and phenolic surface groups from black liquor, a by-product of the paper industry, for the removal of

Pb(II), Cd(II), Zn(II), and Ni(II) ions from water, demonstrating considerable capacity. The work demonstrated the potential dual use in waste valorization and pollutant control. Ajmal et al. (1996) prepared activated carbon from phosphate-treated mango sawdust, which was capable of removing 100% Cr(VI) ions from electroplating wastewater. The adsorbent could be regenerated with NaOH to retain 87% of its adsorption capacity, demonstrating the feasibility of cyclic use in treatment facilities. The performance of activated carbons derived from different natural resources is presented in Table 4.5.

While the performance of activated carbons derived from different sources in heavy metal ion capture varies due to differences in their textural and surface characteristics, other external factors, such as pH, temperature, particle size, and initial metal ion concentration in the reaction medium, also influence their adsorption efficiency. For example, in the adsorption of Pb(II) and Hg(II) ions, a pH range of 4–8 is generally favorable. On the other hand, Cr(VI) ions are effectively adsorbed at lower pH values (pH 2–3). The pH of the medium affects the ionization and speciation of both the adsorbate and the adsorbent surface, altering the adsorption behavior of the adsorbent. At elevated temperatures, the adsorption capacity of adsorbents increases due to the endothermic nature of the process. Additionally, a smaller particle size yields a larger specific surface area, thereby enhancing an adsorbent's adsorption rate and capacity. Finally, a higher initial metal-ion concentration in the probing medium provides a greater driving force for mass transfer, promoting faster adsorption and rapid saturation of the adsorbent surface.

4.3.2 Ion Capture over Carbon Nanotubes and Carbon Nanotube-Based Composites

CNTs are cylindrical, one-dimensional structures with nanometer-scale diameters composed of carbon atoms arranged in a hexagonal pattern, resembling rolled-up graphene sheets. The CNTS can be as long as a few millimeters. CNTs can be classified as single-walled (SWCNTs) or multi-walled (MWCNTs) and are utilized in various applications, including composite production, drug delivery, sensor development, and wastewater treatment. CNTs rank among the stiffest and strongest fibers identified, with Young's moduli as high as 1 TPa and tensile strengths of up to 63 GPa. Depending on the atomic arrangement at their edges, they can behave as metals or semiconductors. Both SWCNTs and MWCNTs have demonstrated their potential as adsorbents for heavy metal ions due to their high specific surface area, hollow structure, and ease of surface functionalization. Bare CNTs usually contain few surface functional groups and tend to aggregate, which limits their capacity to adsorb metal ions. However, surface functionalizations such as oxidation (adding carboxyl, hydroxyl, or carbonyl groups), functionalization with ligands (e.g., amide, polyelectrolytes), or decoration with nanoparticles (usually metal oxides) significantly enhance their affinity for metal ions. The mechanism of heavy-metal binding with CNTs includes electrostatic attraction to negatively charged functional groups over their surface, ion-exchange (as electrodes in electrosorption systems), complexation or chelation with surface ligands, and occasionally precipitation or sorption-precipitation on or near the CNT surface (De Silva et al. 2025). As stated earlier, due to the very low quantity of surface functional

TABLE 4.5
Adsorption of Heavy Metals by Active Carbons Derived from Different Sources

Feedstock	Activation method	BET surface area ($m^2\ g^{-1}$)	Micropore volume ($cm^3\ g^{-1}$) and pore diameter (nm)	Removed heavy metal	Performance of the adsorbent	Reference
Coconut shell	Phosphoric acid, 800 °C, 2 h	2648	V_p: 0.762 r_p:1.4–4.0 nm	-	Utilized for benzene uptake	Lewis (1982)
Spent coffee	Steam-activated at 800 °C	464	V_p: NA, r_p:0.57 nm	Ni^{2+}	51.91 mg/g (at pH 6.0)	Hernández Rodríguez et al. (2018a)
Coffee husk	Steam-activated at 800 °C	383	V_p: NA, r_p:0.57–1.4 nm	Ni^{2+}	57.14 mg/g (at pH 6.0)	Hernández Rodríguez et al. (2018a)
Cocoa seed husk	Steam-activated at 850 °C	428	0.204, 0.57 nm	Ni^{2+}	104.99 mg/g	Hernández Rodríguez et al. (2018b)
Mangrove (*Rhizophora mucronate*) fruit	KOH activated at 900 °C	1061.5	0.5325, 2.0 nm	Pb^{2+} Cd^{2+}	99% removal for 10 ppm of Pb^{2+} and Cd^{2+} ions	Sherugar et al. (2022)
Coffee grounds	Phosphoric acid, 450 °C, 1h	925	0.7, < 3 nm	____	Utilized for dye (adsorption	Reffas et al. (2010)
Bean dreg	Steam-activation at 1000 °C under He gas	1004	0.5, NA	Cr(VI)	3.30 mmol/g in acidic conditions	Chu et al. (2020)
Pomelo peel	Phosphoric acid (85 wt%), 250 °C, 2h, in air	____	____	Ag(I) Pb(II)	137.4 and 88.74 mg/g (at pH 4.0)	Zhao et al. (2018)
Sewage sludge	i) 500 °C, 2h, in N_2 ii) 10M HNO_3 at 90 °C, 4h	202.1 and 721.2	____	Pb(II)	1.25 mg/g, and 4.05 mg/g (at pH 3)	Li et al. (2019)

groups, bare or non-functionalized CNTs are poor adsorbers of metal ions. However, their adsorption capacity increases significantly upon functionalization or when conjugated with other materials in the form of nanocomposites.

Several CNT composites have demonstrated very high heavy metal removal efficiencies and capacities. For example, a composite made of CNT, starch, and magnetite nanoparticles (CNT/starch/Fe_3O_4) has shown excellent capacity for Pb(II) and Cd(II) capture, removing 99.74% of Pb(II) and 98.76% of Cd(II) from aqueous solution under mild conditions (pH 6, room temperature), with maximum Langmuir adsorption capacities (q_m factor) of about 50.36 mg/g for Pb(II9 and ~ 44.72 mg/g for Cd(II) (M. Zhang 2025). The mean free energy of the adsorption process (E_{ads}) was determined to be <8 kJ/mol, which indicated that the process is physical, that is, through physisorption. As another example, we can consider the work of (Yang et al. 2020), who prepared a chestnut-shell biochar-supported CNT composite. This composite achieved a notable Pb(II) sorption capacity of approximately 1641 mg/g, which is significantly higher than that of biochar alone. This suggests that adding a small amount of CNTs enhances the adsorption capacity of biochar. A magnetic MWCNT composite decorated with maghemite (Fe_2O_3) was prepared by in situ precipitation at 325K and pH 10, using ammonia solution as the precipitating agent (Gatabi et al. 2016). The nanocomposite demonstrated good Cd(II) removal capacity (maximum 78.8 mg/g at 298K), with rapid kinetics. A list of CNT-containing nanocomposites used in heavy metal ion removal, along with their performances, is provided in Table 4.6.

As can be noticed, CNTs and their composites are strong candidates for removing heavy metal ions from water, especially after proper functionalization, composite formation, or decoration with magnetic or non-magnetic nanostructures, offering high capacity and rapid kinetics. However, practical challenges, such as production costs, robustness under real wastewater conditions (with mixed ions and varying pH levels), reusability, regeneration, and environmental safety issues, remain.

4.3.3 Ion Capture over Graphene and Graphene-Based Nanocomposites

Graphene is a 2D material composed of sp^2-hybridized carbon atoms arranged in a honeycomb lattice. It exhibits extraordinary surface area and versatile surface chemistry. Graphene derivatives, such as graphene oxide (GO) and reduced graphene oxide (rGO), provide abundant oxygen-containing groups and have tunable properties. Research to date has revealed that these materials can serve as high-performance adsorbents for heavy metal ions, including Pb^{2+}, Cd^{2+}, Hg^{2+}, Cr(VI), U(VI), and actinides (Siti et al. 2020; Rout et al. 2023; Dervin et al. 2016). While pristine graphene offers an exceptional theoretical surface area, it tends to aggregate and restack in aqueous environments, thereby reducing the number of accessible adsorption sites. Therefore, in most of the studies, it serves as a baseline material rather than an adsorbent.

GO is the most extensively studied member of the graphene family in water purification. Its abundant oxygen groups, such as carboxyl, hydroxyl, and epoxy, provide hydrophilicity and chemical reactivity, making GO an efficient platform for electrostatic interactions and surface complexation with metal cations (Wang and Chen 2015). On the other hand, rGO features partially restored graphitic domains,

TABLE 4.6
Performance of Bare CNTs and CNT-Based Composites in Heavy Metal Ion Removal from Water

Material/composite	Target metal ions	Adsorption capacity (mg/g)	Removal efficiency/performance/experimental conditions	References
Bare/oxidized CNTs	Cd(II) Pb(II), Cd(II), Cu(II), Zn(II)	13.5 (for oxidized CNTs, 6 h functionalization)	Moderate. Oxidation improves the removal efficiency significantly. Functionalization time affects the performance. In the mixture, Pb(II) shows the highest adsorption among other ions. Desorption depends on pH and the presence of competing ions. pH and initial concentration affect kinetics and efficiency.	Sheraz et al. (2024) Ma et al. (2015)
CNT-Hydroxiapatite (MWCNT-HAP) composites	Pb(II), Cu(II), Cd(II)	~720 for Pb(II), ~107 for Cu(II) and ~105 for Cd(II)	Pb(II) shows the highest adsorption. The effect of ionic strength is pronounced in Cd(II) case, less pronounced for Pb(II) and Cu(II). Optimum pH: 4.0–6.0. The main adsorption mechanisms of Pb(II) and Cu(II) include decomposition-precipitation of HAP, exchange reaction between Pb^{2+} and Ca^{2+}, and surface complexation. The main mechanisms of Cd(II) adsorption are the exchange reaction between Cd^{2+} and Ca^{2+} and electrostatic adsorption.	Li et al. (2021)
NiO/CNT nanocomposite	Pb(II)	100% removal of Pb(II) by the composite. Bare CNT achieves ~43% under the same conditions and dose.	At pH 7, using 0.1 g of adsorbent in 50 mL of 20 ppm Pb(II), the composite achieves full removal in ~10 minutes (much faster than bare CNTs).	Diva et al. (2017)
–COO(–), –OH, and $-CONH_2$ Functionalized CNTs (MD simulation study)	Pb(II), Cu(II), Cd(II), Hg(II)	1.28–2.04 mmol/g	Adsorption capacity improves significantly by functionalization. Adsorption capacity increases with increasing metal-ion concentration. CNT-COO(–) surface effectively adsorbs over 150–230% more metal ions than the bare CNT surface. –OH and $-CONH_2$ functionalization improves 10–47% adsorption efficiency. Order of adsorption: Pb(II) > Cu(II) > Cd(II) > Hg(II)	Anitha et al. (2015)

enhancing electrical conductivity necessary for electrical and electrochemical applications. rGO has often been used to fabricate scaffolds and to prepare hybrid or composite materials for heavy metal ion capture.

For metal-ion capture, functionalization has proven essential for improving the selectivity and binding capacity of graphene, GO and rGO. For example, thioleted GO showed remarkable affinity for Hg^{2+}, while carboxylated GO effectively captures Pb^{2+} and U(VI) (Jalilian et al. 2025). The adsorption of heavy metal ions on graphene-based adsorbents results from multiple mechanisms (often occurring simultaneously), including (i) Electrostatic attraction and ion-exchange, (ii) surface complexation, (iii) Redox reactions, (iv) π-metal interactions, and (v) physical entrapment. In electrostatic attraction and ion-exchange processes, deprotonated functional groups on the GO surface attract cationic species (the process is pH-dependent). In surface complexation, carboxyl, hydroxyl, amino, and thiol groups form coordination bonds with heavy metal ions, creating inner-sphere complexes. In some instances, the π electrons of graphitic regions interact with soft metals like mercury through π-electron interaction. Finally, in the case of 3D graphene aerogels and foams, metal ions are entrapped and diffusion-limited entrapment occurs within porous networks.

As an example of metal ion capture by functionalized graphene, we can consider the work by Madadrang et al. (2012), who demonstrated that EDTA-functionalized GO exhibits outstanding Pb^{2+} adsorption capacity, highlighting the role of chelating ligands grafted onto GO in boosting metal ion affinity. To increase the capacity of graphene, GO, and rGO, they are often combined with other materials to form composites. Depending on the material combinations, different composites with distinct properties and names, such as magnetic graphene composites, metal oxide/graphene hybrids, polymer/graphene hybrids, MOF/graphene composites, and graphene-based aerogels, have been created. For example, Liu et al. (2013) developed a porous Fe_3O_4 hollow microspheres/graphene oxide (Fe_3O_4/GO) composite by self-assembling amine-functionalized porous Fe_3O_4 microspheres prepared via a hydrothermal process. This magnetic composite adsorbs Cr(VI) ions from aqueous solution, with a maximum capacity of 32.33 mg/g, which is significantly higher than that of bare Fe_3O_4 microspheres. Adding GO to the composite not only prevented aggregation of magnetite microspheres but also enhanced their dispersibility in water, thereby enabling their recovery and reuse after ion adsorption. Several other non-magnetic metal oxides, such as MgO, TiO_2, and MnO_2, have been combined with GO or rGO to improve their adsorption capacities. For example, Mohan et al. (2017) synthesized a GO/MgO nanocomposite to remove Pb(II) ions from water. The composite, synthesized via a simple chemical precipitation method, exhibited a high Pb(II) uptake capacity of 190 mg/g even at a low dose of 0.4 g/L, within 30 minutes at pH 6.5.

Functionalized GO has shown promising results in capturing several other metals, such as actinides, including U(VI), Eu(III), and Np(V), beyond the common heavy metals. Through density functional theory (DFT) calculations, Chen et al. (2025) demonstrated that carboxylated GO is an effective adsorbent for actinides such as Pa(V), U(VI), Np(V), and Pu(VI). By functionalizing with –COOH, –O–COOH, and –NH–COOH groups, they explored the mechanism of actinide capture, finding that the –NH group serves as an electron donor to adjacent –COOH groups, thereby enhancing electrostatic attraction to hydrated actinide cations. Xie et al. (2016) modified their GO samples with carboxylic sulfonate functional groups to study the adsorption mechanism of actinides. The functionalized GO demonstrated an adsorption capacity of 298 mg/g for U(VI). Surface complexation modeling indicated that sorption of Eu(III) and U(VI) to carboxylate groups is the primary mechanism of their sorption to GO.

4.4 ION CAPTURE OVER METAL OXIDES

Metal oxide nanostructures have shown great promise for removing metal ions from wastewater due to their large surface area, porous structures, and strong adsorption capacity. The surface properties of these oxides can be easily tailored by introducing functional groups to enhance their affinity toward specific metal ions. The efficiency of metal oxides in capturing metal ions also depends on their composition. The principal mechanism of ion capture on metal oxide surfaces is adsorption through electrostatic interactions, which is directly influenced by the surface charge of the adsorbents. In the following subsections, the origin of surface charge on metal oxides and their ion-capture performances have been discussed.

4.4.1 Surface Charge over Metal Oxides

Metal oxides exhibit positive or negative surface charges depending on various physicochemical factors. These surface charges arise from structural substitutions, disorders or defects, and interactions with ionic species in aqueous environments (Sposito 1998). Surface charge over metal oxides can be categorized into three main components: (i) structural surface charge (σ_o), originating from intrinsic features such as aliovalent ion substitutions; (ii) protonic surface charge (σ_H), associated with H^+ and OH^- ions forming complexes at the surface of oxide of hydroxides; and (iii) ionic adsorption charge (Δq), derived from the adsorption of other ions (excluding H^+ and OH^-) (Sposito 1998). For example, when trivalent Al^{3+} ions in Al_2O_3 are selectively substituted by divalent cations like Mg^{2+}, Cu^{2+}, or Zn^{2+}, a net negative structural charge (σ_o) is developed at the surface. Protonic surface charge (σ_H) is associated with the adsorption of H^+ or OH^- groups. In contrast, Δq (ionic adsorption charge) covers all other ion adsorption processes, including inner-sphere surface complexes (σ_{IS}), outer-sphere surface complexes (σ_{OS}), and diffuse layer adsorption (σ_d). The ionic adsorption charge can be expressed as:

$$\Delta q \equiv \sigma_{IS} + \sigma_{OS} + \sigma_d$$

where σ_{IS} is the inner sphere charge, σ_{OS} is the outer sphere charge, and σ_d is the dynamic of dipole contribution. The overall charge balance of a metal oxide in contact with an electrolyte solution is given by:

$$\sigma_0 + \sigma_H + \Delta q = 0$$

The net surface charge depends strongly on the solution pH. One key parameter describing this relationship is the point of zero net proton charge (pznpc), the pH at which the net protonic charge (σ_H) equals zero.

Another important reference point is the point of zero charge (PZC) or the pH at which the total charge at the oxide surface is zero. PZC is commonly determined by potentiometric mass titration. Since it depends on the solution's ionic strength, it is generally controlled using inert electrolytes such as KCl, $KClO_4$, KNO_3, and $NaNO_3$. These salts are selected due to the low binding energy of the K^+, Na^+, Cl^-, ClO_4^-, and NO_3^- ions at the oxide surface. Figure 4.8 illustrates a theoretical surface charging model with a PZC of 8.5 under varying ΔpK and Ns values at 0.1 M ionic strength, along with experimental data of surface charge density versus pH for NiO in different $NaNO_3$ concentrations.

The PZC and isoelectric point (IEP) are two standard parameters used to characterize metal oxides in their aqueous dispersions. Both parameters are critical in understanding ion adsorption phenomena in metal oxide surfaces (Kosmulski 2016). Table 4.7 presents a comparative overview of the PZC and IEP values for a range of metal oxides under different experimental conditions.

4.4.2 Ion Capture over Metal Oxides and Their Composites

Metal oxides such as ZnO (Gu et al. 2020), TiO_2 (Huang et al. 2012), ZrO_2 (Zhao et al. 2022), MgO (Wang et al. 2020a), and MnO_2 (Wang et al. 2020) exhibit moderate adsorption capacities for heavy metals in aqueous media. Their performance can be enhanced through surface modification or functionalization, or by forming composites with materials containing amines (Li et al. 2014) or carboxylic functional groups (Liu et al. 2008).

Polymer with low isoelectric points (IEP) or points of zero charge (PZC), such as chitosan (Liu et al. 2008), polydopamine (Zhao et al. 2022), polystyrene (Zhao et al. 2022), are often combined with metal oxides to enhance their heavy metal adsorption (Chen et al. 2018). Because these composites carry a negative surface charge even at acidic pH values (as low as 2), they effectively attract and bind positively charged heavy metal ions (Zhao et al. 2022). In contrast, when polymers with high PZC values (e.g., polypyrrole) are used, adsorption occurs primarily on the metal oxide phase. For example, in a TiO_2–polypyrrol composite (PZC = 8.05 for TiO_2 and 10.4 for polypyrrole), adsorption of heavy metals occurs mainly on the TiO_2 surface (Chen et al. 2018).

Composites containing metal oxides often exhibit higher solubility and hydrophilicity than their individual components, properties that are beneficial for the adsorption of heavy metals (Chen et al. 2018). In general, metal oxide composites exhibit superior performance in heavy metal removal in comparison to bare metal oxides. For example, $MnFe_2O_4$@Mn–Co oxide exhibits maximum adsorption capacities of 481.2, 386.2, and 345.5 mg g^{-1} for Pb(II), Cu(II), and Cd(II), respectively. These values are significantly higher than those of pristine $MnFe_2O_4$ (Ma et al. 2013). Composites prepared by incorporating biochar into metal oxide systems also exhibit higher adsorption capacity than bare metal oxides. For example, Tho et al. (2021) found that the efficiency of As(III), Cd(II), Pb(II), and Cr(VI) removal of a 3% ZnO loaded biochar composite is higher than bare ZnO or biochar. Chen et al. (2018) demonstrated that a TiO_2-polypyrrole composite exhibits higher adsorption capacities for Zn(II) and Pb(II) ions than pure TiO_2. Additionally, Zhao et al. (2020) fabricated

TABLE 4.7
PZC and IEP of Selected Metal Oxides

Oxide	Electrolyte	Instrument	PZC	IEP	References
γ-Al_2O_3	0.001–0.1 M KNO_3	Malvern Zetasizer 5000	8.2	8.6	Bourikas et al. (2005); Bourikas et al. (2003)
Bi_2O_3	0.01 M KCl	electroosmosis	-	9.4	Nechaev EA (1979)
CdO	0.001–0.1 M $NaClO_4$/ 0.01 M KCl 0.001	electroosmosis	10.1	10.6	Nechaev and Shein (1979); Janusz (1991)
CeO_2	0.01 M $NaNO_3$	Pen Kem Zee Meter 501	-	6	Janusz (1991)
CeO_2	0.1–1 M $NaClO_4$		8.6		Nabavi et al. (1993)
Co_3O_4	0.005–0.2 M KNO_3	Mark II	8.2	8.2	Nabavi et al. (1993)
CuO	0.1 M $NaClO_4$	Zeta-Meter Coulter		9.4	Nechaev and Shein (1979)
Fe_3O_4	0.001–0.1 M KNO_3	-	6.9	6.8	Regazzoni et al. (1983); Blesa et al. (1984)
ϒ-Fe_2O_3	0.001–0.1 M KNO_3	Delsa 400 Coulter	6.6	6.6	Garcell et al. (1998); Serna and Morales (2004)
α-Fe_2O_3	0.01 M KCl	Malvern Zetasizer		9.0	Schudel et al. (1997)
Ga_2O_3	0.0001 M $NaNO_3$	Malvern Zetasizer 3000		9.0	Schudel et al. (1997)
HfO_2	0.0001–0.1 M KCl,	Zeta Probe Colloidal Dynamics	7.1	8.1	Zhang et al. (2012)
La_2O_3	0.001–0.1 M NaCl	Electroosmosis	9.6	9.6	Zhang et al. (2012)
MgO	0.001–0.1 M KNO_3	Malvern Zetasizer 5000	12	10	Bourikas et al. (2003)
MnO	0.001 M NaCl	Zeta PALS Brookhaven	-	5.1	Škapin et al. (2015)
Mn_3O_4	0.01 M NaCl	Malvern Mastersizer	-	5.7	Ocaña (2000)
Mn_2O_3	0.001–0.1 M NaCl,	Malvern Zetasizer 3000	6.2	3.5	Janusz and Gałgan (2001)
MnO_2	0.001 M $NaClO_4$	Zeta Plus Brookhaven	-	4.4	Cristiano et al. (2011)
MnO_2	0.001–0.1 M KNO_3	-	-	6.7	Prélot et al. (2003)
Nb_2O_5	0.002 M $NaClO_4$	Zeta Plus Brookhaven		4.1	Kosmulski (1997)
NiO	0.001 M $NaNO_3$	Pen Kem 501		8.1	Naklicki et al. (2002)
PbO	0, 0.001 M KCl	Rank Mark II	-	10.7	Taha et al. (1990)
RuO_2	0.01–0.25 M KNO_3	Rank Brothers Mark II	5.8	5.5	Taha et al. (1990)
SnO_2	0.0001–1M KNO_3	Rank Brothers Mark II	4.3	4.5	Houchin and Warren (1985)
TiO_2	0.001–0.1 M $NaClO_4$	Malvern Zetasizer 3000	5.5	5.2	Houchin and Warren (1985)
WO_3	0, 0.001 M KCl	Rank Mark II	-	2	Taha et al. (1990)
Y_2O_3	0.01 M NaCl	Malvern Zetasizer II	-	8.2	Wei et al. (1999)
ZnO	0.0001 M NaCl	Malvern Zetasizer Nano ZS	-	8.9	Wei et al. (1999)
ZnO		0.005–0.5 M KNO_3	-	9.2	Szekeres and Tombácz (2012)
ZrO_2	0.01 M NaCl	Malvern Zetasizer 2000	6.1	-	Megías-Alguacil et al. (2000)

Source: Kosmulski (2016).

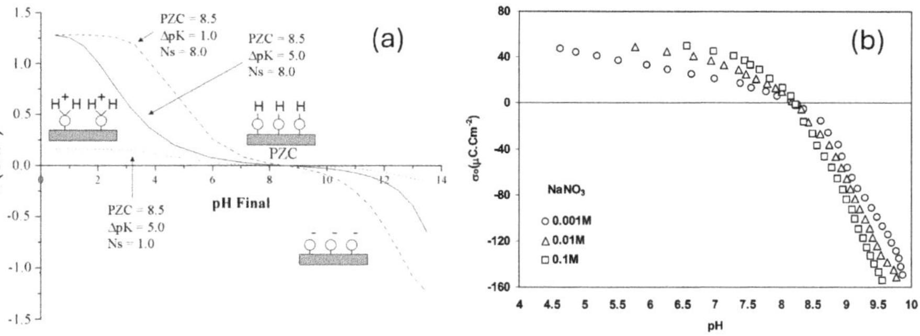

FIGURE 4.8 (a) Simulated surface charging behavior of a metal oxide with a PZC = 8.5, showing the influence of ΔpK and surface site density (Ns). The simulation assumes a surface loading of 50 m^2/L and a background ionic strength of 0.1 M. Reproduced with permission from Schreier et al. (2010; Copyright Elsevier, 2010). (b) Experimental surface charge density (σ_o) of NiO as a function of pH in $NaNO_3$ solutions at concentrations and 303 K. Reproduced with permission Mahmood et al. (2011; Copyright ACS, 2011).

a nanocomposite comprising Fe_3O_4 and graphitic carbon nanosheets for use as electrodes, which can remove Cd(II) and Pb(II) from water with high efficiency. The mechanisms of interaction of heavy metals with these composites include ion-exchange, electrostatic attraction, surface complexation, inner-sphere complexation, and, in some cases, reduction of the heavy metal species followed by adsorption (Wang et al. 2020a).

Metal oxyhydroxides also serve as effective adsorbents for removing heavy metals. A well-known example is goethite (α-FeOOH), used either directly or in combination with membranes, which has demonstrated excellent performance in removing heavy metals from water (Kosmulski and Mączka 2004; Shindel et al. 2023). It is important to note that adsorption capacity can be negatively impacted by the presence of competing ions such as Ca(II), Mg(II), or Zn(II) (Zhao et al. 2022; Huang et al. 2012). Also, the adsorption performance of a metal oxide can vary depending on the type of heavy metal ions. For example, sodium titanates exhibit an adsorption capacity trend of Cd(II) > Zn(II) > Ni(II) (Huang et al. 2012). The kinetics for adsorption frequently follow a pseudo-second-order model (Huang et al. 2012; Gu et al. 2020). Table 4.8 summarizes the adsorption capacities of various metal oxides and their composites. Since adsorption is often reversible, many of these adsorbents can be regenerated and reused.

4.5 ION CAPTURE OVER MOFS

4.5.1 Structure and Porosity of Metal-Organic Frameworks

MOFs are crystalline, porous materials composed of inorganic nodes consisting of metal ions or metal clusters, which are coordinated to multitopic organic linkers, forming extended, three-dimensional lattices. They are characterized by high chemical and structural tunability, ultrahigh porosity, and very large surface areas (Li et al. 2024).

TABLE 4.8
Adsorption Performance of Common Metal Oxides and Their Composites for Heavy Metal Removal from Aqueous Solutions

Adsorbent	Heavy metal	Adsorption isotherm model and performance: q_m, Langmuir model	K_F, Freundlich model	Reference
Amorphous phosphate titanium oxide (APTO)	Pb(II) Cd(II) Cr(III)	Pb(II): 761.3 mg·g^{-1} Cd(II): 193.3 mg·g^{-1} Cr(III): 255.2 mg·g^{-1}	-	Wang et al. (2020b)
Humic acid-coated Fe_3O_4	Cu(II) Cd(II) Pb(II) Hg(II)	Cu(II): 46.3 mg·g^{-1} Cd(II): 50.4 mg·g^{-1} Pb(II): 92.4 mg·g^{-1} Hg(II): 97.7 mg·g^{-1}	-	Liu et al. (2008)
Polydopamine adhered sulfonated polystyrene/ZrO_2	Pb(II)	Pb(II): 428.41 mg·g^{-1}	Pb(II): 176.6 mg·g^{-1}	Zhao et al. (2022)
Sodium titanate nanoflowers	Pb(II)	Pb(II): 1.47x10^{-3} mol·g^{-1}	Pb(II): 7.2x10^{-3} mol·g^{-1}	Huang et al. (2012)
Mg-doped lithium manganese oxide	Pb(II):	Pb(II): 49.46 mg·g^{-1}	-	Saravaia et al. (2018)
ZnO	Cr(III)	Cr(III): 88.57 mg·g^{-1}	Cr(III): 33.55 mg·g^{-1}	Gu et al. (2020)

MOFs stem from polyhedral or small-cluster coordination polymers, which can be progressively extended into one-dimensional chains, two-dimensional layers, and ultimately three-dimensional frameworks (Kobielska et al. 2018). Figure 4.9 illustrates a segment of $Zn_4(BTC)_2(\mu_4\text{-}O)(H_2O)_2$ framework, consisting of 2D layers formed by $Zn_4(\mu_4\text{-}O)$ subunits bridged with 1,3,5-benzenetricarboxylate (BTC^{3-}) ligands. MOFs typically exhibit high surface area (1,000–10,000 m^2 g^{-1}) and possess highly ordered structures (Shamim et al. 2022; Zhao et al. 2021a). Such high surface areas provide abundant active sites for the adsorption of molecules or ions. Moreover, the presence of multiple functional groups enables anchoring of molecules or nanostructures, making MOFs attractive scaffolds for hosting catalysts, photocatalysts, sensing elements, enzymes, and numerous other functional nanomaterials (Kung et al. 2020). MOFs can be classified according to the net charge of the host framework as cationic, neutral, or anionic types (Zhao et al. 2021b). Another widely used classification distinguishes between carboxylate-based MOFs (e.g., HKUST-1, Cu-BTC, MOF-5, IRMOF-3, UiO-66) and imidazolate-based MOFs (e.g., ZIF-7 and its derivatives).

Stable MOFs across a wide range of pH and temperature conditions can be synthesized by combining a hard Lewis acid (typically metal ions with high valence states and small ionic radii, such as Al^{3+}, Cr^{3+}, Fe^{3+}, Zr^{4+}, and Ti^{4+}) with a hard Lewis base, or soft Lewis acids (e.g. Fe^{2+}, Co^{2+}, Ni^{2+}, Mn^{2+}, and Zn^{2+}) with soft Lewis bases. Carboxylate-based ligands present a common class of hard Lewis bases. Soft Lewis

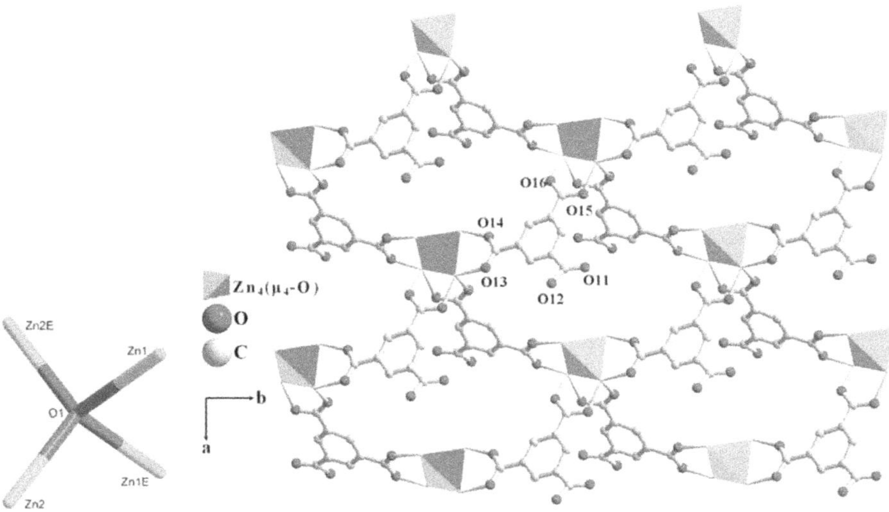

FIGURE 4.9 (a) Structural subunit of $Zn_4(\mu_4\text{-}O)$; (b) 2D layer of the $[Zn_4(BTC)_2(\mu_4\text{-}O)(H_2O)_2]$ MOF, formed through the linkage of $Zn4(\mu_4\text{-}O)$ subunits with BTC^{3-} ligands. Adapted from Xu et al. (2008; Copyright Elsevier, 2008).

TABLE 4.9
Classification of Hard, Broadline, and Soft Acids

Hard acids	Borderline acids	Soft acids
H^+, Li^+, Na^+, K^+,		
Be^{2+}, Mg^{2+}, Ca^{2+}, Sr^{2+}		
BF_3, BCl_3, $B(OR)_3$ Al^{3+}, $Al(CH3)_3$, $AlCl_3$, AlH_3	$B(CH)_3$	BH_3, Tl^+, $Tl(CH_3)_3$
Cr^{3+}, Mn^{2+}, Fe^{3+}, Co^{3+}	Fe^{2+}, Co^{2+}, Ni^{2+}, Cu^{2+}, Zn^{2+}, Rh^{3+}, Ir^{3+} Ru^{3+}, Os^{2+}	Cu^+, Ag^+, Au^+, Cd^{2+}, Hg_2^{2+}, Hg^{2+}, CH_3Hg^+, $[CO(CN)_5]^{2-}$, Pd^{2+}, Pt^{2+}, Pt^{4+}, Br_2, I_2
Zr^{4+}, Ti^{4+}, Si^{4+}, Th^{4+}, U^{4+}, WO^{4+}, Mo^{3+}, etc. (Ions with formal oxidation states of 4 or higher.		Metals with zero oxidation state
HX (hydrogen-bonding molecules).		π acceptors: e.g., trinitrobenzene, quinines, tetracyanoethylene

Source: Pearson (1968).

acids can form stable MOFs when coordinated with soft Lewis bases such as azolate-based ligands (Lam et al. 2023). Soft Lewis acids are generally characterized by lower valence states and medium to high ionic radii. Classifications of hard and soft acids and bases are summarized in Tables 4.9 and 4.10, respectively.

In addition to the aforementioned cations, other commonly employed metal ions in MOF synthesis are Ca^{2+}, Mg^{2+}, Bi^{3+}, Ag^+, La^{3+}, and Pr^{3+}. Among these, Zr-based

TABLE 4.10
Classification of Hard, Broadline, and Soft Bases

Hard bases	Borderline bases	Soft bases
HF^-, Cl^-, H_2O, OH^-, O_2^-, ROH, RO^-,R_2O,CH_3COO^-,NO_3^-,ClO_4^-	Br-, NO_2^-, N_3^-,	I^-, H^-, H_2S, SH^-, S_2^-, RSH, RS^-, R_2S, SCN^-, CN^-, RNC, CO,
CO_3^{2-}, SO_4^{2-}, PO_4^{3-}	SO_3^{2-}	$S_2O_3^{2-}$,
NH_3, RNH_2, N_2H_4	$C_6H_5NH_2$, C_5H_5N, N_2	PR_3, $P(OR)_3$, AsR_3, C_2H_4, C_6H_6

Source: Pearson (1968).

MOFs are regarded as the most water-stable due to the high coordination number of their secondary building units (SBUs). This high stability arises from the strong Zr–O bonds and the ability of the inner Zr_6-cluster to undergo reversible rearrangements upon the removal or addition of μ_3-OH groups, without altering the connectivity of the carboxylate linkers (Cavka et al. 2008).

The pore sizes of MOFs can be finely tuned by varying the length of the organic linkers. Furthermore, the preferred coordination geometry of the metal center (e.g., octahedral, tetrahedral, or square planar) plays a critical role in determining both pore size and shape. The resulting porous networks provide accessible channels that facilitate target (guest molecules and ions) diffusion and transport. A wide variety of MOFs can thus be designed by combining metal nodes with organic linkers through coordinative bonds (Lam et al. 2023).

MOFs are often constructed from highly symmetrical inorganic secondary units, also referred to as "bricks," such as the $Cu_2(OH)_2(CO_2)_4$ brick of HKUST-1, the $Zn_4O(CO_2)_6$ brick of MOF-5, and the $Cr_3O(OH)_3(CO_2)_6$ brick of MIL-88 (Cavka et al. 2008). The organic linkers typically contain two or more coordinating groups, most commonly carboxylates. Representative examples include 1,4-benzenedicarboxylic acid (BDC), (Lam et al. 2023) 1,3,5-benzenetricarboxylate (BTC^{3-}), 2,5-dimercaptoterephthalic acid (H_2DMBD) (Yee et al. 2013), and isophthalic acid, among others. The structures of several linkers used in the synthesis of MOFs bearing carboxylic acid groups are presented in Figure 4.10. Beyond carboxylates, azolate linkers containing two or more nitrogen atoms are also employed for MOF synthesis (Figure 4.11). Other organic linkers used in MOF synthesis include cyanide, pyridyl, sulfonates, phosphonates, and hydroxyl moieties.

Common examples of MOFs include MOF-5 [$Zn_4O(BDC)_3(DMF)_8(C_6H_5Cl)$] (D. Li et al. 2024), MOF-808 [$Zr_6O_5(OH)_3(BTC)_2(HCOO)5(H_2O)_2$] (Khosroshahi et al. 2022), and zeolitic imidazolate framework ZIF-8 (Tanihara et al. 2021). Many MOFs are named after the institutions where they were first synthesized, such as UiO-66 (University of Oslo), HKUST (Hong Kong University of Science and Technology), MIL (Materials Institute Lavoisier), SNU (Seoul National University), JUC (Jilin University, China), ITQMOF (Instituto de Tecnología Química MOF), POST (Pohang University of Science and Technology), and CUK (Cambridge University KRICT) (Li et al. 2024).

The synthesis of MOFs typically involves mixing a metal salt, which provides the cations for the inorganic nodes, with an organic linker in a suitable solvent,

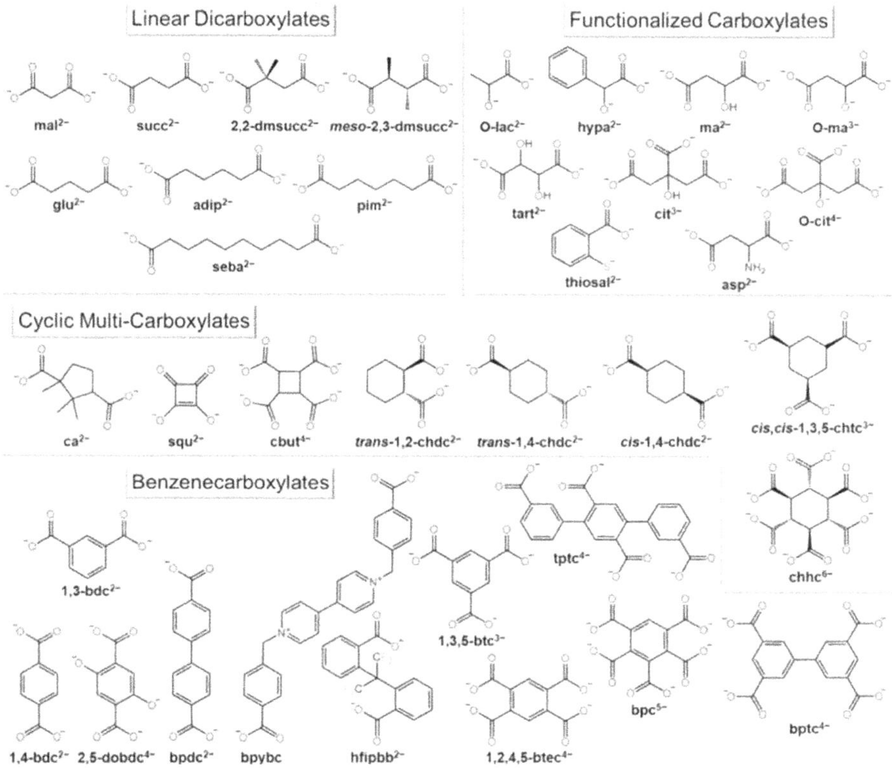

FIGURE 4.10 Structures of some multitopic carboxylate linkers commonly employed in MOF synthesis. Adapted from Thorarinsdottir and Harris (2020; Copyright ACS, 2020).

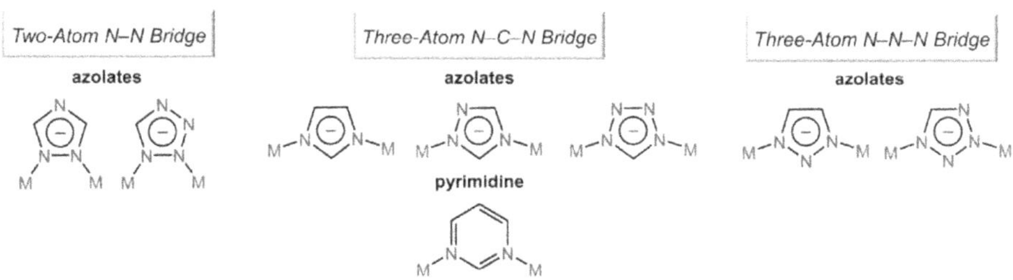

FIGURE 4.11 Examples of short bridging pathways between two metal (M) centers provided by N-heterocyclic ligands. From top to bottom, the structures shown are a two-atom N–N bridge, a three-atom N–C–N bridge, and a three-atom N–N–N bridge. Reproduced from Thorarinsdottir and Harris (2020; Copyright ACS, 2020).

such as *N,N*-dimethylformamide (DMF). Key parameters influencing MOF formation include the choice of solvent, reaction temperature, and counterions. The strategies used for MOF synthesis include solvothermal, hydrothermal, ionothermal, microwave-assisted, sonochemical, electrochemical, and mechanochemical methods, as illustrated in Figure 4.12. Additionally, MOFs can be prepared via slow evaporation techniques or using templates.

Figure 4.13a illustrates a representative procedure for synthesizing a Zr-based MOF by reacting $ZrCl_4$ with H_2DMBD in DMF. Figures 4.13b and 4.13c illustrate a tetracarboxylic acid linker and demonstrate how one of the carboxylic acid groups can interact with and adsorb Cd(II) ions.

Figure 4.14 illustrates the structures of several common MOFs, highlighting both the metal clusters (based on Zr, Fe, Cr, Zn, and Mg elements) and the corresponding

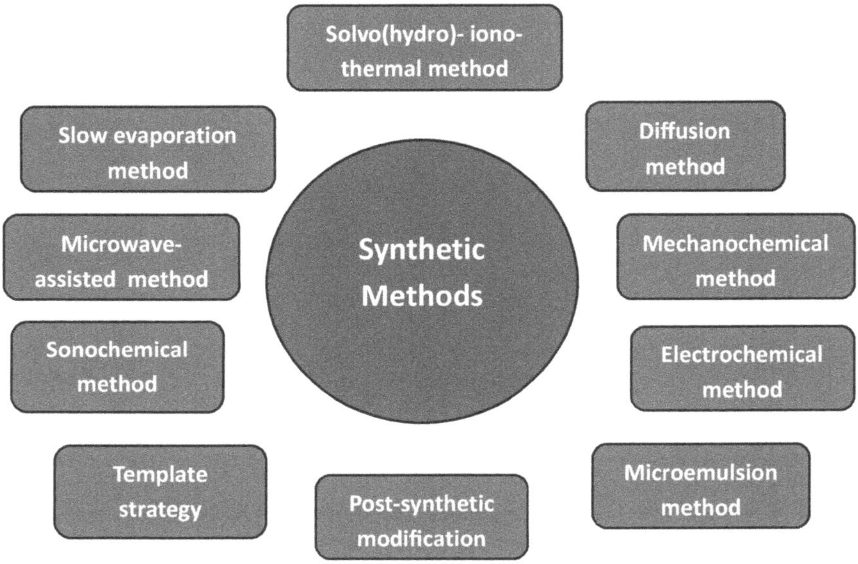

FIGURE 4.12 Common synthesis approaches for MOFs. Reproduced from Raptopoulou (2021; Copyright MDPI, 2021).

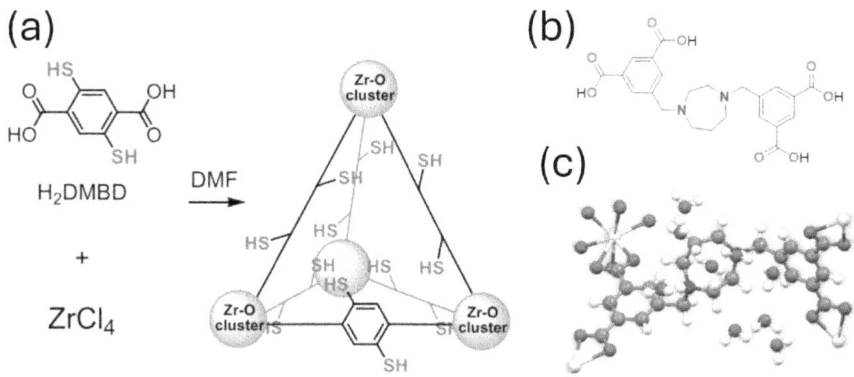

FIGURE 4.13 (a) Zr-DMBD network in a Zr-based MOF Yee et al. (2013), (Copyright ACS, 2013). (b) Tetracarboxylic acid H_4L (L= bis-isophthalic acid). (c) Asymmetric unit of Cd-MOF, poly-[Cd(H_2L)]·9H_2O. Reproduced from Ang et al. (2024; Copyright ACS, 2024).

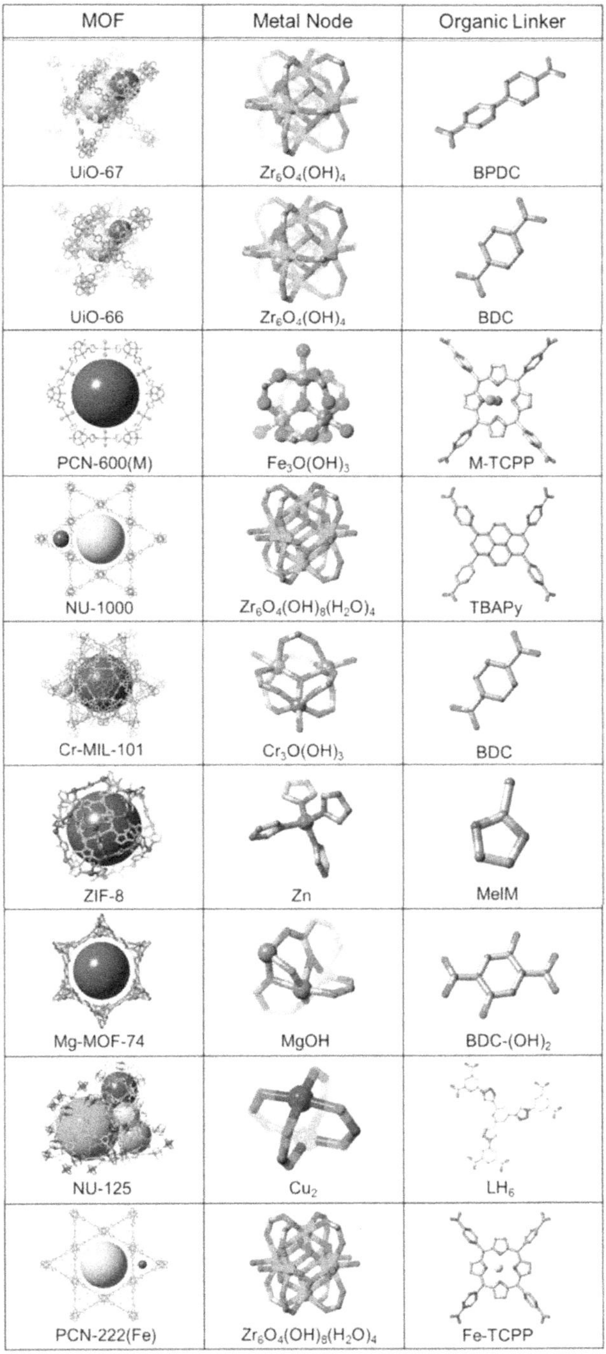

FIGURE 4.14 Representative structures of MOFS, showing metal cluster and organic linkers. Reproduced with permission from Howarth et al. (2017; Copyright ACS, 2017).

organic linkers. Metal clusters composed of transition metal elements (TMEs), such as $Zr_6O_4(OH)_4$, $Zr_6O_4(OH)_8(H_2O)_4$, $Fe_3O(OH)_3$, $Cr_3O(OH)_3$, typically contain multiple metal centers per cluster. These clusters frequently coordinated to the carboxylate groups of the organic linkers, forming the extended 3D frameworks characteristic of MOFs. Such 3D architectures can encapsulate nanoparticles within the MOF cavities. The mechanisms underlying interactions between MOFs and heavy metal ions include diffusion, ion-exchange, electrostatic interactions, van der Waals forces, hydrogen bonding, Lewis acid-base interactions, chelation, and coordination, as summarized in Figure 4.15.

4.5.2 Ion Capture over MOFs

MOFs exhibit high adsorption capacities for heavy metals due to their large surface areas and the presence of functional groups such as carboxylates, amines, thiols, and others, which can effectively bind heavy metal ions (Kobielska et al. 2018; Yuan et al. 2025). Additionally, MOFs can be functionalized post-synthesis by incorporating amine-containing molecules, such as ethylenediamine and quinoline, to enhance their adsorption performance. For example, Zhao et al. reported adsorption capacities of Hg^{2+} for UiO-66 and NH_2-functionalized UiO-66 were 50.3 and 223.8 mg g^{-1}, respectively (Zhao et al. 2021b). Generally, functionalizing MOFs with sulfur-containing molecules (e.g., thiol, thiourea, isothiocyanate) enhances their performance for Hg^{2+} adsorption (Li et al. 2025). However, unfunctionalized MOFs, which do not adsorb Hg^{2+} ions, can still capture other heavy metal ions, such as Pb^{2+}, Cr^{3+}, and Cd^{2+} (Zou

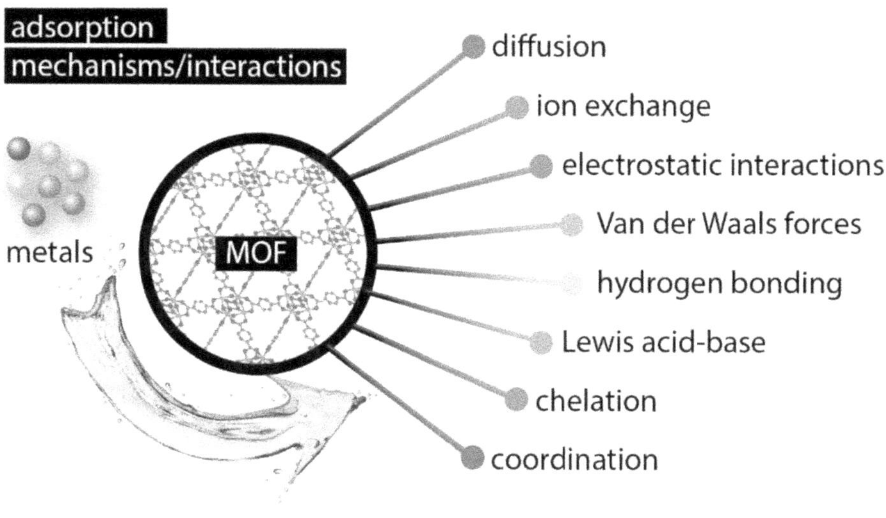

FIGURE 4.15 Schematic representation of the interactions and mechanisms involved in the adsorption of metals by MOFs. Reproduced with permission from Manousi et al. (2019; Copyright MDPI, 2019).

TABLE 4.11
Heavy Metal Adsorption Performance of Metal-Organic Frameworks

MOF	Initial concentration of the heavy metal ($mg \cdot L^{-1}$)	Adsorption isotherm model and performance: q_m, Langmuir model	Adsorption isotherm model and performance: K_F, Freundlich model	Reference
ZIF-8	Pb(II): 1000 Cu(II): 1000	Pb(II): 629 $mg \cdot g^{-1}$ Cu(II): 726 $mg \cdot g^{-1}$	-	Tanihara et al. (2021)
MIL-101(Cr)	Cr(VI)	Cr(VI): 39.5 $mg \cdot g^{-1}$	Cr(VI): 4.3 $mg \cdot g^{-1}$	Liu et al. (2020b)
UiO-66 functionalized with ethylene diamine	Pb(II): 300 Cd(II): 300 Cu(II): 300	Pb(II): 243.9 $mg \cdot g^{-1}$ Cd(II): 217.4 $mg \cdot g^{-1}$ Cu(II): 208.3 $mg \cdot g^{-1}$	Pb(II): 69.0 $mg \cdot g^{-1}$ Cd(II): 35.0 $mg \cdot g^{-1}$ Cu(II): 26.1 $mg \cdot g^{-1}$	Ahmadijokani et al. (2021)
MIL-125 functionalized with quinoline	Pb(II): 500 Cd(II): 500 Cu(II): 500 Cr(III): 500	Pb(II): 262.1 $mg \cdot g^{-1}$ Cd(II): 102.8 $mg \cdot g^{-1}$ Cu(II): 66.9 $mg \cdot g^{-1}$ Cr(III): 53.9 $mg \cdot g^{-1}$	Pb(II): 68.4 $mg \cdot g^{-1}$ Cd(II): 57.6 $mg \cdot g^{-1}$ Cu(II): 30.7 $mg \cdot g^{-1}$ Cr(III): 42.4 $mg \cdot g^{-1}$	Abdelhameed et al. (2019)

et al. 2013). The adsorption capacities of several MOFs for various heavy metals are presented in Table 4.11.

4.5.3 Ion Capture over MOF-Based Nanocomposites

MOFs have been widely used to prepare nanocomposites with polymers, such as chitosan, or with metal oxides to enhance their adsorption performance. Additionally, combining MOFs with magnetic materials facilitates their easy recovery and separation from aqueous solutions (Houchin and Warren 1985; Kobielska et al. 2018). Incorporation of inorganic polyoxometalates, such as phosphotungstate, into MOF-based nanocomposites further improves their stability in water, as these species can be embedded within the MOF's channels (Zou et al. 2013). Performance of some MOF-based composites in heavy metal removal has been listed in Table 4.12.

4.6 ION CAPTURE OVER NON-CARBONACEOUS 2D MATERIALS

4.6.1 Ion Capture over Layered Double Hydroxides

The heavy metal removal efficiency of adsorbents depends on several factors, including the contact time between the adsorbent and the solution, agitation speed, solution pH, adsorbent particle size, initial metal ion concentration, reaction temperature, and ionic strength. Therefore, a careful control of all these factors is essential to optimize the adsorption process. To determine the maximum adsorption capacity and evaluate the kinetics of metal-ion uptake, estimating the equilibrium time is particularly

TABLE 4.12
Performance of a Few MOF-Based Nanocomposites in the Adsorption of Heavy Metals

MOF-based nanocomposite	Heavy metal	Adsorption isotherm model and performance		Reference
		q_m, Langmuir model	K_F, Freundlich model	
MOF808/ polyacrylonitrile	Cd(II) Zn(II)	Cd(II): 225 mg·g^{-1} Zn(II): 287 mg·g^{-1}	Cd(II): 225 mg·g^{-1} Zn(II): 287 mg·g^{-1}	Efome et al. (2018)
ZIF-8/polysulfone	Pb(II) Cd(II)	Pb(II): 220 mg·g^{-1} Cd(II): 161 mg·g^{-1}		Roy et al. (2021)
MIL-101(Cr)/ polyethyleneimine	Cr(VI)	Cr(VI): 62 mg·g^{-1}	Cr(VI): 11.5 mg·g^{-1}	Liu et al. (2020b)
Bi-I-Fe_3O_4@SiO_2@ HKUS-1	Hg(II)	Hg(II): 264 mg·g^{-1}		Huang et al. (2015)
PAN/chitosan/ UiO-66-NH_2	Pb(II)	Pb(II): 441 mg·g^{-1}		Huang et al. (2015)
MIL-100(Fe)/ rGO/δ-MnO_2	As(V)	Pb(II): 162 mg·g^{-1}		Huang et al. (2015)

Note: MOF, metal organic framework.

important (Zou et al. 2013). The solution pH influences the adsorbent surface charge. At pH values below the point of zero charge (pH_{PZC}), the adsorbent surface becomes positively charged, which hinders the uptake of cationic species such as Pb(II), Cd(II), Cr(III), and Hg(II). At pH values higher than the pH_{PZC}, the adsorbent surface acquires a negative charge, promoting the adsorption of positively charged heavy metal ions. On the other hand, reducing the particle size of adsorbents enhances the specific surface area and the number of available active sites increases, which enhances the adsorption capacity of the adsorbent. Similarly, a higher sorbent dosage in solution results in a greater percentage of heavy metal removal. The initial concentration of the metal ions also plays a decisive role, as higher concentrations provide the driving force needed to overcome mass transfer resistance between the metal ions and the adsorbent surface.

LDHs are ionic, clay-like 2D materials with a hydrotalcite-type structure (Abd El-Monaem et al. 2023). Their framework is composed of divalent and trivalent cations coordinated by six hydroxyl groups, forming octahedral units. These octahedra share edges to generate positively charged brucite-like layers (host layers), whose charge is neutralized by exchangeable anions located in the interlayer region (Abd El-Monaem et al. 2023). Figure 4.16 illustrates the brucite-type layers of LDHs, where divalent and trivalent cations occupy the octahedral sites. The general chemical formula of LDHs is: $[M_{1-x}{}^{2+}M_x{}^{3+}(OH)_2]^{x+}(A^{n-})_{x/n}mH_2O$ where

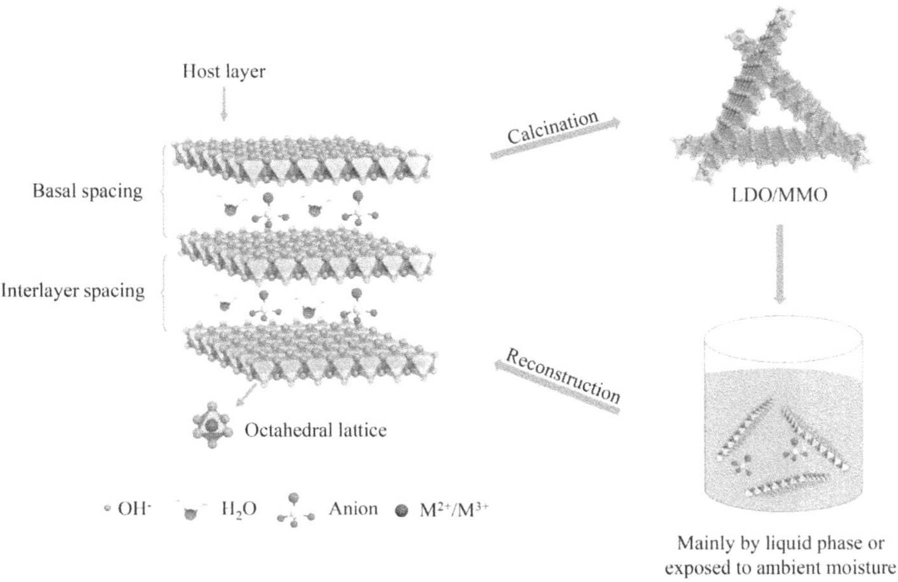

FIGURE 4.16 The characteristic structure and "memory effect" of LDHs. The brucite-type layers, composed of divalent and trivalent cations in octahedral coordination with hydroxyl groups, are balanced by interlayer anions and water molecules. The memory effect describes the reversible transformation in which LDHs are converted into mixed metal oxides or layered double oxides upon calcination, and subsequently regenerate their layered structure when rehydrated in aqueous solutions or steam environments. Reproduced with permission Cui et al. (2023; Copyright Taylor & Francis, 2023).

M^{2+} and M^{3+} are divalent (Zn^{2+}, Mg^{2+}, Co^{2+}, Ni^{2+}, Cu^{2+}, Mn^{2+}) and trivalent (Fe^{3+}, Mn^{3+}, Al^{3+}, Cr^{3+}, V^{3+}, Ga^{3+}) metal cations, respectively. A^{n-} represents charge-balancing interlayer anions, which can be inorganic or organic species such as nitrates, chlorates, sulphate, carbonates, phosphates, chromates, dichromates, carboxylates, surfactant anions, polyoxometalates, or amino acids anions (Kaur et al. 2021a). The parameter x (typically ranging from 0.2 to 0.33) reflects the molar fraction of M^{3+} relative to M^{2+} and defines the net layer charge of LDH, which is directly linked to key properties such as the anion-exchange capacity (Kaur et al. 2021a; Di Bitetto et al. 2017). The anions residing at the interlayer sites increase the hydrophilicity of LDHs, enhancing their structural stability. On calcination under controlled ambient, the LDHs are transformed into mixed metal oxides, that is, layered double oxides (LDOs) through the removal of water and interlayer ions (Figure 4.16). Interestingly, this process is reversible: exposure of LDOs to aqueous solutions containing anions restores the layered LDH structures, a phenomenon known as "memory effect" (Cui et al. 2023). LDHs generally have a higher charge density than other clay minerals, such as montmorillonite, hectorite, and laponite (Wang & O'Hare 2012).

LDHs are highly effective materials for removing heavy metal ions, primarily via ion-exchange. In this process, the interlayer anions of LDH are exchanged with heavy metal anions. To further enhance their adsorption capacity, small organic molecules containing functional groups such as sulfhydryl, carboxyl, or amine moieties (e.g., L-cysteine, EDTA, and mercaptocarboxylic acid) can be incorporated either within the interlayer space or onto the LDH surface. These small molecules act as additional active sites, enabling strong chelation with heavy metal ions (Liu et al. 2023). In addition, LDHs intercalated with MoS_4^{2-} anions have demonstrated improved adsorption performance toward both oxoanions, such as $[HSeO_3]^-$, $[SeO_3]^{2-}$, and $[SeO_4]^{2-}$ oxoanions and cations, including Cr^{3+}, Cd^{2+}, Hg^{2+}, and Cu^+ (Ma et al. 2017a; Ma et al. 2017b). Another strategy involves combining LDHs with polymers, such as polypyrrole, thereby enhancing overall adsorption efficiency. In these composites, the polymers interact with the LDH layers through hydrogen bonding, thereby providing additional stability and functionality. Table 4.13 lists selected LDH-based adsorbents and their performance in heavy-metal removal.

4.6.2 Ion Capture over 2D Chalcogenides

Molybdenum disulfide (MoS_2) is a widely studied 2D chalcogenide material, capable of adsorbing heavy metal ions from aqueous solutions due to exposed S atoms on its surface. Structurally, MoS_2 consists of two sulfur layers sandwiching a molybdenum layer, forming a S-Mo-S configuration. MoS_2 monolayers stack via weak van der Waals interactions to form bulk MoS_2. The adsorption performance of MoS_2 for heavy metals has been extensively reported, as summarized in Table 4.14. Moreover, the development of MoS_2-based nanocomposites has been shown to enhance the adsorption capacity of the chalcogenide. For example, Aghagoli and Shemirani (2017) observed that the adsorption capacity for Ni^{2+} over MoS_2 increases from 98 to 294 mg g^{-1} when MoS_2 is grown over graphene oxide (GO). Similarly, MoS_2 grown over activated carbon was seen to exhibit an adsorption capacity of 22.15 mg g^{-1} for Cd^{2+} (Yin et al. 2022). Functionalizing a MoS_2/activated carbon composite with polydopamine enhanced its Pb(II) adsorption capacity up to 396 mg g^{-1} (Yin et al. 2022). The Pb(II) adsorption mechanism in the functionalized MoS_2/activated carbon composite is mainly attributed to complexation by MoS_2 and chelation through hydroxyl, carbonyl, and amino groups. The nanocomposite can be effectively regenerated using 5 mM EDTA-2Na solution, which acts as a desorbing agent.

The high adsorption capacity of MoS_2 is attributed to soft-soft interactions between the sulfur atoms (soft bases) and heavy metal ions (soft acids), as discussed in Table 4.8 (Wang et al. 2018a; Ai et al. 2016). The adsorption performance of MoS_2 also depends on solution pH. For example, Li et al. (2018) found that the maximum adsorption of Th^{4+} and U^{4+} occurs in the pH range of 8–11. The morphology of MoS_2 also influences its removal efficiency. AI Momani et al. (2024) synthesized MoS_2 nanospheres, nanoplatelets, and nanosheets, which were combined with chitosan to form nanocomposite membranes (Al Momani et al. 2024). Among them, the

TABLE 4.13
Performance of Some LDHs and Their Nanocomposites in Heavy Metals Removal

LDH	Heavy metal	Adsorption isotherm model and performance: q_m, Langmuir model	Adsorption isotherm model and performance: K_F, Freundlich model	Reference
Ca/Al-LDHs	Pb(II) Zn(II)	Pb(II): 76.3 mg·g^{-1} Zn(II): 3.55 mg·g^{-1}	Pb(II): 1886.4 mg·g^{-1} Zn(II): 65.76 mg·g^{-1}	Fan et al. (2025)
MgAl-LDH with intercalated MoS_4^{2-} anion.	Se(VI) Se(IV)	Se(VI): 93 mg·g^{-1} Se(IV): 291 mg·g^{-1}	-	Ma et al. (2017b)
Polypyrrole-modified Layered MgAl-LDH	Cr(VI)	Cr(VI): 75.90 mg·g^{-1}	Cr(VI): 3.67 mg·g^{-1}	Sahu et al. (2019)
MgAl-LDH	As(V)	As(V): 114 mg·g^{-1}	As(V): 390 mg·g^{-1}	Goh et al. (2009)
MgAl-LDH	As(V)	As(V): 0.78 mmol·g^{-1}	-	Yoshida et al. (2015)
LDH/graphene oxide grafted with NH_2 groups.	Cd(II)	Cd(II): 32.78 mg·g^{-1} Cu(II): 182.82 mg·g^{-1}	Cd(II): 8.46 mg·g^{-1} Cu(II): 27.95 mg·g^{-1}	Yoshida et al. (2015)
Fe_3O_4/MgAl-LDH	Cu(II) Cd(II) Pb(II)	Cu(II) 64.66 mg·g^{-1} Cd(II) 74.06 mg·g^{-1} Pb(II) 266.6 mg·g^{-1}	Cu(II) 15.78 mg·g^{-1} Cd(II) 15.86 mg·g^{-1} Pb(II) 54.30 mg·g^{-1}	Yoshida et al. (2015)

TABLE 4.14
Reported Adsorption Capacity of MoS_2 for Various Heavy Metals

Heavy metal	Adsorption capacity, mg/g	Reference
Ag(I)	4000	Wang et al. (2018c)
Ni(II)	98	Aghagoli and Shemirani (2017)
Pb(II)	107	Aghagoli and Shemirani (2017)
Cd(II)	585	Q. Wang et al. (2018)
Hg(II)	2563	Ai et al. (2016)
Th(IV)	455	Li et al. (2018)
Cr(VI)	238	Gao et al. (2016)

membrane containing MoS_2 nanoplatelets exhibited the highest removal efficiency (90%) for Zn^{2+} ions.

Other 2D chalcogenides have also demonstrated ion-capture capabilities. For instance, $(NH_4)_4In_{12}Se_{20}$ exhibits high adsorption capacities for Ag^+, Hg^{2+}, and Pb^{2+} ions via an ion-exchange mechanism involving NH_4^+, although this compound is less stable than MoS_2 (Manos et al. 2007).

4.7 CONCLUSION

A wide variety of inorganic nanomaterials, used in pristine, functionalized, or composite form, have been investigated for removing heavy metal ions from water. Each class of materials offers distinct advantages and limitations in terms of cost, adsorption capacity, selectivity, and reusability. For example, clays, zeolites, and bio-derived porous carbons are low-cost, environmentally friendly options, though their adsorption capacities are often modest. Notable exceptions include sepiolite/polypyrrole composites and Zeolite A-Na, which exhibit outstanding Cr(VI) (>300 mg/g) and Pb(II) (>800 mg/g) uptake, respectively.

By contrast, MOFs, although more expensive than clays, zeolites, or biocarbon materials, generally exhibit higher adsorption performance due to their tunable porosity and surface functionalization. For instance, ZIF-8 shows a Pb(II) adsorption capacity of 629 mg/g and a Cu(II) uptake of 72 mg/g, while ethylenediamine-functionalized UiO-66 captures Pb(II) at 244 mg/g. However, key challenges remain regarding their recyclability and structural stability, limiting their practical use.

Metal oxides and their composites are typically more expensive than MOFs. However, they exhibit a broader affinity for multiple heavy metal ions with moderate adsorption capacities. When combined with CNTs to form nanocomposites, their capacities for adsorbing Pb(II), Cu(II), and Cd(II) ions are enhanced due to synergistic effects.

GO, rGO, and their composites, although not the cheapest, exhibit unique adsorption behavior. They are especially promising for removing actinides and conventional heavy metals such as Pb(II) and Hg(II). High surface area, π-electron interactions, and ease of surface modification make them attractive for advanced applications.

Among non-carbonaceous 2D materials, MoS_2 and $(NH_4)_4In_{12}Se_{20}$ have shown moderate performance in heavy-metal adsorption. On the other hand, LDHs, despite their relatively complex and costly synthetic routes, exhibited some of the highest reported adsorption capacities, owing to their ion-exchange capability and large interlayer spacing.

In summary, while low-cost clays, zeolites, and biocarbons are suitable for large-scale water treatment, advanced materials such as MOFs, graphene composites, and LDHs offer superior adsorption capacities and tunable selectivity. The choice of material should therefore depend on the balance of cost, adsorption efficiency, selectivity, regeneration potential, and, of course, on the application context. A selective integration of these materials into scalable composite systems or hybrid adsorbents, while ensuring recyclability and structural stability, will be critical for real-world applications.

REFERENCES

Abd El-Monaem, Eman M., Hala M. Elshishini, Sara S. Bakr, Hisham G. El-Aqapa, Mohamed Hosny, Gangadhar Andaluri, Gehan M. El-Subruiti, Ahmed M. Omer, and Abdelazeem S. Eltaweil. 2023. "A Comprehensive Review on LDH-Based Catalysts to Activate Persulfates for the Degradation of Organic Pollutants." *NPJ Clean Water* 6 (1). https://doi.org/10.1038/s41545-023-00245-x.

Abdelhameed, Reda M., Rabab A. Ismail, Mohamed El-Naggar, Eman S. Zarie, Ramzy Abdelaziz, and Mardia T. El Sayed. 2019. "Post-Synthetic Modification of MIL-125 with Bis-Quinoline Mannich Bases for Removal of Heavy Metals from Wastewater." *Microporous and Mesoporous Materials* 279 (August 2018): 26–36. https://doi.org/10.1016/j.micromeso.2018.12.018.

Adebowale, Kayode O., Iyayi E. Unuabonah, and Bamidele I. Olu-Owolabi. 2006. "The Effect of Some Operating Variables on the Adsorption of Lead and Cadmium Ions on Kaolinite Clay" 134: 130–39. https://doi.org/10.1016/j.jhazmat.2005.10.056.

Aghagoli, Mohammad Javad, and Farzaneh Shemirani. 2017. "Hybrid Nanosheets Composed of Molybdenum Disulfide and Reduced Graphene Oxide for Enhanced Solid Phase Extraction of Pb(II) and Ni(II)." *Microchimica Acta* 184 (1): 237–44. https://doi.org/10.1007/s00604-016-2000-7.

Ahmadijokani, Farhad, Shima Tajahmadi, Addie Bahi, Hossein Molavi, Mashallah Rezakazemi, Frank Ko, Tejraj M. Aminabhavi, and Mohammad Arjmand. 2021. "Ethylenediamine-Functionalized Zr-Based MOF for Efficient Removal of Heavy Metal Ions from Water." *Chemosphere* 264: 128466. https://doi.org/10.1016/j.chemosphere.2020.128466.

Ai, Kelong, Changping Ruan, Mengxia Shen, and Lehui Lu. 2016. "MoS_2 Nanosheets with Widened Interlayer Spacing for High-Efficiency Removal of Mercury in Aquatic Systems." *Advanced Functional Materials* 26 (30): 5542–49. https://doi.org/10.1002/adfm.201601338.

Ajmal, M., R.A.K. Rao, and B.A. Siddiqui. 1996. "Studies on Removal and Recovery of Cr(VI) from Electroplating Wastes." *Water Research* 30 (6): 1478–82.

Ang, Jade Nadine S., Ali Y. Chahine, Thomas J. Raeber, Stuart R. Batten, and David R. Turner. 2024. "Amine-Based Mof for Precious Metal Remediation." *Inorganic Chemistry* 63 (2): 1258–65. https://doi.org/10.1021/acs.inorgchem.3c03654.

Anitha, K, Sadanandam Namsani, and Jayant K Singh. 2015. "Removal of Heavy Metal Ions Using a Functionalized Single-Walled Carbon Nanotube: A Molecular Dynamics Study." *The Journal of Physical Chemistry A* 119 (30): 8349–58.

Araujo, Leandro Goular, Vinicius Litrenta Medeiros, Guilherme de Paula Guarnieri, Danilo Antonio da Silva, Tamires Watanabe, Júlio Takehiro Marumo, and José Geraldo Nery. 2023. "An Adsorption Agent Based on Chitosan-Zeolite Composite: Environmental and Radioactive Liquid Waste Remediation." *Environmental Science: Advances* 2 (3): 484–94. https://doi.org/10.1039/d2va00148a.

Bansal, R.P., J.P. Donnet, and F. Stoeckli. 1988. *Active Carbon*. New York: Marcel Dekker Inc.

Biblioteca, Ilario, Matteo Sambucci, and Marco Valente. 2023. "Zeolite-Clinoptilolite Conditioning for Improved Heavy Metals Ions Removal: A Preliminary Assessment." *Ceramics International* 49 (23): 39649–56. https://doi.org/10.1016/j.ceramint.2023.09.319.

Bish, David L., and J. William Carey. 2001. "Thermal Behavior of Natural Zeolites." *Reviews in Mineralogy and Geochemistry* 45: 403–52. https://doi.org/10.2138/rmg.2001.45.13.

Biswas, Shanta, Jannatun Fatema, Tonmoy Debnath, and Taslim Ur Rashid. 2021. "Chitosan–Clay Composites for Wastewater Treatment: A State-of-the-Art Review." *ACS ES and T Water* 1 (5): 1055–85. https://doi.org/10.1021/acsestwater.0c00207.

Bitetto, Arnaud Di, Erwan André, Cédric Carteret, Pierrick Durand, and Gwendal Kervern. 2017. "Probing the Dynamics of Layered Double Hydroxides by Solid-State ^{27}Al NMR Spectroscopy." *Journal of Physical Chemistry C* 121 (13): 7276–81. https://doi.org/10.1021/acs.jpcc.6b13106.

Blesa, Miguel A., Nestor M. Figliolia, Alberto J.G. Maroto, and Alberto E. Regazzoni. 1984. "The Influence of Temperature on the Interface Magnetite-Aqueous Electrolyte Solution." *Journal of Colloid And Interface Science* 101 (2): 410–18. https://doi.org/10.1016/0021-9797(84)90052-3.

Bolotina, Nadezhda B., Anatoly N. Sapozhnikov, Nikita V. Chukanov, and Marina F. Vigasina. 2023. "Structure Modulations and Symmetry of Lazurite-Related Sodalite-Group Minerals." *Crystals* 13: 768. https://doi.org/10.3390/cryst13050768.

Borhade, Ashok V., Tushar A. Kshirsagar, Arun G. Dholi, and Jyoti A. Agashe. 2015. "Removal of Heavy Metals Cd^{2+}, Pb^{2+}, and Ni^{2+} from Aqueous Solutions Using Synthesized Azide Cancrinite, $Na_8[AlSiO_4]_6(N_3)_{2.4}(H_2O)_{4.6}$." *Journal of Chemical and Engineering Data* 60: 586–93. https://doi.org/10.1021/je500698x.

Bourikas, Kyriakos, Christos Kordulis, and Alexis Lycourghiotis. 2005. "Differential Potentiometric Titration: Development of a Methodology for Determining the Point of Zero Charge of Metal (Hydr)Oxides by One Titration Curve." *Environmental Science and Technology* 39 (11): 4100–08. https://doi.org/10.1021/es048139n.

Bourikas, Kyriakos, John Vakros, Christos Kordulis, and Alexis Lycourghiotis. 2003. "Potentiometric Mass Titrations: Experimental and Theoretical Establishment of a New Technique for Determining the Point of Zero Charge (PZC) of Metal (Hydr) Oxides." *Journal of Physical Chemistry B* 107 (35): 9441–51. https://doi.org/10.1021/jp035123v.

Campos, V., L.C. Morais, and P.M. Buchler. 2007. "Removal of Chromate from Aqueous Solution Using Treated Natural Zeolite." *Environmental Geology* 52 (8): 1521–25. https://doi.org/10.1007/s00254-006-0596-3.

Cavka, Jasmina Hafizovic, Søren Jakobsen, Unni Olsbye, Nathalie Guillou, Carlo Lamberti, Silvia Bordiga, and Karl Petter Lillerud. 2008. "A New Zirconium Inorganic Building Brick Forming Metal Organic Frameworks with Exceptional Stability." *Journal of the American Chemical Society* 130 (42): 13850–51. https://doi.org/10.1021/ja8057953.

Chen, Jie, Mengting Yu, Caiyun Wang, Jiangtao Feng, and Wei Yan. 2018. "Insight into the Synergistic Effect on Selective Adsorption for Heavy Metal Ions by a Polypyrrole/TiO_2 Composite." *Langmuir* 34 (34): 10187–96. https://doi.org/10.1021/acs.langmuir.8b01987.

Chen, Jun, Xiaoqin Hong, Qingdong Xie, Diankai Li, and Qianfeng Zhang. 2014. "Sepiolite Fiber Oriented-Polypyrrole Nano Fibers for Efficient Chromium(VI) Removal from Aqueous Solution." *Journal of Chemical and Engineering Data* 59: 2275–2282.

Chen, Yiming, Guofeng Jin, Yutian Li, Xueqi Zhao, Yingying Wang, Yang Gao, Zhipan Ma, Yi Geng, and Peng Zhang. 2025. "Investigation into the Adsorption Mechanism of Three Varieties of Carboxylated Graphene Oxide for Pentahydrate Actinide (An = Pa–Pu) Compounds." *ACS Omega* 10 (38): 44706–16. https://doi.org/https://doi.org/10.1021/acsomega.5c07789.

Cheremisinoff, P.N., and F. Ellerbush. 1979. *Carbon Adsorption Handbook*. Michigan: Ann Arbor Science Pub.

Chu, B., Y. Amano, and Machida M. 2020. "Preparation of Bean Dreg Derived N-Doped Activated Carbon with High Adsorption for Cr(VI)." *Colloids and Surfaces A: Physicochemical and Engineering Aspects* 586: 110121.

Chu, Yuting, Sidi Zhu, Fengyun Wang, Wu Lei, Mingzhu Xia, and Chuan Liao. 2019. "Tyrosine-Immobilized Montmorillonite: An Efficient Adsorbent for Removal of Pb2+ and Cu2+ from Aqueous Solution." Research-article. *Journal of Chemical and Engineering Data* 64 (8): 3535–46. https://doi.org/10.1021/acs.jced.9b00304.

Chutia, Pratap, Shigeru Kato, Toshinori Kojima, and Shigeo Satokawa. 2009. "Adsorption of As(V) on Surfactant-Modified Natural Zeolites." *Journal of Hazardous Materials* 162 (1): 204–11. https://doi.org/10.1016/j.jhazmat.2008.05.024.

Cristiano, Elena, Yung Jin Hu, Matthew Siegfried, Daniel Kaplan, and Heino Nitsche. 2011. "A Comparison of Point of Zero Charge Measurement Methodology." *Clays and Clay Minerals* 59 (2): 107–15. https://doi.org/10.1346/CCMN.2011.0590201.

Cui, Shihao, Yutao Peng, Xiao Yang, Xing Gao, Chung Yu Guan, Beibei Fan, Xue Zhou, and Qing Chen. 2023. "Comprehensive Understanding of Guest Compound Intercalated Layered Double Hydroxides: Design and Applications in Removal of Potentially Toxic Elements." *Critical Reviews in Environmental Science and Technology* 53 (4): 457–82. https://doi.org/10.1080/10643389.2022.2070402.

Das, Sankar, Arnab Samanta, Gautam Gangopadhyay, and Subhra Jana. 2018. "Clay-Based Nanocomposites as Recyclable Adsorbent toward Hg(II) Capture: Experimental and Theoretical Understanding." *ACS Omega* 3 (6): 6283–92. https://doi.org/10.1021/acsomega.8b00789.

Das, Triparna, Aditya Kumar, Joel Saji, Ashok Pandey, and Devendra Kumar Patel. 2024. "Remediation of Trace Metals from Aqueous Solution Using a Novel Lignin–Montmorillonite Hydrogel Composite Prepared from Tectona Grandis Sawdust Derived Lignin." *ACS Sustainable Resource Management* 1 (1): 141–53. https://doi.org/10.1021/acssusresmgt.3c00077.

Deb, Amal Kanti, Bhabananda Biswas, Mohammad Mahmudur Rahman, Yunfei Xi, Santosh Kumar Paul, and Ravi Naidu. 2022. "Magnetite Nanoparticles Loaded into Halloysite Nanotubes for Arsenic(V) Removal from Water." *ACS Applied Nano Materials* 5 (9): 12063–76. https://doi.org/10.1021/acsanm.2c00239.

Dervin, Saoirse, Dionysios D. Dionysiou, and Suresh C. Pillai. 2016. "2D Nanostructures for Water Purification: Graphene and Beyond." *Nanoscale* 8: 15115–31. https://doi.org/https://doi.org/10.1039/C6NR04508A.

Diva, T. Navaei, K. Zare, F. Taleshi, and M. Yousefi. 2017. "Synthesis, Characterization, and Application of Nickel Oxide/CNT Nanocomposites to Remove Pb^{2+} from Aqueous Solution." *Journal of Nanostructure in Chemistry* 7: 273–81. https://doi.org/10.1007/s40097-017-0239-0.

Efome, Johnson E., Dipak Rana, Takeshi Matsuura, and Christopher Q. Lan. 2018. "Insight Studies on Metal-Organic Framework Nanofibrous Membrane Adsorption and Activation for Heavy Metal Ions Removal from Aqueous Solution." *ACS Applied Materials and Interfaces* 10 (22): 18619–29. https://doi.org/10.1021/acsami.8b01454.

El-habacha, Mohamed, Youssef Miyah, Salek Lagdali, Guellaa Mahmoudy, Abdelkader Dabagh, Mohamed Chiban, Fouad Sinan, Soulaiman Iaich, and Mohamed Zerbet. 2023. "General Overview to Understand the Adsorption Mechanism of Textile Dyes and Heavy Metals on the Surface of Different Clay Materials." *Arabian Journal of Chemistry* 16 (11): 105248. https://doi.org/10.1016/j.arabjc.2023.105248.

Erdem, E., N. Karapinar, and R. Donat. 2004. "The Removal of Heavy Metal Cations by Natural Zeolites." *Journal of Colloid and Interface Science* 280 (2): 309–14. https://doi.org/10.1016/j.jcis.2004.08.028.

Fan, Chengcheng, Baomin Wang, Hu Feng, Hongyu Zhou, Yihan Zhang, Weigao Ding, and Fei Liu. 2025. "Immobilization and Coordination Chemistry of Friedel's Salt (Ca/Al-LDHs) on Heavy Metals Removal." *Process Safety and Environmental Protection* 193: 262–71. https://doi.org/10.1016/j.psep.2024.11.035.

Fitzer, E., K. Mueller, and W. Schaefer. 1971. "The Chemistry of the Pyrolytic Conversion of Organic Compounds to Carbon." In *The Chemistry and Physics of Carbon*, edited by P.L. Walker, Vol. 237. New York: Marcel Dekker Inc.

Galindo, Valbuena, Hugo Martín, Andrés F. Medina, Julio C. Vargas, and Orlando Hernández Fandiño. 2023. "Synthesis of Zeolites Na-A, Na-X, and Analcime from Crushed Stone Waste and Their Applications in Heavy Metal Removal in Aqueous Media." *Chemical Engineering Research and Design* 197: 159–72. https://doi.org/10.1016/j.cherd.2023.07.018.

Gao, Yang, Changlun Chen, Xiaoli Tan, Huan Xu, and Kairuo Zhu. 2016. "Polyaniline-Modified 3D-Flower-like Molybdenum Disulfide Composite for Efficient Adsorption/Photocatalytic Reduction of Cr(VI)." *Journal of Colloid and Interface Science* 476: 62–70. https://doi.org/10.1016/j.jcis.2016.05.022.

Garcell, L., M.P. Morales, M. Andres-Vergés, P. Tartaj, and C.J. Serna. 1998. "Interfacial and Rheological Characteristics of Maghemite Aqueous Suspensions." *Journal of Colloid and Interface Science* 205 (2): 470–75. https://doi.org/10.1006/jcis.1998.5654.

Gatabi, Maliheh Pashai, Hossain Milani Moghaddam, and Mohsen Ghorbani. 2016. "Efficient Removal of Cadmium Using Magnetic Multiwalled Carbon Nanotube Nanoadsorbents: Equilibrium, Kinetic, and Thermodynamic Study." *Journal of Nanoparticle Research* 18: 189.

Goh, Kok Hui, Teik Thye Lim, and Zhili Dong. 2009. "Enhanced Arsenic Removal by Hydrothermally Treated Nanocrystalline MG/AL Layered Double Hydroxide with Nitrate Intercalation." *Environmental Science and Technology* 43 (7): 2537–43. https://doi.org/10.1021/es802811n.

Gu, Mengqin, Liying Hao, Yigan Wang, Xiyu Li, Yifan Chen, Wei Li, and Li Jiang. 2020. "The Selective Heavy Metal Ions Adsorption of Zinc Oxide Nanoparticles from Dental Wastewater." *Chemical Physics* 534: 110750. https://doi.org/10.1016/j.chemphys.2020.110750.

Guo, Xueyan, Shuzhen Zhang, and Xiao quan Shan. 2008. "Adsorption of Metal Ions on Lignin." *Journal of Hazardous Materials* 151: 134–42. https://doi.org/10.1016/j.jhazmat.2007.05.065.

Hernández Rodríguez, Mónica, Jan Yperman, Robert Carleer, Jens Maggen, Dessalegn Dadi, Grazyna Gryglewicz, Bart Van der Bruggen, José Falcón Hernández, and Alexis Otero Calvis. 2018a. "Adsorption of Ni(II) on Spent Coffee and Coffee Husk Based Activated Carbon." *Journal of Environmental Chemical Engineering* 6: 1161–70.

Hernández Rodríguez, Mónica, Jan Yperman, Robert Carleer, Jens Maggen, Sara Vanderheyden, José Falcón Hernández, Alexis Otero Calvis, and Grazyna Gryglewicz, 2018b. "Evaluation of Activation Parameters of Activated Carbon from Coffee and Cocoa Seed Husk Rests: Carbon Yields and Ni(II) Adsorption Capacity Study." *Desalination and Water Treatment* 104: 175–88. https://doi.org/https://doi.org/10.5004/dwt.2018.21907.

Houchin, Martin R., and Leonard J. Warren. 1985. "Surface Titrations and Electrokinetic Measurements on Australian Cassiterites." *Colloids and Surfaces* 16 (2): 117–26. https://doi.org/10.1016/0166-6622(85)80246-8.

Howarth, Ashlee J., Aaron W. Peters, Nicolaas A. Vermeulen, Timothy C. Wang, Joseph T. Hupp, and Omar K. Farha. 2017. "Best Practices for the Synthesis, Activation, and Characterization of Metal–organic Frameworks." *Chemistry of Materials* 29: 26–39. https://doi.org/10.1021/acs.chemmater.6b02626.

Huang, Jiquan, Yongge Cao, Zhuguang Liu, Zhonghua Deng, Fei Tang, and Wenchao Wang. 2012. "Efficient Removal of Heavy Metal Ions from Water System by Titanate Nanoflowers." *Chemical Engineering Journal* 180: 75–80. https://doi.org/10.1016/j.cej.2011.11.005.

Huang, Lijin, Man He, Beibei Chen, and Bin Hu. 2015. "A Designable Magnetic MOF Composite and Facile Coordination-Based Post-Synthetic Strategy for the Enhanced Removal of Hg^{2+} from Water." *Journal of Materials Chemistry A* 3 (21): 11587–95. https://doi.org/10.1039/c5ta01484k.

Ibrahim, Ahmed H., Xianjun Lyu, and Amr B. ElDeeb. 2023. "Synthesized Zeolite Based on Egyptian Boiler Ash Residue and Kaolin for the Effective Removal of Heavy Metal Ions from Industrial Wastewater." *Nanomaterials* 13: 1091. https://doi.org/10.3390/nano13061091.

IZA-SC. 2017. "Structure Commission of the International Zeolite Association." https://america.iza-structure.org/IZA-SC/search_fw.php.

Jalilian, Milad, Pooya Parvizi, and Mohammad Reza Zangeneh. 2025. "Advances in Graphene-Based Nanomaterials for Heavy Metal Removal from Water: Mini Review." *Water* 97: e70062. https://doi.org/https://doi.org/10.1002/wer.70062.

Janusz, W. 1991. "Adsorption and Precipitation Processes in the Cadmium Oxide/Aqueous NaCl or $NaClO_4$ Solution System." *Journal of Colloid And Interface Science* 145 (1): 119–26. https://doi.org/10.1016/0021-9797(91)90105-H.

Janusz, W., and A. Gałgan. 2001. "Electrical Double Layer at Manganese Oxides/1:1 Electrolyte Solution Interface." *Physicochemical Problems of Mineral Processing* 35: 31–41.

Jiménez, Alejandro, Raquel Trujillano, Vicente Rives, Miguel A. Vicente, and Antonio Gil. 2024. "Removal of Heavy Metals from Aqueous Solutions by Adsorption on Zeolites Synthesized from Aluminum Saline Slags." *Inorganic Chemistry Communications* 170: 113440. https://doi.org/10.1016/j.inoche.2024.113440.

Jiménez-Castañeda, Martha E., and Dora I. Medina. 2017. "Use of Surfactant-Modified Zeolites and Clays for the Removal of Heavy Metals From Water." *Water (Switzerland)* 9 (4): 235. https://doi.org/10.3390/w9040235.

Kaur, Harpreet, Satnam Singh, and Bonamali Pal. 2021a. "A Brief Review on Modified Layered Double Hydroxides for H_2 Production through Photoinduced H_2O Splitting." *Environmental Nanotechnology, Monitoring & Management* 16 (February): 100451. https://doi.org/10.1016/j.enmm.2021.100451.

Kaur, Jaspreet, Anil Kumar Sarma, Poonam Gera, and Mithilesh Kumar Jha. 2021b. "Process Optimization with Acid Functionalised Activated Carbon Derived from Corncob for Production of 4-Hydroxymethyl-2,2-Dimethyl-1,3-Dioxolane and 5-Hydroxy-2,2-Dimethyl-1,3-Dioxane." *Scientific Reports* 11: 8567. https://doi.org/10.1038/s41598-021-87622-z.

Khosroshahi, Negin, Mahnaz Bakhtian, and Vahid Safarifard. 2022. "Mechanochemical Synthesis of Ferrite/MOF Nanocomposite: Efficient Photocatalyst for the Removal of Meropenem and Hexavalent Chromium from Water." *Journal of Photochemistry and Photobiology A: Chemistry* 431: 114033. https://doi.org/10.1016/j.jphotochem.2022.114033.

Kobielska, Paulina A., Ashlee J. Howarth, Omar K. Farha, and Sanjit Nayak. 2018. "Metal–Organic Frameworks for Heavy Metal Removal from Water." *Coordination Chemistry Reviews* 358: 92–107. https://doi.org/10.1016/j.ccr.2017.12.010.

Kosmulski, Marek. 1997. "Attempt to Determine Pristine Points of Zero Charge of Nb_2O_5, Ta_2O_5, and HfO_2." *Langmuir* 13 (23): 6315–19. https://doi.org/10.1021/la970469g.

Kosmulski, Marek. 2016. "Isoelectric Points and Points of Zero Charge of Metal (Hydr) Oxides: 50 Years after Parks' Review." *Advances in Colloid and Interface Science* 238: 1–61. https://doi.org/10.1016/j.cis.2016.10.005.

Kosmulski, Marek, and Edward Mączka. 2004. "Dilatometric Study of the Adsorption of Heavy-Metal Cations on Goethite." *Langmuir* 20 (6): 2320–23. https://doi.org/10.1021/la0356957.

Kumari, Neeraj, and Chandra Mohan. 2021. "Basics of Clay Minerals and Their Characteristic Properties." In *Clays and Clay Minerals*, edited by Gustavo Morari Do Nascimento. IntechOpen. https://doi.org/10.5772/intechopen.95640.

Kumrić, Ksenija R., Ancrossed D. Signelka B. Crossed D Signukić, Tatjana M. Trtić-Petrović, Nikola S. Vukelić, Zoran Stojanović, Jasmina D. Grbović Novaković, and Ljiljana Lj

Matović. 2013. "Simultaneous Removal of Divalent Heavy Metals from Aqueous Solutions Using Raw and Mechanochemically Treated Interstratified Montmorillonite/Kaolinite Clay." *Industrial and Engineering Chemistry Research* 52 (23): 7930–39. https://doi.org/10.1021/ie400257k.

Kung, Chung Wei, Subhadip Goswami, Idan Hod, Timothy C. Wang, Jiaxin Duan, Omar K. Farha, and Joseph T. Hupp. 2020. "Charge Transport in Zirconium-Based Metal-Organic Frameworks." *Accounts of Chemical Research* 53 (6): 1187–95. https://doi.org/10.1021/acs.accounts.0c00106.

Lam, Iris Tsz Yan, Seon Jin Choi, Dong Lu, and Yoonseob Kim. 2023. "Functionalized Metal-Organic Frameworks for Heavy Metal Ion Removal from Water." *Nanoscale* 15 (24): 10189–205. https://doi.org/10.1039/d3nr02250a.

Lazarević, Slavica, Ivona Janković-Častvan, Veljko Djokić, Zeljko Radovanović, Djordje Janaćković, and Rada Petrović. 2010. "Iron-Modified Sepiolite for Ni^{2+} Sorption from Aqueous Solution: An Equilibrium, Kinetic, and Thermodynamic Study." *Journal of Chemical and Engineering Data* 55 (12): 5681–89. https://doi.org/10.1021/je100639k.

Lewis, I.C. 1982. "Chemistry of Carbonization." *Carbon* 20: 519–27. https://doi.org/https://doi.org/10.1016/0008-6223(82)90089-6.

Leyva-Ramos, R., A. Jacobo-Azuara, P.E. Diaz-Flores, R.M. Guerrero-Coronado, J. Mendoza-Barron, and M.S. Berber-Mendoza. 2008. "Adsorption of Chromium(VI) from an Aqueous Solution on a Surfactant-Modified Zeolite." *Colloids and Surfaces A: Physicochemical and Engineering Aspects* 330: 35–41. https://doi.org/10.1016/j.colsurfa.2008.07.025.

Li, Dongxiao, Anurag Yadav, Hong Zhou, Kaustav Roy, Pounraj Thanasekaran, and Chengkuo Lee. 2024. "Advances and Applications of Metal-Organic Frameworks (MOFs) in Emerging Technologies: A Comprehensive Review." *Global Challenges* 8 (2): 1–32. https://doi.org/10.1002/gch2.202300244.

Li, Gang, Jinli Zhang, Yu Lic, Jia Liu, and Zhen Yan. 2021. "Adsorption Characteristics of Pb(II), Cd(II) and Cu(II) on Carbon Nanotube-Hydroxyapatite." *Environmental Technology* 42: 1560–81. https://doi.org/https://doi.org/10.1080/09593330.2019.1674385.

Li, Hui, Feng Zheng, Jing Wang, Jianmin Zhou, Xiaohui Huang, Lai Chen, Pengfei Hu, et al. 2020. "Facile Preparation of Zeolite-Activated Carbon Composite from Coal Gangue with Enhanced Adsorption Performance." *Chemical Engineering Journal* 390: 124513. https://doi.org/10.1016/j.cej.2020.124513.

Li, Jian, Long Yan, Haoyu Li, Weijun Li, Fei Zha, and Ziqiang Lei. 2015. "Underwater Superoleophobic Palygorskite Coated Meshes for Efficient Oil/Water Separation." *Journal of Materials Chemistry A* 3 (28): 14696–702. https://doi.org/10.1039/c5ta02870a.

Li, L.Y., X. Gong, and O. Abida. 2019. "Waste-to-Resources: Exploratory Surface Modification of Sludge-Based Activated Carbon by Nitric Acid for Heavy Metal Adsorption." *Waste Manage.* 87: 375–86.

Li, Wei, Fang Xia, Jin Qu, Ping Li, Dehong Chen, Zhe Chen, Yu Yu, Yu Lu, Rachel A. Caruso, and Weiguo Song. 2014. "Versatile Inorganic-Organic Hybrid WOx-Ethylenediamine Nanowires: Synthesis, Mechanism and Application in Heavy Metal Ion Adsorption and Catalysis." *Nano Research* 7 (6): 903–16. https://doi.org/10.1007/s12274-014-0452-9.

Li, Xiulei, Zhiqian Jia, Hong Wei Tan, Yu Yang, and Li An Hou. 2025. "Enhanced Simultaneous Adsorption and Detection of Mercury (II) Using Functionalized Metal – Organic Framework with Defect Structures." *Separation and Purification Technology* 354: 129110. https://doi.org/10.1016/J.SEPPUR.2024.129110.

Li, Xue, Qian Li, Wensheng Linghu, Runpu Shen, Baoshan Zhao, Lijia Dong, Ahmed Alsaedi, Tasawar Hayat, Jin Wang, and Juan Liu. 2018. "Sorption Properties of U(VI) and Th(IV) on Two-Dimensional Molybdenum Disulfide (MoS_2) Nanosheets: Effects of PH, Ionic Strength, Contact Time, Humic Acids and Temperature." *Environmental Technology and Innovation* 11: 328–38. https://doi.org/10.1016/j.eti.2018.06.001.

Liu, Jing Fu, Zong Shan Zhao, and Gui Bin Jiang. 2008. "Coating Fe_3O_4 Magnetic Nanoparticles with Humic Acid for High Efficient Removal of Heavy Metals in Water." *Environmental Science and Technology* 42 (18): 6949–54. https://doi.org/10.1021/es800924c.

Liu, Lingmei, Ning Wang, Chongzhi Zhu, Xiaona Liu, Yihan Zhu, Peng Guo, Lujain Alfilfil, Xinglong Dong, Daliang Zhang, and Yu Han. 2020a. "Direct Imaging of Atomically Dispersed Molybdenum That Enables Location of Aluminum in the Framework of Zeolite ZSM–5." *Angewandte Chemie* 59 (2): 819–25.

Liu, Mancheng, Tao Wen, Xilin Wu, Changlun Chen, Jun Hu, Jie Li, and Xiangke Wang. 2013. "Synthesis of Porous Fe_3O_4 Hollow Microspheres/Graphene Oxide Composite for Cr(VI) Removal." *Dalton Transactions* 42: 14710–17.

Liu, Wencai, Yang Liu, Zhiqin Yuan, and Chao Lu. 2023. "Recent Advances in the Detection and Removal of Heavy Metal Ions Using Functionalized Layered Double Hydroxides: A Review." *Industrial Chemistry & Materials* 1: 79–92. https://doi.org/10.1039/d2im00024e.

Liu, Zhuan Nian, A. Ping Fan, Xiao Gang Han, Habiba Shapour, and Qing Yun Zhang. 2020b. "Pei-Modified Chromium-Based Metal Organic Framework for Cr(VI) Removal from Aqueous Solution." *Desalination and Water Treatment* 184: 139–49. https://doi.org/10.5004/dwt.2020.24992.

Luo, Peng, Jiang-shan Zhang, Bing Zhang, Jin-hua Wang, Ya-fei Zhao, and Jin-dun Liu. 2011. "Preparation and Characterization of Silane Coupling Agent Modified Halloysite for Cr(VI) Removal." *Industrial and Engineering Chemistry Research* 50: 10246–52. https://doi.org/10.1111/j.1442-200X.1985.tb00686.x.

Lyman, W.J. 1987. "Applicability of Carbon Adsorption to the Treatment of Hazardous Industrial Waste." In *Carbon Adsorption Handbook*, edited by P.N. Cheremisinoff and F. Ellerbusch. Michigan: Ann Arbor Sci. Publ.

Ma, Lijiao, Saiful M. Islam, Hongyun Liu, Jing Zhao, Genban Sun, Huifeng Li, Shulan Ma, and Mercouri G. Kanatzidis. 2017a. "Selective and Efficient Removal of Toxic Oxoanions of As(III), As(V), and Cr(VI) by Layered Double Hydroxide Intercalated with MoS_4^{2-}." *Chemistry of Materials* 29 (7): 3274–84. https://doi.org/10.1021/acs.chemmater.7b00618.

Ma, Lijiao, Saiful M. Islam, Chengliang Xiao, Jing Zhao, Hongyun Liu, Mengwei Yuan, Genban Sun, Huifeng Li, Shulan Ma, and Mercouri G. Kanatzidis. 2017b. "Rapid Simultaneous Removal of Toxic Anions $[HSeO_3]^-$, $[SeO_3]^{2-}$, and $[SeO_4]^{2-}$, and Metals Hg^{2+}, Cu^{2+}, and Cd^{2+} by MoS_4^{2-} Intercalated La." *Journal of the American Chemical Society* 139 (36): 12745–57. https://doi.org/10.1021/jacs.7b07123.

Ma, Xin, Sheng-Tao Yang, Huan Tang, Yuanfang Liu, and Haifang Wang. 2015. "Competitive Adsorption of Heavy Metal Ions on Carbon Nanotubes and the Desorption in Simulated Biofluids." *Journal of Colloid and Interface Science* 448: 347–55. https://doi.org/10.1016/j.jcis.2015.02.042.

Ma, Zichuan, Dongyuan Zhao, Yongfang Chang, Shengtao Xing, Yinsu Wu, and Yuanzhe Gao. 2013. "Synthesis of $MnFe_2O_4$@Mn–Co Oxide Core–Shell Nanoparticles and Their Excellent Performance for Heavy Metal Removal." *Journal of the Chemical Society. Dalton Transactions* 42 (39): 14261–67. https://doi.org/10.1039/c3dt51310f.

Macht, Felix, Karin Eusterhues, Geertje Johanna Pronk, and Kai Uwe Totsche. 2011. "Specific Surface Area of Clay Minerals: Comparison between Atomic Force Microscopy Measurements and Bulk-Gas (N_2) and -Liquid (EGME) Adsorption Methods." *Applied Clay Science* 53 (1): 20–26. https://doi.org/10.1016/j.clay.2011.04.006.

Madadrang, Clemonne J., Hyun Yun Kim, Guihua Gao, Ning Wang, Jun Zhu, Huan Feng, Matthew Gorring, Marc L. Kasner, and Shifeng Hou. 2012. "Adsorption Behavior of EDTA-Graphene Oxide for Pb (II) Removal." *ACS Applied Materials & Interfaces* 4 (3): 1186–93. https://doi.org/https://doi.org/10.1021/am201645g.

Mahmood, Tahira, Muhammad Tahir Saddique, Abdul Naeem, Paul Westerhoff, Syed Mustafa, and Absar Alum. 2011. "Comparison of Different Methods for the Point of Zero Charge Determination of NiO." *Industrial and Engineering Chemistry Research* 50 (17): 10017–23. https://doi.org/10.1021/ie200271d.

Manos, Manolis J., Christos D. Malliakas, and Mercouri G. Kanatzidis. 2007. "Heavy-Metal-Ion Capture, Ion-Exchange, and Exceptional Acid Stability of the Open-Framework Chalcogenide $(NH_4)_4In_{12}Se_{20}$." *Chemistry—A European Journal* 13 (1): 51–8. https://doi.org/10.1002/chem.200600892.

Manousi, Natalia, Dimitrios A. Giannakoudakis, Erwin Rosenberg, and George A. Zachariadis. 2019. "Extraction of Metal Ions with Metal–Organic Frameworks." *Molecules* 24 (24): 1–21. https://doi.org/10.3390/molecules24244605.

Megías-Alguacil, D., J.D.G. Durán, and A.V. Delgado. 2000. "Yield Stress of Concentrated Zirconia Suspensions: Correlation with Particle Interactions." *Journal of Colloid and Interface Science* 231 (1): 74–83. https://doi.org/10.1006/jcis.2000.7121.

Mendoza-Barrón, Jovita, Araceli Jacobo-Azuara, Roberto Leyva-Ramos, Ma Selene Berber-Mendoza, Rosa Ma Guerrero-Coronado, Laura Fuentes-Rubio, and J. Merced Martínez-Rosales. 2011. "Adsorption of Arsenic (V) from a Water Solution onto a Surfactant-Modified Zeolite." *Adsorption* 17 (3): 489–96. https://doi.org/10.1007/s10450-010-9307-1.

Meng, Qingpeng, Hong Chen, Junzhong Lin, Zhang Lin, and Junliang Sun. 2017. "Zeolite A Synthesized from Alkaline Assisted Pre-Activated Halloysite for Efficient Heavy Metal Removal in Polluted River Water and Industrial Wastewater." *Journal of Environmental Sciences (China)* 56: 254–62. https://doi.org/10.1016/j.jes.2016.10.010.

Misaelides, Panagiotis. 2011. "Application of Natural Zeolites in Environmental Remediation: A Short Review." *Microporous and Mesoporous Materials* 144: 15–18. https://doi.org/10.1016/j.micromeso.2011.03.024.

Mohan, Sweta, Vijay Kumar, Devendra Kumar Singh, and Syed Hadi Hasan. 2017. "Effective Removal of Lead Ions Using Graphene Oxide-MgO Nanohybrid from Aqueous Solution: Isotherm, Kinetic and Thermodynamic Modeling of Adsorption." *Journal of Environmental Chemical Engineering* 5: 2259–73.

Momani, Delal E. Al, Fathima Arshad, Inas Taha, Dalaver H. Anjum, and Linda Zou. 2024. "Effect of Different MoS_2 Morphologies on the Formation and Performance of Adsorptive-Catalytic Nanocomposite Membranes." *NPJ Clean Water* 7 (1): 64. https://doi.org/10.1038/s41545-024-00360-3.

Moreno-Sader, Kariana, Alvaro García-Padilla, Alvaro Realpe, María Acevedo-Morantes, and João B.P. Soares. 2019. "Removal of Heavy Metal Water Pollutants (Co^{2+} and Ni^{2+}) Using Polyacrylamide/Sodium Montmorillonite (PAM/Na-MMT) Nanocomposites." *ACS Omega* 4 (6): 10834–44. https://doi.org/10.1021/acsomega.9b00981.

Muttil, Nitin, Saranya Jagadeesan, Arnab Chanda, Mikel Duke, and Swadesh Kumar Singh. 2023. "Production, Types, and Applications of Activated Carbon Derived from Waste Tyres: An Overview." *Applied Sciences (Switzerland)* 13: 257. https://doi.org/10.3390/app13010257.

Nabavi, Minou, Olivier Spalla, and Bernard Cabane. 1993. "Surface Chemistry of Nanometric Ceria Particles in Aqueous Dispersions." *Journal of Colloid And Interface Science* 160 (2): 459–71. https://doi.org/10.1006/jcis.1993.1417.

Naji, Samah Zaki, and Ching Thian Tye. 2022. "A Review of the Synthesis of Activated Carbon for Biodiesel Production: Precursor, Preparation, and Modification." *Energy Conversion and Management: X* 13: 100152. https://doi.org/10.1016/j.ecmx.2021.100152.

Nakamoto, Kohtaroh, Masaru Ohshiro, and Takaomi Kobayashi. 2017. "Mordenite Zeolite—Polyethersulfone Composite Fibers Developed for Decontamination of Heavy Metal Ions." *Journal of Environmental Chemical Engineering* 5: 513–25. https://doi.org/10.1016/j.jece.2016.12.031.

Naklicki, M.L., S.R. Rao, M. Gomez, and J.A. Finch. 2002. "Flotation and Surface Analysis of the Nickel (II) Oxide/Amyl Xanthate System." *International Journal of Mineral Processing* 65 (2): 73–82. https://doi.org/10.1016/S0301-7516(01)00061-8.

Nechaev, E.A., and V.N. Shein. 1979. "Specific Adsorption of Ions on Oxides." *Kolloidn Zh* 41: 361–63.

Nurmalita, Nurmalita, Raudhatul Raihan, Zulkarnain Jalil, Syahrun Nur, and Adi Setiawan. 2022. "The Physical and Chemical Properties of Activated Nanocarbon Produced from Robusta (Coffea Canephora) Coffee Pulp under Slow Pyrolysis Method." *Coffee Science* 17: e172019. https://doi.org/10.25186/.v17i.2019.

Obaid, Shamsan S., D.K. Gaikwad, M.I. Sayyed, Khader Al-Rashdi, and P.P. Pawar. 2018. "Heavy Metal Ions Removal from Waste Water By the Natural Zeolites." *Materials Today: Proceedings* 5 (9): 17930–34. https://doi.org/10.1016/j.matpr.2018.06.122.

Ocaña, M. 2000. "Uniform Particles of Manganese Compounds Obtained by Forced Hydrolysis of Manganese(II) Acetate." *Colloid and Polymer Science* 278 (5): 443–49. https://doi.org/10.1007/s003960050537.

Orhan, Y., and H. Büyükgüngör. 1993. "The Removal of Heavy Metals by Using Agricultural Wastes." *Water Science & Technology* 28 (2): 247–55. https://doi.org/10.2166/wst.1993.0114.

Oubagaranadin, John U. Kennedy, and Z.V.P. Murthy. 2009. "Adsorption of Divalent Lead on a Montmorillonite-Illite Type of Clay." *Industrial and Engineering Chemistry Research* 48 (23): 10627–36. https://doi.org/10.1021/ie9005047.

Pan, Junyao, Binyu Wang, Haoyang Zhang, Yufei Wang, Shuang Liu, and Wenfu Yan. 2025. "Synthesis of MOR Zeolite from Natural Stellerite and Clinoptilolite for Efficient Pb^{2+} and Cd^{2+} Removal from Aqueous Solutions." *Crystal Growth and Design* 25 (8): 2484–97. https://doi.org/10.1021/acs.cgd.4c01710.

Pearson, Ralph G. 1968. "Hard and Soft Acids and Bases, HSAB, Part 1: Fundamental Principles." *Journal of Chemical Education* 45: 581–87. http://dx.doi.org/10.1021/ed045p581%5Cnhttp://pubs.acs.org/doi/abs/10.1021/ed045p581.

Phouthavong, Vanpaseuth, Takeshi Hagio, Jae Hyeok Park, Supinya Nijpanich, Kanchanok Duangkhai, Ratana Rujiravanit, Piyatida Thaveemas, et al. 2024. "Removal of Heavy Metals by BEA Zeolite/Fe_3O_4 Composite Prepared via Dry-Gel Conversion Method Using Agrowaste-Derived Raw Material." *Solid State Sciences* 149: 107473. https://doi.org/10.1016/j.solidstatesciences.2024.107473.

Prélot, Bénédicte, Christiane Poinsignon, Fabien Thomas, Elisabeth Schouller, and Frédéric Villiéras. 2003. "Structural-Chemical Disorder of Manganese Dioxides 1. Influence on Surface Properties at the Solid-Electrolyte Interface." *Journal of Colloid and Interface Science* 257 (1): 77–84. https://doi.org/10.1016/S0021-9797(02)00013-9.

Raptopoulou, Catherine P. 2021. "Metal-Organic Frameworks: Synthetic Methods and Potential Applications." *Materials* 14 (2): 1–32. https://doi.org/10.3390/ma14020310.

Reffas, A., V. Bernardet, B. David, L. Reinert, M.B. Lehocine, M. Dubois, N. Batisse, and L. Duclaux. 2010. "Carbons Prepared from Coffee Grounds by H_3PO_4 Activation: Characterization and Adsorption of Methylene Blue and Nylosan Red N-2RBL." *Journal of Hazardous Materials* 175: 779–88.

Regazzoni, Alberto E., Miguel A. Blesa, and Alberto J.G. Maroto. 1983. "Interfacial Properties of Zirconium Dioxide and Magnetite in Water." *Journal of Colloid And Interface Science* 91 (2): 560–70. https://doi.org/10.1016/0021-9797(83)90370-3.

Rout, Dibya Ranjan, Hara Mohan Jena, Omirserik Baigenzhenov, Ahmad Hosseini-Bandegharaei. 2023. "Graphene-Based Materials for Effective Adsorption of Organic and Inorganic Pollutants: A Critical and Comprehensive Review." *Science of the Total Environment* 863: 160871. https://doi.org/https://doi.org/10.1016/j.scitotenv.2022.160871.

Roy, Debashis, Sudarsan Neogi, and Sirshendu De. 2021. "Adsorptive Removal of Heavy Metals from Battery Industry Effluent Using MOF Incorporated Polymeric Beads: A Combined Experimental and Modeling Approach." *Journal of Hazardous Materials* 403: 123624. https://doi.org/10.1016/j.jhazmat.2020.123624.

Sahu, Sumanta, Padmaja Kar, Nisarani Bishoyi, Laxmi Mallik, and Raj Kishore Patel. 2019. "Synthesis of Polypyrrole-Modified Layered Double Hydroxides for Efficient Removal of Cr(VI)." *Journal of Chemical and Engineering Data* 64 (10): 4357–68. https://doi.org/10.1021/acs.jced.9b00444.

Saravaia, Hitesh, Hariom Gupta, Pooja Popat, Parthrajsinh Sodha, and Vaibhav Kulshrestha. 2018. "Single-Step Synthesis of Magnesium-Doped Lithium Manganese Oxide Nanosorbent and Their Polymer Composite Beads for Selective Heavy Metal Removal." *ACS Applied Materials and Interfaces* 10 (50): 44059–70. https://doi.org/10.1021/acsami.8b17141.

Schreier, Marc, Theresa E. Feltes, Melanie T. Schaal, and John R. Regalbuto. 2010. "The Determination of Oxide Surface Charging Parameters for a Predictive Metal Adsorption Model." *Journal of Colloid and Interface Science* 348 (2): 571–78. https://doi.org/10.1016/j.jcis.2010.04.064.

Schudel, M., S.H. Behrens, H. Holthoff, R. Kretzschmar, and M. Borkovec. 1997. "Absolute Aggregation Rate Constants of Hematite Particles in Aqueous Suspensions: A Comparison of Two Different Surface Morphologies." *Journal of Colloid and Interface Science* 196 (2): 241–53. https://doi.org/10.1006/jcis.1997.5207.

Serna, C.J., and M.P. Morales. 2004. "Maghemite (γ-Fe_2O_3): A Versatile Magnetic Colloidal Material." In *Surface and Colloid Science*, edited by M. Borkovec and E. Matijevic, 27–81. New York: Kluwer.

Shah, Kinjal J., Jiacheng Yu, Ting Zhang, Zhaoyang You, and Hyunook Kim. 2024. "Simultaneous Removal of Cu(II) And Pb(II) From Stormwater Runoff by Y-Type-Zeolite-Modified Bioretention System." *Water, Air, and Soil Pollution* 235 (6): 1–13. https://doi.org/10.1007/s11270-024-07179-1.

Shamim, Mohd Arham, Hammad Zia, Mohd Zeeshan, Mohammad Yasir Khan, and M. Shahid. 2022. "Metal Organic Frameworks (MOFs) as a Cutting-Edge Tool for the Selective Detection and Rapid Removal of Heavy Metal Ions from Water: Recent Progress." *Journal of Environmental Chemical Engineering* 10 (1): 106991. https://doi.org/10.1016/j.jece.2021.106991.

Sheraz, Nashra, Afzal Shah, Abdul Haleem, and Faiza Jan Iftikhar. 2024. "Comprehensive Assessment of Carbon-, Biomaterial- and Inorganic-Based Adsorbents for the Removal of the Most Hazardous Heavy Metal Ions from Wastewater." *RSC Advances* 14: 11284–310.

Sherugar, Prajwal, Mahesh Padaki, Nagaraj S. Naik, and Dharmapura H.K. Murthy Sajan D. George. 2022. "Biomass-Derived Versatile Activated Carbon Removes Both Heavy Metals and Dye Molecules from Wastewater with near-Unity Efficiency: Mechanism and Kinetics." *Chemosphere* 287: 132085.

Shindel, Benjamin, Stephanie M. Ribet, Caroline Harms, Vikas Nandwana, and Vinayak P. Dravid. 2023. "Nano-SCHeMe: Nanomaterial Sponge Coatings for Heavy Metals, an Environmental Remediation Platform." *ACS ES and T Water* 3 (8): 2120–29. https://doi.org/10.1021/acsestwater.2c00646.

Silva, Mahima De, Gaili Cao, and K.C. Tam. 2025. “Nanomaterials for the Removal and Detection of Heavy Metals: A Review.” *Environmental Science: Nano* 12 (4): 2154–76. https://doi.org/10.1039/d4en01041h.

Siti, Zu Nurain Ahmad, Wan Norharyati Wan Salleh, Ahmad Fauzi Ismail, Norhaniza Yusof, Mohd Zamri Mohd Yusop, Farhana Aziz. 2020. “Adsorptive Removal of Heavy Metal Ions Using Graphene-Based Nanomaterials: Toxicity, Roles of Functional Groups and Mechanisms.” *Chemosphere* 248: 126008. https://doi.org/https://doi.org/10.1016/j.chemosphere.2020.126008.

Škapin, Srečo D., Vida Čadež, Danilo Suvorov, and Ivan Sondi. 2015. “Formation and Properties of Nanostructured Colloidal Manganese Oxide Particles Obtained through the Thermally Controlled Transformation of Manganese Carbonate Precursor Phase.” *Journal of Colloid and Interface Science* 457: 35–42. https://doi.org/10.1016/j.jcis.2015.06.041.

Smržová, Darina, Lórant Szatmáry, Petra Ecorchard, Aneta Machálková, Monika Maříková, Petra Salačová, and Martin Straka. 2022. “Carbon and Zeolite-Based Composites for Radionuclide and Heavy Metal Sorption.” *Heliyon* 8 (12): e12293. https://doi.org/10.1016/j.heliyon.2022.e12293.

Solomon, Alemayehu, Dinsefa Mensur Andoshe, Adane Muche Abebe, Tatek Temesgen Terfasa, T. Ganesh, Fekadu Gashaw Hone, Newayemedhin A. Tegegne, and Belay Brehane Tesfamariam. 2023. “Optimization of Preparation Parameters of Ceramic Pot Water Filters for Potential Application of Microbial and Fluoride Removal from Groundwater.” *Heliyon* 9 (2): e13261. https://doi.org/10.1016/j.heliyon.2023.e13261.

Sposito, Garrison. 1998. “On Points of Zero Charge.” *Environmental Science and Technology* 32 (19): 2815–19. https://doi.org/10.1021/es9802347.

Srinivasan, K., and N. Balasubramanian. 2023. “Removal of Cadmium (II), Mercury (II) and Lead (II) by Rice Husk Carbon.” *Oriental Journal of Chemistry* 19 (2).

Szekeres, Márta, and Etelka Tombácz. 2012. “Surface Charge Characterization of Metal Oxides by Potentiometric Acid-Base Titration, Revisited Theory and Experiment.” *Colloids and Surfaces A: Physicochemical and Engineering Aspects* 414: 302–13. https://doi.org/10.1016/j.colsurfa.2012.08.027.

Taha, F., A.M. El-Roudi, A.A. Abd El-Gaber, and F.M. Zahran. 1990. “Adsorption of Cations on Oxides.” *Revue Roumaine de Chimie* 35: 503–9.

Tanihara, Ayane, Kouhei Kikuchi, and Hiroki Konno. 2021. “Insight into the Mechanism of Heavy Metal Removal from Water by Monodisperse ZIF-8 Fine Particles.” *Inorganic Chemistry Communications* 131: 108782. https://doi.org/10.1016/j.inoche.2021.108782.

Tchinda, Aurélien Jieumboué, Emmanuel Ngameni, Ignas Tonlé Kenfack, and Alain Walcarius. 2009. “One-Step Preparation of Thiol-Functionalized Porous Clay Heterostructures: Application to Hg(II) Binding and Characterization of Mass Transport Issues.” *Chemistry of Materials* 21 (18): 4111–21. https://doi.org/10.1021/cm8024022.

Tho, P.T., Huu Tap Van, Lan Huong Nguyen, Trung Kien Hoang, Thi Ngoc Ha Tran, Thi Tuyet Nguyen, Thi Bich Hanh Nguyen, et al. 2021. “Enhanced Simultaneous Adsorption of As(III), Cd(II), Pb(II) and Cr(VI) Ions from Aqueous Solution Using Cassava Root Husk-Derived Biochar Loaded with ZnO Nanoparticles.” *RSC Advances* 11 (31): 18881–97. https://doi.org/10.1039/d1ra01599k.

Thorarinsdottir, Agnes E., and T. David Harris. 2020. “Metal-Organic Framework Magnets.” *Chemical Reviews* 120 (16): 8716–89. https://doi.org/10.1021/acs.chemrev.9b00666.

Tran, Lytuong, Pingxiao Wu, Yajie Zhu, Lin Yang, and Nengwu Zhu. 2015. “Highly Enhanced Adsorption for the Removal of Hg(II) from Aqueous Solution by Mercaptoethylamine/Mercaptopropyltrimethoxysilane Functionalized Vermiculites.” *Journal of Colloid and Interface Science* 445: 348–56. https://doi.org/10.1016/j.jcis.2015.01.006.

Tufail, Muhammad Khurram, Muhammad Ifrahim, Muhammad Rashid, Ihtisham Ul Haq, Rizwana Asghar, U.T. Uthappa, Manickam Selvaraj, and Mahaveer Kurkuri. 2025. "Chemistry of Zeolites and Zeolite Based Composite Membranes as a Cutting-Edge Candidate for Removal of Organic Dyes & Heavy Metal Ions: Progress and Future Directions." *Separation and Purification Technology* 354 (P2): 128739. https://doi.org/10.1016/j.seppur.2024.128739.

Vasconcelos, da Silva, Luiz Thiago, Armando Diego Lima de Freitas, Thaizy de Gois Martins, Antônia Mayza de Morais França, Adonay Rodrigues Loiola, and Ronaldo Ferreira do Nascimento. 2023. "Comparative Study on the Performance of Two Synthetic Zeolite 4A on the Removal of Heavy Metals from Aqueous Solution—Effect of Coal Fly Ash as Al and Si Source." *Desalination and Water Treatment* 310: 157–66. https://doi.org/10.5004/dwt.2023.29948.

Velarde, Lisbania, Dariush Nikjoo, Edwin Escalera, and Farid Akhtar. 2024. "Bolivian Natural Zeolite as a Low-Cost Adsorbent for the Adsorption of Cadmium: Isotherms and Kinetics." *Heliyon* 10 (1): e24006. https://doi.org/10.1016/j.heliyon.2024.e24006.

Vipin, Manviri Rani, Uma Shanker, and Mika Sillanpää. 2023a. "Green Synthesis of N-Doped ZnO@Zeolite Nanocomposite for the Efficient Photo-Adsorptive Removal of Toxic Heavy Metal Ions Cr^{6+} and Cd^{2+}." *Inorganic Chemistry Communications* 157: 111401 Contents. https://doi.org/10.1016/j.inoche.2023.111401.

Vipin, Manviri Rani, Uma Shanker, and Mika Sillanpää. 2023b. "Green Synthesis of N-Doped ZnO@Zeolite Nanocomposite for the Efficient Photo-Adsorptive Removal of Toxic Heavy Metal Ions Cr^{6+} and Cd^{2+}." *Inorganic Chemistry Communications* 157: 111401. https://doi.org/10.1016/j.inoche.2023.111401.

Vishnu, Dhanya, Balaji Dhandapani, K. Senthil Kumar, G. Balaji, and Surianarayanan Mahadevan. 2021. "Removal of Heavy Metals from Mine Waters by Natural Zeolites." *New Trends in Removal of Heavy Metals from Industrial Wastewater* 39 (12): 161–75. https://doi.org/10.1016/B978-0-12-822965-1.00008-8.

Wang, Qiang, and Dermot O'Hare. 2012. "Recent Advances in the Synthesis and Application of Layered Double Hydroxide (LDH) Nanosheets." *Chemical Reviews* 112: 4124–55.

Wang, Jun, and Baoliang Chen. 2015. "Adsorption and Coadsorption of Organic Pollutants and a Heavy Metal by Graphene Oxide and Reduced Graphene Materials." *Chemical Engineering Journal* 281: 379–88. https://doi.org/https://doi.org/10.1016/j.cej.2015.06.102.

Wang, Li, Chengxiang Shi, Lun Pan, Xiangwen Zhang, and Ji Jun Zou. 2020a. "Rational Design, Synthesis, Adsorption Principles and Applications of Metal Oxide Adsorbents: A Review." *Nanoscale* 12 (8): 4790–4815. https://doi.org/10.1039/c9nr09274a.

Wang, Peng, Du Sun, Mingxia Deng, Shaoning Zhang, Qingyuan Bi, Wei Zhao, Fuqiang Huang. 2020b. "Amorphous Phosphated Titanium Oxide with Amino and Hydroxyl Bifunctional Groups for Highly Efficient Heavy Metals Removal." *Environmental Science Nano* 7: 1266–74. https://doi.org/10.1039/C9EN01466G.Environmental.

Wang, Qingmiao, Lang Yang, Feifei Jia, Yanmei Li, and Shaoxian Song. 2018a. "Removal of Cd(II) from Water by Using Nano-Scale Molybdenum Disulphide Sheets as Adsorbents." *Journal of Molecular Liquids* 263: 526–33. https://doi.org/10.1016/j.molliq.2018.04.149.

Wang, Yaquan, Yi Feng, Jinlong Jiang, and Jianfeng Yao. 2019a. "Designing of Recyclable Attapulgite for Wastewater Treatments: A Review." *ACS Sustainable Chemistry and Engineering* 7 (2): 1855–69. https://doi.org/10.1021/acssuschemeng.8b05823.

Wang, Yaquan, Yi Feng, Xiong Fei Zhang, Xingguang Zhang, Jinlong Jiang, and Jianfeng Yao. 2018b. "Alginate-Based Attapulgite Foams as Efficient and Recyclable Adsorbents for the Removal of Heavy Metals." *Journal of Colloid and Interface Science* 514: 190–98. https://doi.org/10.1016/j.jcis.2017.12.035.

Wang, Zhe, Keyan Tan, Jingyi Cai, Shitian Hou, Yue Wang, Peng Jiang, and Minghui Liang. 2019b. "Silica Oxide Encapsulated Natural Zeolite for High Efficiency Removal of Low Concentration Heavy Metals in Water." *Colloids and Surfaces A: Physicochemical and Engineering Aspects* 561: 388–94. https://doi.org/10.1016/j.colsurfa.2018.10.065.

Wang, Zhongying, Sim Alison, Urban Jeffrey, and Mi Baoxia. 2018c. "Removal and Recovery of Heavy Metal Ions by Two-Dimensional MoS_2 Nanosheets: Performance and Mechanisms." *Environmental Science & Technology* 52 (17): 9741–48.

Warchoł, J., P. Misaelides, R. Petrus, and D. Zamboulis. 2006. "Preparation and Application of Organo-Modified Zeolitic Material in the Removal of Chromates and Iodides." *Journal of Hazardous Materials* 137 (3): 1410–16. https://doi.org/10.1016/j.jhazmat.2006.04.028.

Weber, W.J. Jr., and B.M. Vanvliet. 1980. "Fundamental Concepts for Applying Activated Carbon in Water and Wastewater Treatment." In *Activated Carbon Adsorption of Organics from the Aqueous Phase*, edited by I.H. Suffet and M.J. McGuire, Vol. 16. Ann Arbor Sci.

Wei, Wen Cheng J., Sheng Chang Wang, and Fang Yuan Ho. 1999. "Electrokinetic Properties of Colloidal Zirconia Powders in Aqueous Suspension." *Journal of the American Ceramic Society* 82 (12): 3385–92. https://doi.org/10.1111/j.1151-2916.1999.tb02255.x.

Wick, Silvan, Bart Baeyens, Maria Marques Fernandes, and Andreas Voegelin. 2018. "Thallium Adsorption onto Illite." *Environmental Science and Technology* 52 (2): 571–80. https://doi.org/10.1021/acs.est.7b04485.

Xie, Yu, Edward M. Helvenston, Lindsay C. Shuller-Nickles, and Brian A. Powell. 2016. "Surface Complexation Modeling of Eu(III) and U(VI) Interactions with Graphene Oxide." *Environmental Science & Technology* 50: 1821–7. https://doi.org/10.1021/acs.est.5b05307.

Xu, Ling, Eun Young Choi, and Young Uk Kwon. 2008. "Ionothermal Synthesis of a 3D Zn-BTC Metal-Organic Framework with Distorted Tetranuclear [Zn4(μ_4-O)] Subunits." *Inorganic Chemistry Communications* 11 (10): 1190–93. https://doi.org/10.1016/j.inoche.2008.07.001.

Yadav, Vijay Bahadur, Ranu Gadi, and Sippy Kalra. 2019. "Clay Based Nanocomposites for Removal of Heavy Metals from Water: A Review." *Journal of Environmental Management* 232 (December 2018): 803–17. https://doi.org/10.1016/j.jenvman.2018.11.120.

Yang, Xiaotong, Yi Zhou, Yinhang Fan, Guocheng Lv, and Libing Liao. 2025. "Structural Engineering of Clay Minerals for Heavy Metal Remediation: Molecular Mechanisms, Performance Enhancement, and Sustainable Applications." *Progress in Natural Science: Materials International* (March). https://doi.org/10.1016/j.pnsc.2025.03.008.

Yang, Yuewei, Fengfei Sun, Jing Li, Junfeng Chen, and Meizhen Tang. 2020. "The Effects of Different Factors on the Removal Mechanism of Pb(II) by Biochar-Supported Carbon Nanotube Composites." *RSC Advances* 10: 5988–95.

Yee, Ka Kit, Nele Reimer, Jie Liu, Sum Yin Cheng, Shek Man Yiu, Jens Weber, Norbert Stock, and Zhengtao Xu. 2013. "Effective Mercury Sorption by Thiol-Laced Metal-Organic Frameworks: In Strong Acid and the Vapor Phase." *Journal of the American Chemical Society* 135 (21): 7795–98. https://doi.org/10.1021/ja400212k.

Yin, Taozhu, Yongsheng Zhang, Duo Dong, Tao Wang, and Jiawei Wang. 2022. "Highly Efficient Capacitive Removal of Cd^{2+} over MoS_2-Carbon Framework Composite Material in Desulphurisation Wastewater from Coal-Fired Power Plants." *Journal of Cleaner Production* 355: 131814. https://doi.org/10.1016/j.jclepro.2022.131814.

Yoshida, Mari, Paulmanickam Koilraj, Xinhong Qiu, Tsuyoshi Hirajima, and Keiko Sasaki. 2015. "Sorption of Arsenate on MgAl and MgFe Layered Double Hydroxides Derived from Calcined Dolomite." *Journal of Environmental Chemical Engineering* 3 (3): 1614–21. https://doi.org/10.1016/j.jece.2015.05.016.

Yuan, Fang, Dingfan Yan, Shuaibo Song, Jinbo Zhang, Yangyang Yang, Zheng Chen, Jiahao Lu, Simin Wang, and Yongjun Sun. 2025. "Removal of Heavy Metals from Water by Adsorption on Metal Organic Frameworks: Research Progress and Mechanistic Analysis in the Last Decade." *Chemical Engineering Journal* 506 (January): 160063. https://doi.org/10.1016/j.cej.2025.160063.

Yuan, Mingliang, Tangfeng Xie, Guanjie Yan, Qichao Chen, and Long Wang. 2018. "Effective Removal of Pb^{2+} from Aqueous Solutions by Magnetically Modified Zeolite." *Powder Technology* 332: 234–41. https://doi.org/10.1016/j.powtec.2018.03.043.

Yusof, Alias Mohd, and Nik Ahmad Nizam Nik Malek. 2009. "Removal of Cr(VI) and As(V) from Aqueous Solutions by HDTMA-Modified Zeolite Y." *Journal of Hazardous Materials* 162: 1019–24. https://doi.org/10.1016/j.jhazmat.2008.05.134.

Zeng, Yubin, Heesoo Woo, Gwanghun Lee, and Junboum Park. 2010. "Adsorption of Cr(VI) on Hexadecylpyridinium Bromide (HDPB) Modified Natural Zeolites." *Microporous and Mesoporous Materials* 130 (1–3): 83–91. https://doi.org/10.1016/j.micromeso.2009.10.016.

Zhang, Haiyuan, Zhaoxia Ji, Tian Xia, Huan Meng, Cecile Low-Kam, Rong Liu, Suman Pokhrel, et al. 2012. "Use of Metal Oxide Nanoparticle Band Gap to Develop a Predictive Paradigm for Oxidative Stress and Acute Pulmonary Inflammation." *ACS Nano* 6 (5): 4349–68. https://doi.org/10.1021/nn3010087.

Zhang, Meng. 2025. "Lead and Cadmium Decontamination from Water Media by CNT/Starch/Fe_3O_4, as Reclaimable Magnetic Nanocomposite." *Biomass Conversion and Biorefinery* 15: 8487–500. https://doi.org/https://doi.org/10.1007/s13399-024-05594-2.

Zhang, Yunhui, Daniel S. Alessi, Ning Chen, Mina Luo, Weiduo Hao, Md Samrat Alam, Kurt O. Konhauser, Yong Sik Ok, and Abir Al-Tabbaa. 2021. "Spectroscopic and Modeling Investigation of Sorption of Pb(II) to ZSM-5 Zeolites." *ACS Environmental Science and Technology Water* 1 (1): 108–16. https://doi.org/10.1021/acsestwater.0c00010.

Zhao, Cuijiao, Xinlei Wang, Shengbo Zhang, Na Sun, Hongjian Zhou, Guozhong Wang, Yunxia Zhang, Haimin Zhang, and Huijun Zhao. 2020. "Porous Carbon Nanosheets Functionalized with Fe_3O_4 Nanoparticles for Capacitive Removal of Heavy Metal Ions from Water." *Environmental Science: Water Research and Technology* 6 (2): 331–40. https://doi.org/10.1039/c9ew00472f.

Zhao, T., Y. Yao, D. Li, F. Wu, C. Zhang, and B. Gao. 2018. "Facile Low-Temperature One-Step Synthesis of Pomelo Peel Biochar under Air Atmosphere and Its Adsorption Behaviors for Ag(I) and Pb(II)." *Science of the Total Environment*, 640–641: 73–9.

Zhao, Yufeng, Dongfang Wang, Wei Wei, Longzhe Cui, Chul Woong Cho, and Guiping Wu. 2021a. "Effective Adsorption of Mercury by Zr(IV)-Based Metal-Organic Frameworks of UiO-66-NH_2 from Aqueous Solution." *Environmental Science and Pollution Research* 28 (6): 7068–75. https://doi.org/10.1007/s11356-020-11080-9.

Zhao, Xiaoliang, Xuezheng Yu, Xueyao Wang, Shoujuan Lai, Yuanyuan Sun, and Dongjiang Yang. 2021b. "Recent Advances in Metal-Organic Frameworks for the Removal of Heavy Metal Oxoanions from Water." *Chemical Engineering Journal* 407: 127221. https://doi.org/10.1016/j.cej.2020.127221.

Zhao, Zhixue, Yaran Song, Pikky Phanlavong, Xuan Zhao, Qiuming Peng, Tifeng Jiao, Zhaoxiang Zhang, and Qingrui Zhang. 2022. "Efficient Heavy Metal Removal from Water by Polydopamine Confined ZrO_2 Nanocrystals with Improvements in Nanoparticles Utilization and Ion Diffusion." *ACS ES and T Engineering* 2 (5): 794–806. https://doi.org/10.1021/acsestengg.1c00370.

Zhou, Tianzhu, Cuiping Li, Huiling Jin, Yangyang Lian, and Wenmei Han. 2017. "Effective Adsorption/Reduction of Cr(VI) Oxyanion by Halloysite@Polyaniline Hybrid Nanotubes." *ACS Applied Materials and Interfaces* 9 (7): 6030–43. https://doi.org/10.1021/acsami.6b14079.

Zhu, Lin, and Karl Seff. 1999. "Reinvestigation of the Crystal Structure of Dehydrated Sodium Zeolite X." *The Journal of Physical Chemistry B* 103 (44): 9512–18. https://doi.org/https://doi.org/10.1021/jp991817l.

Zou, Fang, Runhan Yu, Rongguan Li, and Wei Li. 2013. "Microwave-Assisted Synthesis of HKUST-1 and Functionalized HKUST-1-@$H_3PW_{12}O_{40}$: Selective Adsorption of Heavy Metal Ions in Water Analyzed with Synchrotron Radiation." *ChemPhysChem* 14 (12): 2825–32. https://doi.org/10.1002/cphc.201300215.

5 Heavy Metal Ion Removal Using BioBased Materials

Varsha Srivastava, Sidra Iftekhar, Hanna Runtti, Amina Tariq, Sherif Hegazy, and Ulla Lassi

5.1 INTRODUCTION

Heavy metals have numerous applications in different industrial activities. The presence of heavy metals in the environment due to inefficient water treatment may pose a significant threat to flora and fauna (O'Connell et al., 2008; Sharma et al., 2007). Several approaches, including membrane filtration, solvent extraction, coagulation, electrochemical methods, adsorption, ion-exchange, chemical precipitation, and biological treatments, have been developed for the removal of heavy metal ions from various types of wastewater (Carolin et al., 2017; Sharma et al., 2009). Nevertheless, most existing methods are associated with significant operational challenges, high costs, and long treatment times. Among available technologies, the adsorption process is preferred for metal removal applications due to its simplicity, efficiency, and cost-effectiveness (Hegazy et al., 2024). All processes, from adsorbent synthesis to recovery, play a vital role (Figure 5.1). Adsorbent properties can significantly affect metal removal efficiency. Additionally, it is also important that adsorbents do not cause secondary pollution during application and can be regenerated for reuse.

A variety of adsorbents, including zeolites, silica-based adsorbents, industrial by-products, polymers, and inorganic-organic hybrid adsorbent materials, have been employed in various studies for heavy metal removal (Hegazy et al., 2024; Srivastava and Lassi, 2024; Srivastava and Sharma, 2013). In recent years, greater emphasis has been placed on the engineering of cost-effective and environmentally friendly adsorbent materials. Biobased adsorbent materials offer a sustainable alternative to commercially available adsorbents, which are typically associated with high costs (Kainth et al., 2024; Sharma et al., 2024). Plant- and algal-based biomass have been investigated for the development of adsorbent materials to remove metallic pollutants (Kainth et al., 2024; Kartik et al., 2021).

Biomass waste such as agricultural by-products (corn stalks, fruit peel-off, nut shells, bagasse, wheat and rice bran, etc.) and forest residues (such as sawdust, paper mill discards, wood waste, etc.) are abundantly available biomass offering versatile solutions for the development of biobased adsorbent materials (Carolin et al., 2017; Makoś-Chełstowska et al., 2024). Additionally, algal biomass-based adsorbents have

 DOI: 10.1201/9781003599098-5

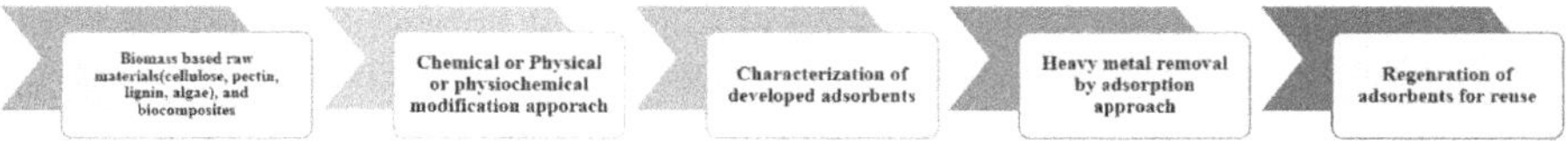

FIGURE 5.1 Schematic flow of overall processes from synthesis to application and recovery of biobased materials in metal adsorption.

potential for metal removal from wastewater (Ramesh et al., 2023). Utilization of biomass waste as a raw material for the development of adsorbent materials can significantly reduce environmental pollution by repurposing biomass waste into affordable adsorbent materials for water purification applications.

5.1.1 Biobased Adsorbents

Various types of adsorbent materials have been evaluated for the removal of heavy metals and other pollutants from wastewater streams (Ávila et al., 2025; Sheraz et al., 2024). It is noteworthy that adsorbents with high efficiency and low environmental impact are always preferable. Activated carbons have been extensively used for metal removal; however, commercially available activated carbons are often associated with high costs (Febrianto et al., 2009). The use of sustainable adsorbent materials in water purification has attracted considerable interest in recent years. Biobased adsorbents are a promising sustainable material for heavy metal removal as they can offer a cost-effective alternative to conventional adsorbents. Biobased adsorbents are usually developed using natural or modified bio-derived materials (Martins et al., 2025). Biomass wastes, such as lignocellulosic residues (e.g., wood and agricultural residues), are potential raw materials for developing adsorbents.

Lignocellulosic wastes mainly contain cellulose (30–50%), hemicellulose (20–30%), lignin (15–20%), and some percentage of pectin and extractives (Makoś-Chełstowska et al., 2024; Srivastava et al., 2024). It is reported that the annual production of agricultural and food waste is approximately 1.3 billion metric tons, which necessitates processing and effective strategies for managing biomass waste (Makoś-Chełstowska et al., 2024). Adsorbents derived from lignocellulosic wastes offer advantages such as inherent functionality, low cost, and worldwide abundance. Different fractions of lignocellulosic biomass can be used directly for adsorbent preparation or after integration of various modification approaches. Biomass fractions such as cellulose, lignin, pectin, etc. have been extensively studied in water purification applications (El Mahdaoui et al., 2024; Kainth et al., 2024; Supanchaiyamat et al., 2019). Biobased adsorbents such as carbonaceous materials have been shown to have high surface area, thermal stability, and mechanical stability. Additionally, due to the abundance of organic functional groups in biobased carbonaceous materials, they have the potential for many modifications. Biobased carbon has also been used to anchor and functionalize biopolymers to produce hybrid biosorbents. For the enhancement of efficiency and recyclability of the adsorbent, the magnetic biosorbents and nanoparticle biosorbents have been developed for metal removal applications (Tee et al., 2022). Furthermore, algae biomass generation has increased in recent years due to more awareness about algae-based products. Due to the availability of different functional groups in algal biomass to bind metallic species, algal biomass has garnered massive attention for heavy metal removal applications. It is noteworthy that using biomass waste in developing adsorbents can also help in the reduction of biomass waste.

5.1.2 Properties of Biobased Adsorbents

Biomass waste has a heterogeneous composition that varies with the biomass species and source (Makoś-Chełstowska et al., 2024). Similarly, the composition of algal biomass varies with the environmental conditions present during the growth stage. Generally, biobased adsorbents derived from biomass possess multiple surface functional groups, including –OH, –C=O, and –COOH, which may significantly affect adsorption efficiency (Kainth et al., 2024; Miretzky and Cirelli, 2010). Furthermore, modifying biobased raw materials to incorporate additional functional groups, such as amine, phosphate, and carboxyl, can enhance heavy metal interactions at the adsorbent surface.

Metal adsorption can be enhanced due to the formation of complexes by the interaction between metal ions and surface ligands of adsorbents. Biosorbents, owing to their highly porous nature, large surface area, and abundant ligand groups, are ideal for adsorption. Furthermore, pH_{zpc} (point of zero charge) is an important parameter that affects heavy metal interactions with the adsorbent surface. It has been demonstrated that the negatively charged surface of biobased adsorbents accelerates the adsorption of positively charged metallic species. It is noteworthy that the negatively charged surface of the adsorbent facilitates the electrostatic attraction toward positively charged metal ions.

Various mechanisms, including hydrogen bonding, electrostatic interactions, chelation, surface complexation, diffusion, and ion-exchange, can be involved in heavy metal removal by biobased adsorbents (Singh et al., 2020). To enhance adsorption properties, raw biosorbents can be modified via physical, chemical, or physicochemical activation (Tee et al., 2022). By integrating a modification approach, biomass waste, such as lignocellulosic biomass, can be converted into carbon/activated carbon or biochar. Furthermore, by employing various modification approaches, algal biomass can be converted into efficient adsorbent materials (Wang et al., 2023). In an oxygen-limited environment during thermal treatment, such as the pyrolysis of biobased raw materials, carbon materials are produced that can be further modified to meet requirements for heavy-metal removal applications (Sharma et al., 2024). During thermal treatment, cellulose decomposes into carbon, and volatile components (e.g., nitrogen, oxygen, and hydrogen) are removed from biomass waste.

The properties of biobased adsorbent materials can also be engineered through chemical activation, including single- or two-step activation, which can affect the adsorbent's functionality, surface area, and pore structure. Chemical activation methods, such as acid, alkali, or metal salt impregnation, have improved the adsorption capacity of biosorbents (Dey et al., 2025; Sharma et al., 2024). Chemical activation typically occurs at low activation temperatures. Chemicals such as NaOH, $NaHCO_3$, $Ca(OH)_2$, HCl, and H_2SO_4 have been extensively studied for the chemical activation of biobased adsorbents. Chemical activation has been shown to generate functional groups that can significantly affect sorption capacity (Sharma et al., 2024). Depending on the metallic pollutant species, biobased adsorbents can be engineered to possess the required characteristics through various modification approaches to enhance their adsorption efficiency. High surface area, porosity, multiple functionalities, surface charge, high adsorption capacity, regeneration potential, cost-effectiveness, and environmental friendliness are important aspects of an effective adsorbent.

5.1.3 Current Status of Biobased Adsorbents

The development of biobased adsorbents for water purification applications has attracted considerable attention owing to their eco-friendly nature, high adsorption capacity, and cost-effectiveness (Kainth et al., 2024). Various biobased adsorbent materials have been examined in different studies for heavy metal removal (Table 5.1).

Biomass-waste-derived activated carbon/modified activated carbon has been extensively used for water treatment applications, including heavy-metal removal, owing to their porous structures and various functional groups (Febrianto et al., 2009). Algal biomass, agricultural residues, forest biomass, and wood waste have been examined using various modification approaches to enhance their adsorption properties for metal interactions. Since chemical functional groups like carboxyl and hydroxyl are the primary sites for metal binding, raw lignocellulosic biosorbents are usually modified chemically to enhance their efficiency for metal adsorption.

To improve the recoverability of the adsorbent after adsorption, magnetic properties can be introduced into the bio-adsorbent material, thereby facilitating its recovery after metal removal (Tee et al., 2022). Different approaches have been developed for synthesizing magnetic adsorbent materials, including co-precipitation, hydrothermal, mechanochemical, and microwave-assisted processes. Notably, the adsorption performance of magnetic biosorbents can be influenced by the properties of the raw biomass and the magnetic modification process. Jing et al. used bamboo pulp-derived carbon to develop a $MnFe_2O_4$-loaded carbon-based aerogel composite with magnetic properties (Jing et al., 2024). Furthermore, sulfone-modified magnetic activated carbon was developed, exhibiting adsorption capacities of 147.05, 151.51, and 119.04 mg/g for Pb(II), As(III), and Cd(II), respectively (Nejadshafiee and Islami, 2019). In another study, algal-mediated magnetic activated carbon was prepared for Cr(VI) removal (Sharma et al., 2023). It has been reported that abundant active sites on carboxymethylated lignin facilitate ion-exchange and hydrogen-bonding interactions for the adsorption of Pb(II) and Cu(II) ions (Zhang et al., 2019). The integration of nanoparticles with biomass-derived raw materials for the fabrication of hybrid adsorbents has also been explored in recent years (Song et al., 2022; Srivastava and Sharma, 2013). Furthermore, hybrid pectin-based biosorbents were synthesized using secondary polysaccharide additives, and the adsorption capacity of the hybrid adsorbent was recorded to be 0.85 mmol/g for Pb(II) and 0.70 mmol/g for Cd(II), respectively (Jakóbik-Kolon et al., 2016).

Cellulose and nanocellulose-based adsorbents have gained significant attention because they possess abundant –OH groups on their surfaces, which facilitate the incorporation of chemical moieties (Pandey et al., 2023). Additionally, high specific

TABLE 5.1
Biobased Adsorbents for Heavy Metal Removal

Adsorbent	Heavy metal	Adsorbent activation/ modification	Adsorption mechanism/ key functional group	Adsorption capacity (Langmuir) q_m (mg/g)	Reference
Biobased aerogels	Cu(II)	Citric acid crosslinking	$(-NH_2)$, (−COOH), (−OH)	205.9	Varamesh et al., (2023)
Lignin-based adsorbent	Pb(II)	Free-radical polymerization	$(-SO_3H)$, $(-NH_2)$, (−COOH), (−OH)	397.3	Xiaohong Wang et al., (2021)
Jackfruit peel biochar	Pb(II)	Hydrothermal activation	(−COOH), (−OH) Ion-exchange, intraparticle diffusion	83.86	Ton-That et al., (2024)
Sugar cane bagasse	Pb(II)	Pyrolysis	Ion-exchange, surface precipitation	86.96	Abdelhafez and Li (2016)
Alkali lignin-based adsorbent	Pb(II)	Chloroacetic acid and carbon disulfide reaction	(−CSS), (−COOH), (−OH)	208.3	Zhang et al., (2021)
Modified jute fiber biosorbents (PANI/JF)	Cd(II)	in-situ aniline polymerization	(−COOH), (=N−)	155.8	Q. Huang et al., (2019)
Nickel-oxide-infused orange peel nanobiocomposite	Cd(II)	Nickel oxide infusion	(−COOH), (−OH) intraparticle diffusion	138.25	Njimou et al., (2025)
CMC-g-HAp/Fe3O4 nanobiocomposite	Cu(II)	Grafting polymerization and Fe_3O_4 embedding	(−COOH), (−OH), $(-PO_4)^{-3}$, (Fe-O)	88.49	Foroutan et al., (2018)
CMC-g-HAp/Fe3O4 nanobiocomposite	Ni(II)	Grafting polymerization and Fe_3O_4 embedding	(−COOH), (−OH), $(-PO_4)^{-3}$, (Fe-O)	86.95	Foroutan et al., (2018)
Corn stalks	Pb(II)	Grafting polymerization and amination	Surface precipitation, complexation, electrostatic attraction	59.10	Zhao et al., (2025)
Corn stalks	Cu(II)	Grafting polymerization and amination	Surface precipitation, complexation, electrostatic attraction	68.92	Zhao et al., (2025)

Corn stalks	Zn(II)	Grafting polymerization and amination	Surface precipitation, complexation, electrostatic attraction	60.79	Zhao et al., (2025)
Sulfated wood (SW-H2SO4)	Ni(II)	Sulfation by sulfuric acid	$(–SO_3H)$	35.6	Huo et al., (2025)
Sulfated wood (SW-ClHSO3)	Ni(II)	Sulfation by chlorosulfonic acid	$(–SO_3H)$	49.9	Huo et al., (2025)
Sulfated wood (SW-NH2SO3H)	Ni(II)	Sulfation by sulfamic acid	$(–SO_3H)$, $(–NH_2)$, complexation, electrostatic attraction	69.6	Huo et al., (2025)
Modified cornstalk cellulose-derived poly(amidoxime) ligand	Co(II)	Grafting polymerization and amidoximation	$(=NOH)NH_2$	249	Rahman et al., (2025)
Klason Lignin based biosorbent	Cd(II)	Acid treatment and lignin extraction	$(–OCH_3)$, (–COOH), (–OH), Ion-exchange	12.5	Yulia et al., (2025)
Pectin modified with β-cyclodextrin	Zn(II)	Crosslinking with β-cyclodextrin	(–COOH), (–OH), Ion-exchange	12.51	Fu et al., (2024)
Pine waste	Cu(II)	Physical treatment	(–NH), (–COOH), (–OH), (–C=O)	27.78	Touihri et al., (2021)

surface area, recyclability, and biodegradability also make these materials attractive for water treatment applications. Saravanan et al. reported that the surface area and pore structure of biobased adsorbents can be modified via ultrasonication (Saravanan et al., 2023). By using surfactants, the metal removal efficiency of adsorbents can be improved, as surfactants can significantly modify the surface charge of bio-adsorbents (Tamjidi et al., 2021).

There are a variety of biobased adsorbents; however, this book chapter focuses on cellulose-based, lignin-based, pectin-based, algae-based, and biocomposite adsorbents.

The trend of publications in biobased adsorbents is shown in Figure 5.2, which clearly indicates that cellulose-based adsorbents have been extensively explored. Studies on lignin and algae-based adsorbents are also increasing. However, studies on pectin- and biocomposite-based adsorbents have been less explored than those on cellulose-, lignin-, or algal-based adsorbents, particularly for metal removal applications.

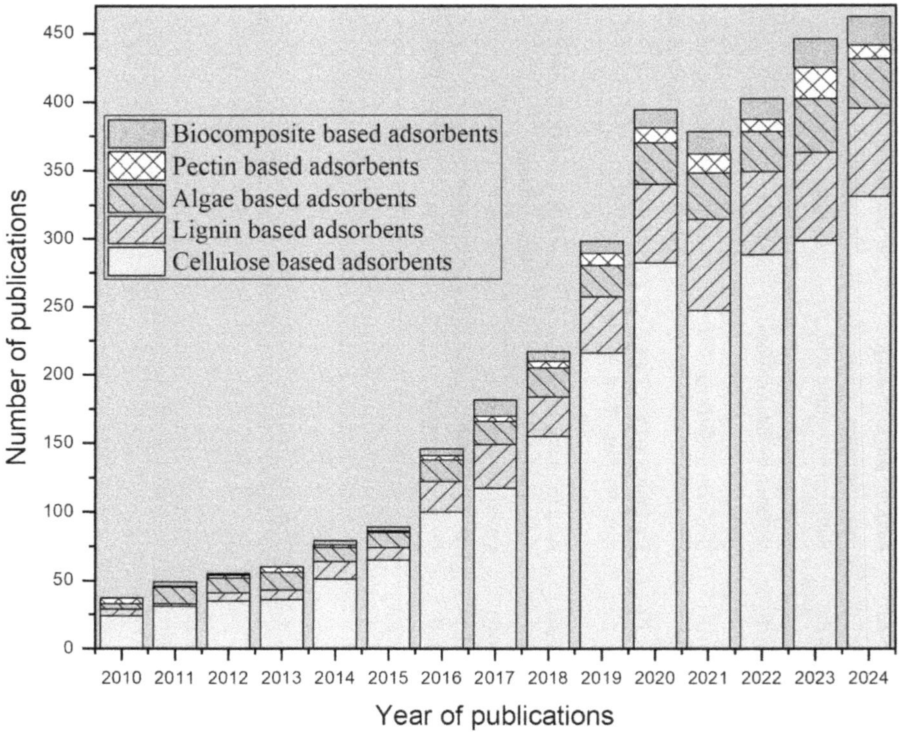

FIGURE 5.2 Publication trend related to biobased adsorbents (Biocomposite adsorbents, Metal removal (Web of Science, date of access 14.05.2025); Pectin adsorbents, Metal removal (Web of Science, date of access 14.05.2025); Algae adsorbents, Metal removal (Web of Science, date of access 14.05.2025); Lignin adsorbents, Metal removal (Web of Science, date of access 14.05.2025); Cellulose adsorbents, Metal removal (Web of Science, date of access 14.05.2025).

5.2 CELLULOSE-BASED ADSORBENTS

Cellulose and cellulose-derived materials have garnered immense attention for the synthesis of adsorbents for water treatment applications owing to their low cost and abundant availability (Farhan et al., 2024; Iftekhar et al., 2021; Jin et al., 2023). Materials at the micro- and nano-scale can be created from cellulose produced from biomass (Collazo-Bigliardi et al., 2018). It has been demonstrated that the cellulose source material and extraction parameters significantly affect the yield of extracted cellulose. Typically, in the pretreatment step, lignin and hemicellulose are solubilized to increase the cellulose mass fraction in the biomass, thereby enhancing accessibility to hydrolyzing catalysts during hydrolysis (Srivastava et al., 2023). After eliminating non-cellulosic components during pretreatment, cellulose can be extracted using various methods, including enzymatic and chemical procedures, as well as mechanical and chemical combinations. Cellulose-based adsorbents are prepared by chemically modifying the structure of natural cellulose to enhance their adsorptive capacity for pollutants, particularly heavy metals, in matrices such as water (Farhan et al., 2024; Iftekhar et al., 2021). Nevertheless, cellulose contains hydrophilic functional groups and possesses a limited number of active adsorption sites; therefore, chemical and physical modifications are required to enhance its adsorption capacity (El Mahdaoui et al., 2024; Tosh and Routray, 2014).

5.2.1 Synthesis of Cellulose-Based Adsorbents

Cellulose-based substrates are promising materials due to the existence of abundant hydroxyl groups on the cellulose's surface, which facilitate metal adsorption (El Mahdaoui et al., 2024). Additionally, the high chemical stability of cellulose makes it a suitable substrate for fabricating a variety of adsorbent materials. Different modification approaches can further improve the selectivity and adsorption capacity of cellulose-based adsorbents. Regeneration and reusability can be improved by selecting appropriate modification approaches. Acid or alkali pretreatments are commonly used to enhance access to reactive hydroxyl groups (Min et al., 2004). For instance, H_2SO_4 and citric acid were used to modify cellulose. Citric acid introduces additional functional groups onto the cellulose chain, whereas pretreatment with H_2SO_4 results in chain breakdown (Iftekhar et al., 2017a). Other mechanical treatments, such as sonication and freeze-drying, increase the porosity and surface area of cellulose. Cellulose is then chemically tailored to incorporate functional groups and to provide active sites that bind metal ions. The basic modification techniques include grafting, crosslinking, and surface modification (El Mahdaoui et al., 2024; Iftekhar et al., 2017b). Grafting is when one or many branches of polymers are grafted onto a cellulose backbone (Kang et al., 2015; M. Li et al., 2021). Usually, vinyl monomers, such as acrylic acid, acrylamide, or methacrylic acid, are used for grafting cellulose by the free-radical method (Ibrahim et al., 2002). As a result, carboxyl, hydroxyl, or amine groups are formed more frequently on the adsorbent surface, groups that have been shown to have a high affinity for metal ions (Gürdağ and Sarmad, 2013).

To enhance the stability and reusability of cellulose-based adsorbent materials, crosslinking agents such as glutaraldehyde, epichlorohydrin, and citric acid can be used (O'Connell et al., 2008). Covalent bond formation stabilizes the cellulose matrix, thus reducing its dissolution and mechanical breakdown during adsorption (Kaur et al., 2022). The modification approach involves altering the surface charge of cellulose fibers and enhancing their functionality. This can be achieved by chemical modification of material surfaces in the presence of charged functional groups, such as sulfonate, phosphate, and amine groups, or by other methods, such as plasma treatment and irradiation (Fakhre and Ibrahim, 2018; Yu et al., 2023). A method such as TEMPO(2,2,6,6-tetramethylpiperidine-1-oxyl) oxidation is one of the examples, where primary hydroxyl groups on cellulose are transformed into carboxyl ones, and, thus, cellulose becomes much more reactive toward metals (Abd El-Aziz et al., 2023; Nieminen et al., 2022; Saito and Isogai, 2005). The corn cob-based cellulose was functionalized under mild conditions by partial esterification of the surface−OH groups using α-bromoisobutyryl bromide (García-Vargas et al., 2025).

The cellulose molecules possess active hydroxyl groups that enable cationic exchange, thereby binding heavy metal ions to their surface. Using chemical or physical methods, these hydroxyl groups can be functionalized with additional substituents. These more reactive functional groups enhance adsorption on the cellulose surface (Rahaman et al., 2021). Cellulose, one of the most abundant naturally occurring biopolymers, has been used as an eco-friendly adsorbent. However, when tailoring cellulose-based adsorbents, it is essential to consider the environmental impact of the synthesis method.

5.2.2 Application of Cellulose-Based Adsorbents in Heavy Metal Removal

Cellulose- and nanocellulose-based adsorbents are gaining popularity in metal recovery applications due to their unique properties, including environmental friendliness, biocompatibility, and mechanical strength (Joshi et al., 2024). By reducing cellulose dimensions to nanometric levels (nanocellulose), the heavy metal adsorption increases significantly. For instance, the Cr(III) and Cr(VI) ion adsorption efficiencies from contaminated water increase by 42.02% and 5.79% with microcellulose and by 62.40% and 5.98% with nanocellulose (Singh et al., 2014). Nanocellulose is hydrophilic and can be functionalized to enhance surface properties for metal adsorption (Surendran and Sherje, 2022; Voisin et al., 2017). Surface modification with functional groups increases sorption capacity, as evidenced by the addition of succinic acid groups to CNC, which significantly improved binding to Pb(II) and Cd(II) (Yu et al., 2013). Similarly, CNFs' carboxylation increases their heavy metal adsorption capacity by 3–10% compared to uncarboxylated CNFs (Srivastava et al., 2012). It is reported that grafting of poly (acrylic acid)-sodium humate and poly (acrylic acid) on the surface of bamboo-derived CNF increased the removal of Cu(II) from the aqueous solution(Zhang et al., 2014).

Thiol-modulated cellulose exhibited adsorption of Cd(II), Pb(II), and Cu(II) with adsorption capacities of 45.9, 22.0, and 49.0 mg/g, respectively (Choi et al., 2020). Conjugation of cellulose with dibenzo-18-crown-6 enhanced the adsorption capacity for heavy metal removal compared with unconjugated cellulose (Fakhre and Ibrahim, 2018). The cellulose hydrogel modified with acrylamide and acrylic acid exhibited maximum capacities of 157.5, 289.9, and 393.2 mg/g for Cu(II), Cd(II), and Pb(II), respectively (Zhao et al., 2019). In another study, higher removal of Hg(II) and Cu(II) was achieved using modified porous cellulose beads and modified cellulose powder compared with unmodified cellulose (Li et al., 2019). The hyperbranched polyamide functionalized cellulose (HPFC) exhibited the highest adsorption capacity of 138 mg/g for Cu(II) ions (Yu et al., 2019). The adsorption capacities of CMC and polyacrylamide (CMC/PAM) hydrogels were studied for Pb(II), Cd(II), and Cu(II) in single- and multi-ion adsorption systems (Godiya et al., 2019).

The adsorption capacities of cellulose/chitosan composite synthesized at pH 5.0 were determined to bind heavy metals, and the adsorbed quantities were found to be 0.16, 0.28, and 0.53 mmol/g for Pb (II), Cd (II), and Cu (II), respectively(Xiao and Hu, 2017). Guo et al. used cellulose nanofibrils and carbon dot-based nanogels to adsorb heavy metals, achieving adsorption capacities of 1246 mg/g for Cu(II) and 2056 mg/g for Pb(II) (Guo et al., 2019). At pH 4.0, the adsorption potential of chitosan-g-poly (acrylic acid) hydrogel for Cu (II) and Pb (II) was 325.5 and 818.4 mg/g, respectively (Rodrigues et al., 2019). Functionalizing cellulose nanofibrils with an aldehyde group can enhance metal removal efficiency (Yao et al., 2016). In another study, an iron-based phenolic metal network coating was examined to modify a cellulosic adsorbent for Cr(VI) removal, and it was demonstrated that surface complexation via phenolic hydroxyl groups was responsible for Cr(VI) adsorption (Lunardi et al., 2024). Notably, the adsorbent retained more than 50% of its adsorption efficiency even after four adsorption-desorption cycles. In amidoxime-functionalized cellulose-based adsorbents, the involvement of amidoxime group chelation with metal ions has been reported (Yue et al., 2022). Furthermore, the efficiency of carboxymethylated cellulose fiber for removing Cu(II) and Ni(II) ions from wastewater was investigated (J. Wang et al., 2019). The roles of surface complexation and electrostatic interactions in metal adsorption have been reported. Cellulose-based adsorbents are promising materials for addressing challenges of heavy metal pollution.

5.3 LIGNIN-BASED ADSORBENTS

Lignin is the second most abundant biopolymer after cellulose and is typically found in plant cell walls, particularly in wood and bark, providing structural rigidity (Supanchaiyamat et al., 2019; Shorey et al., 2024). A high content of lignocellulosic biomass can be found in many industrial wastes, such as agricultural and forest waste (Ruthran et al., 2022). The paper and pulp industry generates a significant amount of lignin as a by-product (Shorey et al., 2024). Lignin has a three-dimensional structure containing monomeric units (monolignols): p-coumaryl, coniferyl, and sinapyl alcohols(Srivastava et al., 2023; Suhas et al., 2007). The proportion of these units varies with the biomass type and its origin. The dominant bond in all wood types (softwood and hardwood species) is the β-O-4 linkage (Zakzeski et al., 2010). Lignin is low-cost, biodegradable, environmentally friendly, and abundant in nature. Due to these properties, many researchers have drawn attention to lignin for removing heavy metals from wastewaters (Ruthran B et al., 2022).

5.3.1 Synthesis of Lignin-Based Adsorbents

Pure lignin or lignocellulosic biomass can be utilized as an adsorbent for heavy metal removal from aqueous phases. If pure lignin is utilized as an adsorbent or precursor for the preparation of adsorbents, the first step is to separate lignin from the lignocellulosic biomass, which consists of lignin, cellulose, and hemicellulose (Ruthran et al., 2022). Different extraction methods have been developed to isolate lignin from the lignocellulosic biomass. Kraft lignin, lignosulfonate, soda, and organosolv lignin are the major types of technical lignin, categorized by pulping technology. At an industrial scale, Kraft lignin and lignosulfonates are the two most prominent lignin types; however, Kraft lignin accounts for approximately 85% of global lignin production (Alam et al., 2024; Ruthran et al., 2022). Conventional extraction processes often employ toxic reagents, such as sulfuric acid and sodium hydroxide. Therefore, more environmentally safe and sustainable extraction methods, such as the organosolv technique, supercritical fluid, non-thermal plasma, ionic liquids, microwave-assisted extraction, and deep eutectic solvents, have also been developed (Alam et al., 2024). However, the development of these "sustainable techniques" is still in its early stages, and they are not yet used at an industrial scale. When extracted, pure lignin can be utilized as an adsorbent.

Pure lignin has generally been reported to be inefficient in removing heavy metals from liquid phases (Suhas et al., 2007). Therefore, different activation and modification methods are needed to enhance the removal efficiency of pure lignin. In addition to pure lignin, lignocellulosic biomass, for example, rice husk, coconut waste, barley straw, fruit shells, and coffee residue, can also be utilized as a lignin source to be used as an adsorbent (Ruthran B et al., 2022). One option to enhance the adsorption efficiency of lignin is to prepare lignin-based activated carbons or lignin-based char. The synthesis of activated carbon comprises two steps: carbonization and activation. Carbonization of lignin precursors can be achieved via pyrolysis (600–900 °C), during which small aromatic products, such as aldehydes, toluols, styrenes, and guaiacyl hydroxyls, can form. During pyrolysis, β-O-4 linkages are broken, and free-radical species are formed, leading to chain propagation, random repolymerization, and the formation of polyaromatic biochar. The carbonization step is typically performed before activation in many activated carbon processes. Activation can be done physically or chemically. Physical activation leads to the formation of a micropore structure. In chemical activation, carbonized char is impregnated with chemicals such as KOH, NaOH, or H_3PO_4. Subsequently, the slurry is heated under a nitrogen flow at 450–900 °C. The activating agent removes oxygen and hydrogen from the lignocellulosic raw material as water, thereby introducing porosity into the carbon (Suhas et al., 2007; Wang et al., 2022).

Microwave heating is also an approach that can be used for the preparation of lignin-based activated carbon (Ao et al., 2018). Microwave heating offers several advantages, including uniform/selective heating and short treatment times, which reduce energy consumption. A higher surface area (2,500–3,000 $m^2\ g^{-1}$) of activated carbon can be achieved by microwave heating under optimized conditions. The use of microwave heating in the preparation of activated carbon remains limited (Ao et al., 2018). Lignin can be modified, for example, by adding oxygen-, nitrogen-, or sulfur-functional groups (de Quadros Melo et al., 2016; Wang et al., 2022). Other modification methods, such as graft copolymerization, also increase the specific surface area of lignin (Wang et al., 2022). In addition to the aforementioned activation and modification methods, lignin-based composite adsorbents, such as lignin-graphene, lignin-nanoparticles, lignin-carbon nanotubes, and lignin-chitosan, can also be prepared (Wang et al., 2022). Lignin-based and composite adsorbents have been successfully employed for water purification.

5.3.2 Application of Lignin-Based Adsorbents in Heavy Metal Removal

Because lignin has high structural porosity and numerous aromatic rings, it can be used, without modification, to remove impurities from liquid phases. However, unmodified lignin is generally neither practical nor selective as an adsorbent for heavy metals (Suhas et al., 2007; Wang et al., 2022). For example, Sciban and Klasnja (2004) studied wood, pulp, and Kraft lignin as adsorbents for the removal of Cu(II). They reported that the maximum adsorption capacities across all cases ranged from 1.77 to 9.35 mg/g (Sciban and Klasnja, 2004). Due to relatively low adsorption capacities of unmodified lignin, researchers have employed various activation and modification strategies to enhance the performance of lignin-based adsorbents. Producing carbon-based lignin adsorbents is one option. Carbonization of pure lignin or lignin biomass, followed by physical or chemical activation, increases the material's surface area. Generally, chemical activation is more common than physical activation. Interestingly, the adsorbent produced by chemical activation has a higher surface area than that produced by physical activation (Wang et al., 2022). For example, Gao et al., (2013) prepared highly porous activated carbon from paper-making black liquor via KOH activation, which exhibited a specific surface area of 2,943 m^2/g.

Functionalization of lignin using O-, N-, and S-containing functional groups can enhance the metal binding capacity of developed adsorbents because, after modification, the structure of the adsorbent consists of atoms with free valence electrons, which are conductive to metal ion bonding (Wang et al., 2022). For example, a simultaneously aminated and sulfomethylated modified lignin-based adsorbent showed the maximum adsorption capacities of 45.4 mg/g and 53.8 mg/g for Cu(II) and Pb(II), respectively. Lignin-based composite adsorbents have gained researchers' attention in recent years (Wang et al., 2022). These materials are composed of different materials (e.g. lignin and graphene) with different physical and chemical properties, making them promising adsorbent materials for removing heavy metals. For example, Sun et al. prepared a lignosulfonate-modified graphene hydrogel that exhibited an ultra-high adsorption capacity (1743.9 mg/g) for Cr(VI). Yulia et al., (2025) used Klason lignin-based biosorbents derived from sugar palm fruit shells (Figure 5.3a). The developed adsorbent was employed for the removal of Pb(II) and Cd(II). The presence of various functionalities, including phenolic, hydroxyl, carboxyl, and methoxy groups, in KLSPFS provides potential sites for metal adsorption (Figure 5.3b).

Currently, lignin-based adsorbents are not commercialized. However, the development of effective lignin-based adsorbents is ongoing. The objective is to develop adsorbents that selectively remove metals across a wide pH range. Therefore, various

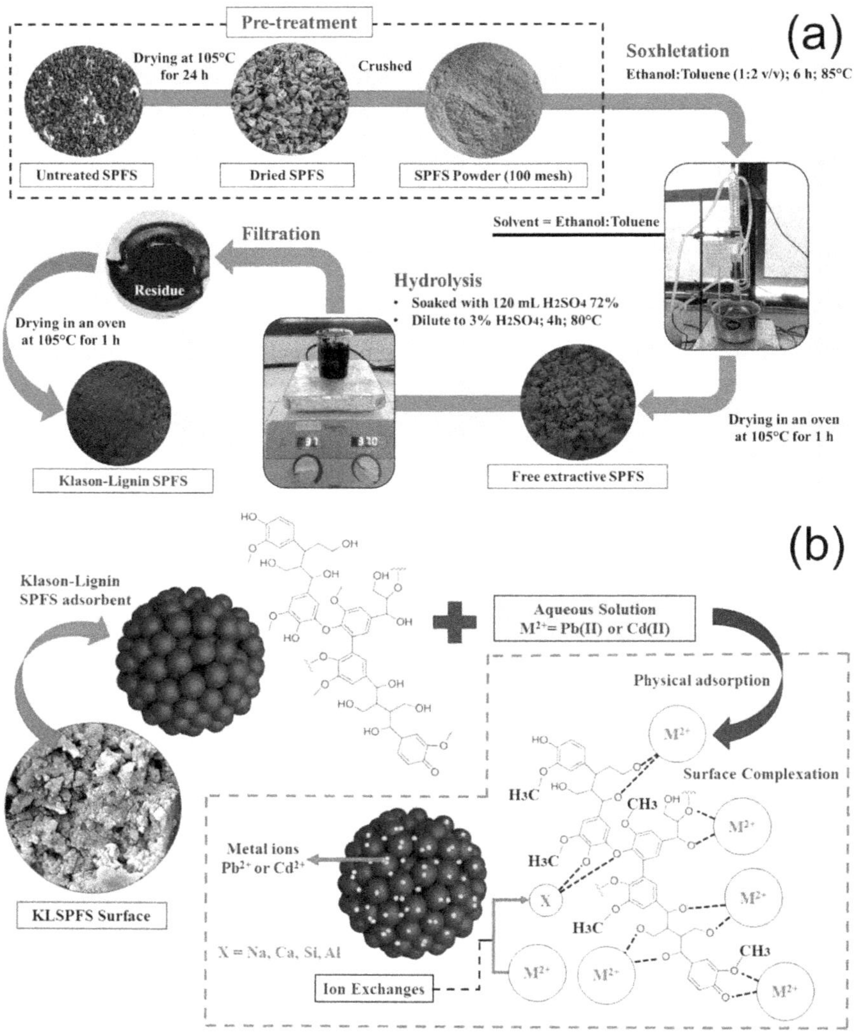

FIGURE 5.3 (a) Lignin extraction from sugar palm fruit shells(SPFS) Klason-method; (b) adsorption mechanism of Pb(II) and Cd(II) on KLSPFS. Reproduced from Yulia et al., (2025; Copyright Elsevier, 2025).

surface modification techniques need to be developed. In the literature, several studies on lignin-based adsorbents for heavy metal removal from liquid phases have been reported (Ge et al., 2014; Suhas et al., 2007).

5.4 PECTIN-BASED ADSORBENTS

Pectin is a naturally occurring polymer that commonly exists in plant cell walls and is known to have nontoxic properties. Pectin, a heteropolysaccharide, is the most abundant constituent of the middle lamella of plant cell walls, where it helps hold cells together (Thakur et al., 2019). The non-woody parts of terrestrial and earthly plants have excessive amounts of pectin; however, the amount of pectin may differ in every portion of the plant. Plant-based materials, including fruit peels and pulp, sugar beet, and sunflower heads, are potential sources for pectin extraction. It is noteworthy that industrial by-products, apple pomace and citrus peels from cider or juice industries are often used for the commercial production of pectin. In pectin, (l–4)-linked-α-d-galacturonic acid residues are branched with neutral sugars via different glycosidic bonds (Khotimchenko et al., 2007). Pectin consists of different polysaccharides, viz., homogalacturonan (HG), rhamnogalacturonan-I(RG-I), and rhamnogalacturonan-II (RG-II)(Li et al., 2021; Wang et al., 2019).

5.4.1 Synthesis of Pectin-Based Adsorbents

Pectin can be extracted under acidic conditions, followed by chemical modification to produce high- or low-methoxy pectin (Thibault and Ralet, 2003). Pectin molecules are anionic polysaccharides with abundant electron-rich functional groups such as acylamino, carboxyl, and hydroxyl groups, which can interact electrostatically with metal ions during adsorption. Pectin is an effective raw material for developing adsorbents owing to its diverse functional groups. Extraction approaches can affect pectin structure; therefore, pectin extracted from the same source using different approaches can exhibit pronounced differences in adsorbent properties.

Fares et al. synthesized poly(acrylic acid) grafted pectin hydrogel using glutaraldehyde crosslinker under N_2 atmosphere (Fares et al., 2011). It was demonstrated that additional carboxyl groups on pectin networks can be introduced via graft copolymerization of poly(acrylic acid), making it a viable candidate for metal-binding and water-treatment applications. Sisal waste is primarily the lignocellulosic residue left over from Agave sisalana (sisal) after separating stiff fibers (Zhu et al., 2021). Zhu et al. used a sequential enzymatic-ultrasonic process to prepare pectin-rich dietary fiber (P-SF) adsorbent from sisal waste (Zhu et al., 2021). During sonic cavitation, hydrolysis of carbohydrate structural components and disruption of the cellulosic network led to a slight increase in pectin content in P-SF.

5.4.2 Application of Pectin-Based Adsorbents in Heavy Metal Removal

It is noteworthy that pristine pectin, modified pectin, and pectin-based composite materials have all been shown to be effective for metal removal (Li et al., 2022). The presence of functionalities in pectin or pectin-based adsorbent significantly impacts heavy metal removal (Figure 5.4) (Wang et al., 2019). Pectin-based hybrid adsorbents or composite materials can be prepared via crosslinking or physical blending to enhance their effectiveness for metal adsorption (Li et al., 2022). To improve pectin's metal adsorption capacity, a range of modification techniques can be employed, including crosslinking, hydrolysis, esterification, amination, and free-radical reactions (Thakur et al., 2019). Electron-rich functional groups can be introduced into pectin via various modification approaches (Thakur et al., 2019). In pectin-based adsorbents, firstly available carboxyl groups can interact with the metal ions via

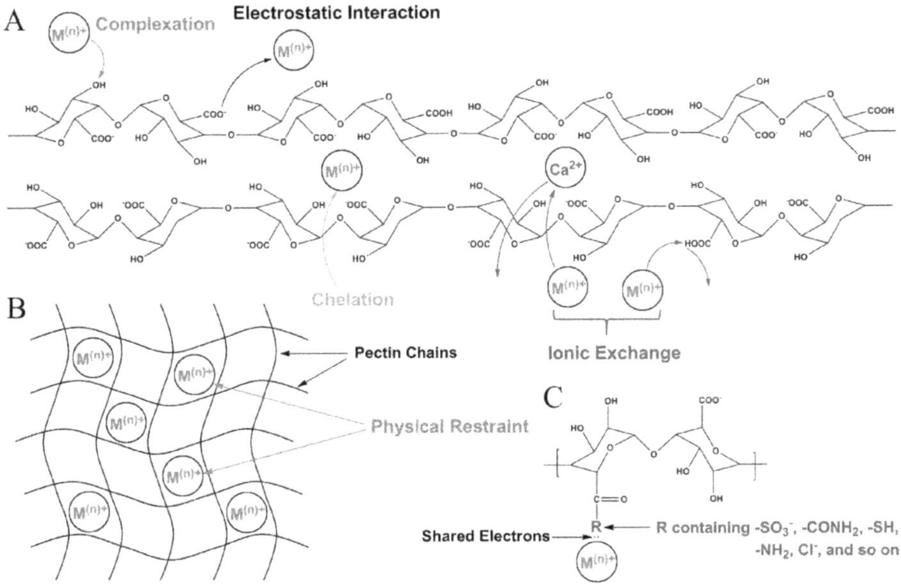

FIGURE 5.4 (a) Depiction of removal mechanisms for pectin; (b) Pectin-based hydrogels; (C) Chemically modified pectin. Reproduced from R. Wang et al. (2019; Copyright Elsevier, 2019).

electrostatic attraction, while electron-rich groups such as hydroxyl and dissociated carboxyl groups can serve as coordination sites for metal ion binding (Kodoth and Badalamoole, 2020). It is clear from Figure 5.4 that several mechanisms can be involved in the removal of heavy metals ($M^{(n)+}$) by pectin-based adsorbents, which makes pectin-based adsorbents quite effective.

Citrus pectin was developed for Pb(II) removal, exhibiting an adsorption capacity of 300 mg/g (Khotimchenko et al., 2007). Lessa et al. synthesized pectin/cellulose composite beads for Cu(II) and Cd(II) removal in a batch system, which showed 192.3 and 88.5 mg/g adsorption capacities for Cd(II) and Cu(II), respectively (Lessa et al., 2020). The involvement of functional groups in metal adsorption on pectin/cellulose composite beads was demonstrated using FTIR analysis. Shao et al. incorporated pectin into chitosan to develop a bead-like structure, which was subsequently evaluated for heavy metal removal. Interestingly, adding pectin to chitosan improved the adsorption capacity and resulted in an adsorbent with high porosity and stability (Shao et al., 2021). Chemisorption was involved in the adsorption of selected heavy metals over chitosan-pectin gel beads. In another study, a pectin/poly(m-phenylenediamine)-based microsphere was developed for Pb(II) adsorption, and it was reported that ion-exchange and chelation were the main mechanisms involved in Pb(II) adsorption (Xue-dong Wang et al., 2021). It was also demonstrated that these microspheres are effective up to five consecutive adsorption/desorption cycles. A sisal waste-derived adsorbent was employed for Pb(II) adsorption from aqueous solution, achieving a maximum adsorption capacity of 184 mg/g (Zhu et al., 2021).

In another study, a pectin-based hydrogel was synthesized to adsorb Cu(II) and Pb(II) ions from water and wastewater (Guilherme et al., 2010). The developed hydrogel exhibited adsorption capacities of 120 mg/g and 130 mg/g for Cu(II) and Pb(II), respectively, at pH 5.5. Zhang et al. synthesized a pectin-based adsorbent from grapefruit peel waste for Cu(II) removal (Zhang et al., 2020). In this study, biochar, pectin, and alginate were crosslinked to produce a composite bead with high chemical and mechanical stability. It is clear from the above discussion that pectin can produce both chemically and mechanically stable adsorbents by integrating with other biobased components, such as alginate, chitosan, and biochar. Furthermore, chemical modification approaches, such as the use of crosslinkers, can provide additional functionality to the pectin-based adsorbent, thereby enhancing metal adsorption. Given the abundance of pectin-based waste, it can serve as a sustainable raw material for developing efficient adsorbents for heavy metal adsorption.

5.5 ALGAE-BASED ADSORBENTS

A variety of algae species can be found in marine and freshwater environments (Wang et al., 2023). Based on their morphology, they can be classified as microalgae or macroalgae (Kumar et al., 2020; Lee et al., 2022). More than 20,000 species of microalgae (green algae) and 18,000 species of macroalgae (red, green, and brown algae) have been reported (Lee et al., 2022). Algae have simpler structures compared to higher plants (Ramesh et al., 2023). They are abundant and diverse, ranging from microscopic phytoplankton to large seaweeds (Ramesh et al., 2023). Microalgae and macroalgae usually consist of different chemical contents (moisture content, volatile matter, fixed carbon, and ash content) and biochemical contents (lipid, protein, and carbohydrate content). In addition, they also have different physical properties. Microalgae can grow within the size range between 1 and 400 μm, while macroalgae can grow up to 60 m. The physical, chemical, and biochemical properties of algal species depend on cultivation conditions and environmental factors (Lee et al., 2022).

Both macroalgae and microalgae are widely researched biomasses worldwide due to their potential to develop low-cost, environmentally friendly, renewable, and recyclable adsorbent materials (Lee et al., 2022; Ramesh et al., 2023). The core of an alga's cellular membrane contains proteins, lipids, and polysaccharides. In addition, algae contain various chemical/functional groups, such as carbonyl, carboxyl, ester, hydroxyl, amino, and sulfhydryl, which play a vital role in metal adsorption because they can bind metal species.

5.5.1 Synthesis of Algae-Based Adsorbents

Numerous synthesis techniques, including chemical, physical, mechanical, and other modification approaches, have been used to develop algae-based adsorbents (Lee et al., 2022). Typically, washing and drying are required to remove water from the algal biomass feedstock and impurities from its surface. Recently, studies have reported cases in which dewatering steps are bypassed. This type of synthesis reduces the overall cost of adsorbent synthesis. Following possible dewatering, algae are mechanically treated (grinding and crushing), which can affect the pore size, specific surface area, and surface morphology of the biomass. Among mechanical techniques, the ultra-centrifugal and cutting mills are commonly used (Lee et al., 2022). Further, algae-based biomass can be converted into biochar, which can be used as an adsorbent for metal removal from wastewater. Usually, biomass can be converted to biochar via pyrolysis, gasification, torrefaction, hydrothermal carbonization, or flash carbonization (Lee et al., 2022; Wang et al., 2023). The removal efficiency of algae-based biochar can be low toward some pollutants. This is due to the small surface area, low porosity, and lack of active sites (Wang et al., 2023). Therefore, algae-based biochars require activation to enhance their adsorption efficiency.

Physical and chemical activations are the most common methods for converting algae-based biochars into porous activated carbon (Lee et al., 2022). In physical activation, algae-based biochar is activated by an oxidizing mixture of gaseous reactants, such as CO_2, steam, or air, at elevated temperatures (>800 °C) (Lee et al., 2022). For chemical activation, algal feedstock is usually impregnated with chemical reagents, such as acids (e.g., HCl, HNO_3, H_2SO_4, H_3PO_4), bases (e.g., KOH, K_2CO_3, NaOH, Na_2CO_3), or metal salts (e.g., $ZnCl_2$, $AlCl_3$). After impregnation, the produced algae biomass slurry is directly subjected to pyrolysis (<600 °C) (Lee et al., 2022). Algae-based adsorbents can also be synthesized via other modification methods. Modification differs from activation methods in that it involves no high-temperature heat treatment. Chemical modification can be done using acids, alkalis, halides, metal salts, metal oxides, H_2O_2, or UV/H_2O_2 advanced oxidation processes (AOPs). In addition, biochar can be modified by doping with nitrogen-containing functional groups (Wang et al., 2023). This can be achieved, for example, by using polyethyleneimine beads to form a stable adsorbent (Ramesh et al., 2023). The choice of modification method depends on which impurities are to be removed. For example, acidic activators generate negative charges, whereas alkaline activators facilitate the generation of positive charges (Tan et al., 2023). Depending on the required surface functionalities to accelerate metal removal, the selection of a suitable modification approach is crucial.

5.5.2 Application of Algae-Based Adsorbents in Heavy Metal Removal

Algae-based adsorbents can be used to treat industrial wastewater because of their high surface area and the presence of various functional groups that facilitate metal binding. In literature, several studies have employed algae-based adsorbents for metal removal. To mention a few studies, for example, Katiyar et al. (2021) synthesized a magnetic kelp biochar with an adsorption capacity of 223 mg/g for Cu(II) removal from wastewater (Katiyar et al., 2021). Liu et al. (2021) developed blue algae-based biochar to remove Cd(II) from wastewater, which exhibited a maximum adsorption capacity of 135.7 mg/g (Liu et al., 2021). In another study, an adsorbent based on microalgal biochar immobilized a complex and was studied for Cd(II) removal, achieving an adsorption capacity of 217.41 mg/g (Shen et al., 2017).

Currently, most studies on algae-based adsorbents have been conducted at the laboratory scale and have not been evaluated at the industrial scale. One possible industrial-scale utilization method is to use reactor setups. The goal of these reactor setups is to handle large volumes of wastewater. At the same time, pollutant removal should be high, and algal growth should remain optimal. It is possible to use various reactor types, including photobioreactors, sequencing batch reactors, upflow anaerobic sludge blanket reactors, membrane bioreactors, continuous-flow reactors, and high-rate algal ponds (Ramesh et al., 2023). In the literature, the challenges associated with scaling up of algae-based adsorbents for commercial-scale use are widely reported. However, progress in developing green and sustainable methods for producing engineered algal adsorbents appears promising (Lee et al., 2022).

5.6 BIOCOMPOSITE-BASED ADSORBENTS

The fabrication of hybrid biocomposites involves blending organic components with either organic or inorganic matrices that are synthetic or chemically modified for the adsorption of pollutants, including heavy metals, critical metals, and organic compounds (El Mahdaoui et al., 2024; Zheng et al., 2024). This concept leverages the properties of biopolymers and inorganic or organic additives to enhance adsorption capacity, selectivity, and stability (El Mahdaoui et al., 2024; Iftekhar et al., 2018a; Zare et al., 2021). Extensive efforts have been made to develop biocomposite-based adsorbents for metal removal in recent years (Iftekhar et al., 2021; Kang et al., 2015; Li et al., 2021).

5.6.1 Synthesis of Biocomposite-Based Adsorbents

The first step in synthesizing bionanocomposite-based adsorbents involves the selection of suitable biopolymers. Commonly used biopolymers include cellulose, chitosan, alginate, starch, and others. To improve the properties of bionanocomposite adsorbents, various matrices can be incorporated, like inorganic (metal oxides, layered double oxides, silica-based, metal-organic frameworks (MOFs), etc.), nanomaterials (graphene oxide, carbon nanotubes, metal nanoparticles) and organic (ionic liquids, ligands) or chemical (acids, bases, or surfactants) modifiers (Iftekhar et al., 2025, 2021). This mixing of two or more matrices has been used to modify the surface properties of the biopolymer and to increase the number of active sites for adsorption (El Mahdaoui et al., 2024; Zheng et al., 2024). Several methods can be employed to synthesize biocomposite-based adsorbents, including coprecipitation, grafting, crosslinking, hydrothermal and thermal analyses, and freeze-drying.

A widely used method for synthesizing bionanocomposites is coprecipitation. The process involves the formation of precipitates by using basic solutions (Figure 5.5a) (Iftekhar et al., 2018c). In an organic-inorganic combination, the organic part is mainly the biopolymer, and the inorganic part is to enhance the properties of the bionanocomposite by ligand exchange using organic functional groups or surface-active agents after preparation (ex-situ method) or during the synthesis process in the reaction mixture (in-situ method). CL-Ca/Al LDH was synthesized by adding Ca and Al solutions to the CL solution and thereafter adjusting the pH to 10. The resulting precipitates were aged for 18 h and then dried and used for As(V) removal (Figure 5.5b,c) (Bessaies et al., 2021, 2020). Using a one-pot synthesis approach, XG-LDH was prepared by simultaneous coprecipitation, and freeze-drying was used to dry the slurry (Figure 5.5d,e) (Iftekhar et al., 2018b).

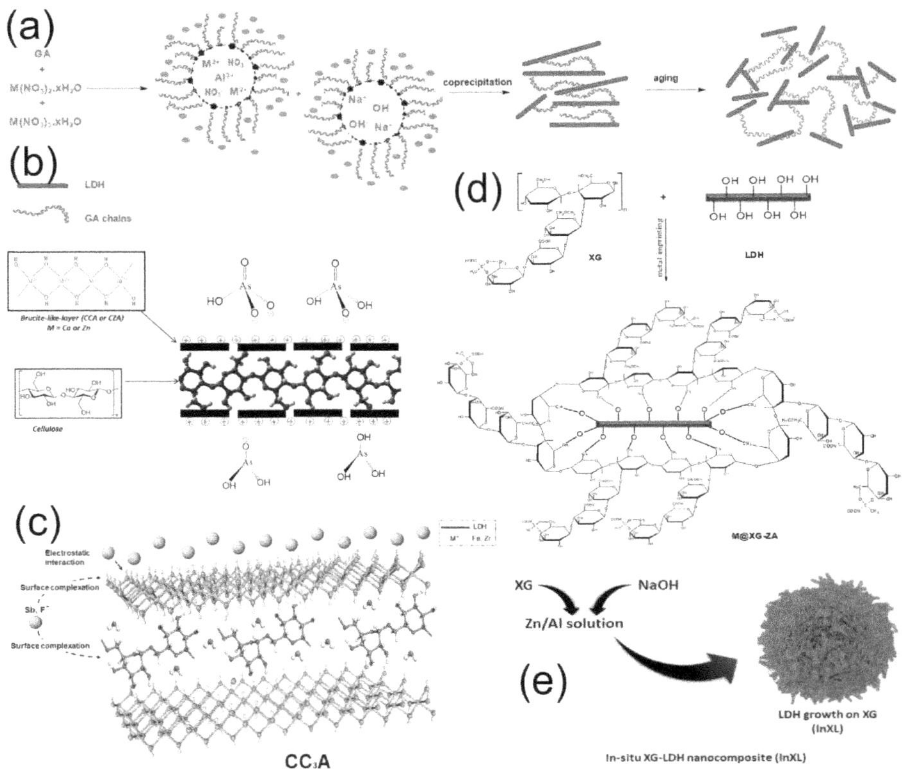

FIGURE 5.5 (a) Exfoliated GA-LDH (Iftekhar et al., (2018c). Reproduced with permission from Iftekhar et al., (2018c; Copyright Elsevier, 2018); (b) Cellulose intercalated LDH Bessaies et al., (2020). Reproduced with permission from Bessaies et al., (2020; Copyright Elsevier, 2020); (c) Cellulose with Ca/Al LDH. Reproduced with permission from Bessaies et al., (2021; Copyright Elsevier, 2020); (d) Metal ions anchored XG-LDH. Reproduced with permission from Iftekhar et al., (2018b; Copyright Elsevier, 2018); (e) Synthesis route of in-situ xanthan nanocomposites. Reproduced with permission from Iftekhar et al., (2020; Copyright Elsevier, 2020).

Grafting involves the chemical attachment of polymer chains onto the backbone of a biopolymer. This technique can enhance the functionality and adsorption capacity of the biocomposite. For instance, cellulosic resin was prepared by copolymerization of acrylic acid/acrylamide using N,N-methylenediacrylamide (MBA) as a crosslinker. It was observed that under microwave irradiation in the presence of initiators such as sodium thiosulfate/potassium persulfate, the reaction time decreased to a few minutes from 10 h with conventional heating (Bao-Xiu et al., 2006).

Ding et al. synthesized the bentonite/cellulose-grafted polyacrylic acid (Ben-PAA@MCC) material using ammonium persulfate as a catalyst (Ding et al., 2023). The biopolymer can be chemically cross-linked using agents like EDTA, glutaraldehyde, or epichlorohydrin. This process improves the mechanical stability and reusability of the adsorbent. Naturally occurring biopolymers, that is, chitosan and cellulose, were crosslinked using EDTA as a chelator (Chitosan@EDTA@CMC) (Manzoor et al., 2019). Epichlorohydrin and diethylenetriamine served as etherifying and crosslinking agents, respectively, to develop hydroxypropyl methylcellulose (HOPropMCell) functionalized with amino groups and to extract Cr(VI) (H. Huang et al., 2019). Likewise, to facilitate crosslinking between microcrystalline cellulose and tetraethylenepentamine (TEPA), epichlorohydrin was also examined(Xue et al., 2019). The performance of cellulose adsorbents can be enhanced by integrating nanomaterials, for example, metal nanoparticles, carbon nanotubes, or graphene oxide. These hybrid materials combine the high surface area and reactivity of nanomaterials with the mechanical and environmental benefits of cellulose. Nanocomposites such as CNC doped with zero-valent iron (nZVI) have demonstrated exceptional Cr(VI) adsorption capacity (Hu et al., 2023). Similarly, a one-pot synthesis approach was used to incorporate nZVI into cellulose nanofibrils (CNF) for the removal of Ni(II) (Song et al., 2022).

5.6.2 Application of Biocomposite-Based Adsorbents in Heavy Metal Removal

Applying biocomposite-based adsorbents is a promising approach to addressing environmental contamination with effective, sustainable materials. By combining the unique properties of biopolymers with additives and employing various synthesis techniques, researchers can develop tailored adsorbents capable of efficiently removing heavy metals and other pollutants from aqueous solutions. Naeimi et al. examined the effect of 3-aminopropyl triethoxysilane (APTES) modification on natural cellulose and plant-based cellulose. The former contains a chelating Schiff base, demonstrating higher Cd(II) adsorption capacity at pH 4.0 than plant-based cellulose (extracted from *Sesbania sesban)* (Naeimi and Amini, 2018). An imidazole-functionalized glycidyl methacrylate-modified cellulose (gGMA-Imidazole@Cell) composite was fabricated to remove Ni(II) (O'Connell et al., 2006).

The adsorption of Hg(II) on epichlorohydrin (Epi) pre-grafting cellulose (Epi-DTPA@DMCell) was 3.5 times lower than glycidyl methacrylate grafted cellulose (GM-DTPA@DMCell) (M. Li et al., 2021). Bentonite and cellulose were selected as base materials for the preparation of microcrystalline cellulose/bentonite grafted polyacrylic acid biocomposite that attained a 242.5 mg/g of adsorption capacity for Cd(II) (Ding et al., 2023). Furthermore, a carboxymethyl cellulose (CMC) and Attapulgite (APT) hydrogel composite was prepared by polymerizing acrylic acid (PAA) (Liu et al., 2010). The adsorption mechanism was complex, and the results indicated that the addition of APT improved Pb(II) removal, attributable to the presence of –OH groups and exchangeable cations on the surface. The adsorption equilibrium was reached faster, and the adsorption capacity was higher with 5% APT (952.4 mg/g) than with 20% (847.5 mg/g) (Liu et al., 2010). The most abundantly occurring biopolymers, i.e., cellulose and chitosan, were crosslinked via EDTA to produce a biocomposite (Chitosan@EDTA@CMC) which was effective in the adsorption of Cu(II)(142.95 mg/g) and attained equilibrium in 40 min at pH 5.5(Manzoor et al., 2019). The biocomposite not only exhibited potential for reuse for five cycles but also, when covered with soil, began to degrade within 20 days (Manzoor et al., 2019). Inorganic matrices such as metal silica, magnetic nanoparticles, and so forth can be incorporated to further enhance the performance of cellulose adsorbents. A cellulose silica composite carrying amino groups (SiO_2@Cell-TEPA) was employed for the Cu(II) removal and detection from wastewater. At pH 5.9 and in 2 min, the SiO_2@Cell-TEPA demonstrated a maximum adsorption of 166.6 mg/g for Cu(II) (Yousif et al., 2019).

The CCN@Fe_3O_4 showed 63.8 mg/g adsorption capacity for Pb(II) at 25 °C after 240 min, and the reaction was endothermic(Lu et al., 2016). The addition of magnetic nanoparticles to cellulose composites facilitated magnetic separation. Likewise, the Fe_3O_4@MCC composite was coated with manganese oxides, which showed an adsorption capacity of 446.4 mg/g for Pb(II) at pH 5(Jin et al., 2023). The carboxyl groups of Fe_3O_4@MCC composite shared a significant role in Pb(II) adsorption. In a study by Anirudhan et al., cellulose was grafted onto TiO_2 and subsequently functionalized with an amino group at the surface (TiO_2@CellPG$^+$Et) (Anirudhan et al., 2013). The TiO_2@CellPG$^+$Et employed in batch mode for Cr(VI) removal indicated that at pH 4.5 and in 60 min, the adsorption capacity attained was 123.6 mg/g (Anirudhan et al., 2013). The TiO_2@CellPG$^+$Et can be reused up to five cycles. In another study, MOFs and cellulose-based aerogel were developed for Cr(VI) removal. Compared to ZIF-8 (8.3 mg/g) and cellulose aerogel (5.0 mg/g), the hybrid composite ZIF-8@CellAr (41.8 mg/g) showed enhanced adsorption for Cr(VI) (Bo et al., 2018). By using iron scrap waste and alginate, an efficient adsorbent (green magnetic carbon/alginate biocomposite beads) comprising microporous and mesoporous structures was developed for the adsorption of Pb(II) ions (Saning et al., 2024). EDX confirmed the presence of high oxygen and carbon contents in the biocomposite beads. In a batch process, the efficiency of polyhydroxybutyrate biocomposites for Pb(II) adsorption was examined (Hungund et al., 2018).

In another study, a biocomposite was developed using lignin extracted from barley straw, exhibiting an adsorption capacity of 344.83 mg/g for Pb(II) (Mohammadabadi and Javanbakht, 2020). The thermodynamic study revealed that Pb(II) adsorption is endothermic. Furthermore, rice husk-derived lignin–TEOS biocomposites were employed for Ni(II) removal (Shweta and Jha, 2015).

In biobased composites, the feasibility of improving functionalities facilitates the metal interactions on the adsorbent surface and within pores, which can significantly impact metal adsorption. However, continued research and development in this field will contribute to the development of effective, sustainable, advanced materials for environmental remediation and sustainable practices.

5.7 PILOT-SCALE STUDIES OF BIOSORBENTS FOR HEAVY METAL REMOVAL

Pilot-scale studies link laboratory-scale investigations of biosorbents to their real-world applications in wastewater treatment. This helps in estimating the effectiveness of adsorbents in terms of cost, feasibility in operation, and performance of materials under dynamic flow conditions. Pilot-scale demonstrations of cellulose and lignin-based biosorbents for heavy metal ion removal remain limited, with most studies confined to laboratory or bench-scale investigations. Available pilot-scale evidence, including fixed-bed column systems using cellulose biomass and activated lignin composites, demonstrates technical feasibility for continuous operation but reveals significant economic and regulatory validation gaps (Escudero-Oñate et al., 2017; Islam et al., 2023). Conventional chemical precipitation remains more cost-effective for bulk metal removal than biosorption, as highlighted by pilot-plant studies. These studies reveal that while biosorption offers eco-friendly alternatives, it faces significant challenges in capital and operating expenditures, particularly in feedstock preparation, regeneration costs, and the longevity of adsorbents in multi-metal effluent scenarios (El Messaoudi et al., 2025).

Using cost analysis, the economic feasibility of the cellulose/chitosan composite relative to activated carbon was assessed (Zeng et al., 2021). The findings indicated that the cost of removing 1g of Cr using AC is 58.8 times higher (CNY 36.48) than that of the cellulose/chitosan composite (CNY 0.62). In another study, a techno-economic analyses (TEA) showed that operational costs for cellulose-based biosorbent treatment systems are on the same order of magnitude of \$1.5–2.0/m^3 of treated wastewater, making these systems economically competitive in comparison to conventional activated carbon processes for which costs are in the order of \$3–4/m^3 of treated wastewater (Ighalo et al., 2022). Cellulose-based systems show broader pilot deployment with established engineering precedents for hydrogel and nanocellulose formulations (capacities ranging 2.3–2,240 mg/g), while lignin-based adsorbents, despite strong intrinsic polyphenolic coordination sites and modification potential, lack extensive pilot economic validation (Banerjee et al., 2018; Kumari et al., 2025; Persano et al., 2024).

Biosorbents are most suitable for tertiary/polishing applications following primary precipitation and secondary clarification, as their performance in replacing primary treatment is not economically justified (Younas et al., 2021). Critical scale-up barriers include feedstock heterogeneity, mechanical attrition in packed beds, mass-transfer limitations, and the absence of systematic regulatory-compliance documentation against EPA or WHO discharge standards from pilot trials (Kumari et al., 2025). While cellulose-derived engineered adsorbents demonstrate higher near-term commercialization prospects due to clearer engineering pathways, and lignin offers sustainability advantages through industrial waste stream utilization, both technologies require robust pilot validation with long-term performance data, multi-cycle regeneration efficiency, and comprehensive techno-economic assessments before widespread industrial adoption can be realized (Banerjee et al., 2018; Kumari et al., 2025).

It is well known that biobased adsorbents and composites offer low production costs; however, the cost of large-scale production is very high, which restricts their application to pilot scale. Therefore, most studies in the literature on the application of biobased adsorbents and composites are limited to laboratory-scale batch-type experiments, and a significant gap remains in the development of pilot- and industrial-scale systems for the treatment of real wastewater samples (Rocky et al., 2023). Further research is needed to address this gap and support a shift from conducting investigations at laboratory scales to outlining large-scale implementation.

5.8 CHALLENGES AND FUTURE DIRECTIONS

Although biobased adsorbents have been extensively investigated for removing heavy metals from various waste streams, numerous challenges remain that must be addressed for large-scale applications of these materials. The following sections discuss these challenges and the future direction of biobased adsorbents.

5.8.1 Challenges in the Application of Biobased Adsorbents for Heavy Metal Removal

The efficiency of biobased adsorbent materials strongly depends on the source of biomaterials (agricultural residues and industrial waste) and on the techniques used for their preparation and functionalization. For example, the composition of the biomass may differ, which will also affect its interactions with metal ions. This unpredictability makes it difficult to decide which type of material to use in a particular application and hampers the formulation of protocols for their use (Iftekhar et al., 2021; Joseph et al., 2021; Saravanan et al., 2023). Several biobased adsorbents with low mechanical strength are synthesized, which pose challenges in cyclic processes such as wastewater treatment plants. Conditions such as prolonged stationary time under flow conditions or mechanical stress can damage the adsorbent's structure and reduce its adsorption capacity. Therefore, researchers must design and synthesize biobased materials with high adsorption capacities and durability.

In real wastewater scenarios, biobased adsorbents may compete with ions and other compounds in solution. Interference from other pollutants with target metallic species reduces the adsorption capacity of the developed material for various metal ions, as they may share adsorption sites with the adsorbent. Hence, the synthesis of biobased adsorbents with an affinity for specific metal ions, or the modification of biobased materials to achieve preferred selectivity for metal ions over competing ions, is required. Recycling of the powdered biobased adsorbent material is challenging. Furthermore, the use of harmful regeneration solutions imposes an additional burden on the environment. Regulatory and standardization issues are also crucial aspects of biobased adsorbent materials. Among the challenges facing biobased adsorbents, the absence of standard test methods and standards or guidelines for their use is the primary barrier to their commercial adoption. As guidelines on their effectiveness, safety, and environmental impact have not been established, it is often difficult to obtain permission to use them in industry. Biobased adsorbents are likely more environmentally friendly than traditional materials, but they can be expensive to develop, particularly if for commercial use. Factors such as material procurement, fabrication methods, and whether further changes are required can affect the total cost. The issues that arise in improving economic feasibility include focusing on workmanship to develop more effective processes and on identifying suitable, as well as renewable, feedstocks.

5.8.2 Future Directions in the Development of Biobased Adsorbents

In light of the existing challenges and limitations of biobased adsorbents, the following options will be pursued. Proper treatment and recovery of resources from exhausted adsorbents that cannot be reused for metal removal applications are also important for reducing waste generation. Improving the chemical and physical treatment approach can improve the selectivity and performance of biobased adsorbents. To reduce environmental impact, green chemicals (e.g., biobased ones) can be used to modify biobased raw materials during the development of adsorbents. Adjusting the size or incorporating nanomaterials can also enhance adsorbent performance.

Genetic and metabolic engineering offer considerable potential for developing biobased adsorbents with characteristics tailored to their utility in removing heavy metals. This finding indicates that integrating agricultural and industrial residues may enhance the sustainability and cost-effectiveness of biobased adsorbent materials. This approach also has implications for waste management, as surplus items can be remodeled to add value. Field trials and pilot-scale examinations will therefore be crucial for determining the field capabilities of biobased adsorbents to enhance their uptake in treatment systems (Asif et al., 2021; Iftekhar et al., 2022). Furthermore, the use of advanced tools, such as artificial intelligence, in the synthesis and application of biobased adsorbent materials can help elucidate key aspects of material synthesis and their application (Lakshmi et al., 2021). To bring the biobased material to market, it is imperative to understand its cost-effectiveness and environmental impact; hence, technoeconomic analysis (TEA) and life cycle assessment (LCA) should be considered.

5.9 CONCLUSION

The presence of heavy metals in the environment, even at low concentrations, can have harmful effects on flora and fauna. Hence, it is necessary to recover and reuse them in different applications. This approach will not only have a positive impact on the environment but also provide a secondary source of metals for reuse. Among the available approaches for heavy metal removal from aqueous phases, adsorption is the most suitable approach which requires a sustainable and effective adsorbent material. Biobased adsorbents are potential materials for water treatment applications. Cellulose, lignin, pectin, and algal biomass can be efficiently utilized to develop adsorbent materials for heavy metal removal.

Biomass waste is a promising raw material for preparing adsorbents due to its abundance and inherent functionalities. Biomass composition varies with source and origin; therefore, developing an adsorbent material with a similar profile is challenging. It is important to design and modify the synthesis methodology based on the raw material, biomass waste. By employing modification approaches, the properties of adsorbents can be tailored to target heavy metals. Recycling of adsorbents for reuse should also be considered. The use of advanced tools, such as artificial intelligence, in the development of biobased adsorbent materials and their application can help address synthesis-related challenges. Although biobased adsorbent materials have been extensively employed in water treatment, there remains a need to advance material synthesis methodologies to ensure effectiveness and sustainability for large-scale applications in heavy metal removal from diverse wastewater streams.

REFERENCES

Abd El-Aziz, M.E., Morsi, S.M.M., Salama, D.M., Elwahed, M.S.A., Shaaban, E.A., Abou-Zeid, R.E., 2023. Carbohydr. Polym. Technol. Appl. 6, 100388.

Abdelhafez, A.A., Li, J., 2016. J. Taiwan Inst. Chem. Eng. 61, 367–375.

Alam, M.M., Greco, A., Rajabimashhadi, Z., Esposito Corcione, C., 2024. Clean. Mater. 13, 100253.

Anirudhan, T.S., Nima, J., Divya, P.L., 2013. Appl. Surf. Sci. 279, 441–449.

Ao, W., Fu, J., Mao, X., Kang, Q., Ran, C., Liu, Y., Zhang, H., Gao, Z., Li, J., Liu, G., Dai, J., 2018. Energy Rev. 92, 958–979.

Asif, M.B., Zhang, Z., Iftekhar, S., Lehto, V.-P., 2021. 21—in: Kanwar, S., Kumar, A., Nguyen, T.A., Sharma, S., Slimani, Y.B.T.-B.N. (Eds.), Micro and Nano Technologies. Elsevier, pp. 447–469.

Ávila, F.G., Cabrera-Sumba, J., Valdez-Pilataxi, S., Villalta-Chungata, J., Valdiviezo-Gonzales, L., Alegria-Arnedo, C., 2025. Clean. Eng. Technol. 24, 100879.

Banerjee, M., Basu, R.K., Das, S.K., 2018. Process Saf. Environ. Prot. 116, 693–702.

Bao-Xiu, Z., Peng, W., Tong, Z., Chun-Yun, C., Jing, S., 2006. J. Appl. Polym. Sci. 99(6), 2951–2956.

Bessaies, H., Iftekhar, S., Asif, M.B., Kheriji, J., Necibi, C., Sillanpää, M., Hamrouni, B., 2021. J. Environ. Sci. 102, 301–315.

Bessaies, H., Iftekhar, S., Doshi, B., Kheriji, J., Ncibi, M.C., Srivastava, V., Sillanpää, M., Hamrouni, B., 2020. J. Environ. Sci. 91, 246–261.

Bo, S., Ren, W., Lei, C., Xie, Y., Cai, Y., Wang, S., Gao, J., Ni, Q., Yao, J., 2018. J. Solid State Chem. 262, 135–141.

Carolin, C.F., Kumar, P.S., Saravanan, A., Joshiba, G.J., Naushad, M., 2017. J. Environ. Chem. Eng. 5, 2782–2799.

Choi, H.Y., Bae, J.H., Hasegawa, Y., An, S., Kim, I.S., Lee, H., Kim, M., 2020. Carbohydr. Polym. 234, 115881.

Collazo-Bigliardi, S., Ortega-Toro, R., Chiralt Boix, A., 2018. Carbohydr. Polym. 191, 205–215.

de Quadros Melo, D., de Oliveira Sousa Neto, V., de Freitas Barros, F.C., Raulino, G.S.C., Vidal, C.B., do Nascimento, R.F., 2016. J. Appl. Polym. Sci. 133.

Dey, S., Veerendra, G.T.N., Phani Manoj, A. V., Anjaneya Babu, P.S.S., Jagadeesh, G.V., Swarup Kumar, J.N.V.R., 2025. Water-Energy Nexus 8, 73–92.

Ding, W., Liang, H., Zhang, H., Sun, H., Geng, Z., Xu, C., 2023. J. Water Process Eng. 51, 103414.

El Mahdaoui, A., Radi, S., Elidrissi, A., Faustino, M.A.F., Neves, M.G.P.M.S., Moura, N.M.M., 2024. J. Environ. Chem. Eng. 12, 113870.

El Messaoudi, N., Miyah, Y., Georgin, J., Franco, D.S.P., Ergan, B.T., Mbanga, B.G.F., Knani, S., Messali, M., 2025. React. Funct. Polym. 217, 106501.

Escudero-Oñate, C., Villaescusa, I., Poch, J., Fiol, N., 2017. in: Tumuluru, J.S. (Ed.),. IntechOpen, London.

Fakhre, N.A., Ibrahim, B.M., 2018. J. Hazard. Mater. 343, 324–331.

Fares, M.M., Tahboub, Y.R., Khatatbeh, S.T., Abul-Haija, Y.M., 2011. J. Polym. Environ. 19, 431–439.

Farhan, A., Khalid, A., Maqsood, N., Iftekhar, S., Sharif, H.M.A., Qi, F., Sillanpää, M., Asif, M.B., 2024. Progress in layered double hydroxides (LDHs): Sci. Total Environ. 912, 169160.

Febrianto, J., Kosasih, A.N., Sunarso, J., Ju, Y.-H., Indraswati, N., Ismadji, S., 2009. J. Hazard. Mater. 162, 616–645.

Foroutan, R., Ahmadlouydarab, M., Ramavandi, B., Mohammadi, R., 2018. J. Environ. Chem. Eng. 6, 6049–6058.
Fu, M., Tuo, X., Yan, X., Li, D., Zhu, H., Gao, S., Han, X., Zhou, J., Mou, D., Xiu, J., 2024. Int. J. Biol. Macromol. 274, 133563.
Gao, Y., Yue, Q., Gao, B., Sun, Y., Wang, W., Li, Q., Wang, Y., 2013. Chem. Eng. J. 217, 345–353.
García-Vargas, M., Munguía-Quintero, M.F., Alcaraz-Cienfuegos, J., Rosas-Aburto, A., Valdivia-López, M. de los Á., Hernández-Luna, M.G., Vivaldo-Lima, E., 2025. Int. J. Biol. Macromol. 284, 137986.
Ge, Y., Xiao, D., Li, Z., Cui, X., 2014. J. Mater. Chem. A 2, 2136–2145.
Godiya, C.B., Cheng, X., Li, D., Chen, Z., Lu, X., 2019. J. Hazard. Mater. 364, 28–38.
Guilherme, M.R., Reis, A. V, Paulino, A.T., Moia, T.A., Mattoso, L.H.C., Tambourgi, E.B., 2010. J. Appl. Polym. Sci. 117, 3146–3154.
Guo, X., Xu, D., Yuan, H., Luo, Q., Tang, S., Liu, L., Wu, Y., 2019. J. Mater. Chem. A 7, 27081–27088.
Gürdağ, G., Sarmad, S., 2013. Cellulose Graft Copolymers: in: Kalia, S., Sabaa, M.W. (Eds.),. Springer Berlin Heidelberg, Berlin, Heidelberg, pp. 15–57.
Hegazy, S., Manninen, M., Tuomikoski, S., Lassi, U., Srivastava, V., 2024. Sep. Purif. Technol. 329, 125163.
Hu, X., Song, M., Li, S., Chu, Y., Zhang, W., Deng, Z., 2023. Chemosphere 343, 140212.
Huang, H., He, D., Tang, Y., Guo, Y., Li, P., Qv, W., Deng, F., Lu, F., 2019. J. Chem. Eng. Data 64, 98–106.
Huang, Q., Hu, D., Chen, M., Bao, C., Jin, X., 2019. J. Mol. Liq. 285, 288–298.
Hungund, B.S., Umloti, S.G., Upadhyaya, K.P., Manjanna, J., Yallappa, S., Ayachit, N.H., 2018. Mater. Today Proc. 5, 21023–21029.
Huo, H., Xu, J., Hou, H., Yu, Y., Feng, W., Lu, M., Wang, S., Min, D., 2025. Int. J. Biol. Macromol. 309, 142835.
Ibrahim, M.M., Flefel, E.M., El-Zawawy, W.K., 2002. Polym. Adv. Technol. 13, 548–557.
Iftekhar, S., Haluska, O., Rahmani, A., Thapa, R., Riikonen, J., Lappalainen, R., Vepsäläinen, J., Lehto, V.-P., 2025. Chemosphere 379, 144404.
Iftekhar, S., Heidari, G., Amanat, N., Zare, E.N., Asif, M.B., Hassanpour, M., Lehto, V.P., Sillanpaa, M., 2022. Environ. Chem. Lett. 20, 3697–3746.
Iftekhar, S., Srivastava, V., Abdul Wasayh, M., Hezarjaribi, M., Sillanpää, M., 2020. Chem. Eng. J. 388.
Iftekhar, S., Srivastava, V., Casas, A., Sillanpää, M., 2018a. J. Clean. Prod. 170, 251–259.
Iftekhar, S., Srivastava, V., Hammouda, S. Ben, Sillanpää, M., 2018b. Carbohydr. Polym. 194, 274–284.
Iftekhar, S., Srivastava, V., Ramasamy, D.L., Naseer, W.A., Sillanpää, M., 2018c. Chem. Eng. J. 347, 398–406.
Iftekhar, S., Srivastava, V., Sillanpää, M., 2017a. Chem. Eng. J. 320, 151–159.
Iftekhar, S., Srivastava, V., Sillanpää, M., 2017b. Chem. Eng. J. 309, 130–139.
Iftekhar, S., Wasayh, M.A., Asif, M.B., Lehto, V.-P., Sillanpää, M., 2021. 2—in: Kanwar, S., Kumar, A., Nguyen, T.A., Sharma, S., Slimani, Y.B.T.-B.N. (Eds.), Micro and Nano Technologies. Elsevier, pp. 13–35.
Ighalo, J.O., Omoarukhe, F.O., Ojukwu, V.E., Iwuozor, K.O., Igwegbe, C.A., 2022. Clean. Chem. Eng. 3, 100042.
Islam, M.S., Rahaman, M.S., Barbeau, B., 2023. J. Environ. Chem. Eng. 11, 111181.
Jakóbik-Kolon, A., Andrzej K., M., Krzysztof, K., and Bok-Badura, J., 2016. Sep. Sci. Technol. 51, 2604–2611.
Jin, G., Gu, P., Qin, L., Li, K., Guan, Y., Su, H., 2023. Int. J. Biol. Macromol. 239, 124277.

Jing, W., Yang, C., Lin, X., Tang, M., Lian, D., Yu, Y., Liu, D., 2024. RSC Adv. 14, 39995–40005.
Joseph, J., Iftekhar, S., Srivastava, V., Fallah, Z., Zare, E.N., Sillanpää, M., 2021. Chemosphere 284.
Joshi, N.C., Joshi, A., Mitra, D., Gururani, P., Kumar, N., Joshi, H.K., 2024. Environ. Nanotechnology, Monit. Manag. 21, 100942.
Kainth, S., Sharma, P., Pandey, O.P., 2024. Appl. Surf. Sci. Adv. 19, 100562.
Kang, H., Liu, R., Huang, Y., 2015. Polymer (Guildf). 70, A1–A16.
Kartik, A., Akhil, D., Lakshmi, D., Panchamoorthy Gopinath, K., Arun, J., Sivaramakrishnan, R., Pugazhendhi, A., 2021. Bioresour. Technol. 329, 124868.
Katiyar, R., Patel, A.K., Nguyen, T.-B., Singhania, R.R., Chen, C.-W., Dong, C.-D., 2021. Bioresour. Technol. 328, 124829.
Kaur, J., Sengupta, P., Mukhopadhyay, S., 2022. Ind. Eng. Chem. Res. 61, 1921–1954.
Khotimchenko, M., Kovalev, V., Khotimchenko, Y., 2007. J. Hazard. Mater. 149, 693–699.
Kodoth, A.K., Badalamoole, V., 2020. Polym. Bull. 77, 541–564.
Kumar, M., Sun, Y., Rathour, R., Pandey, A., Thakur, I.S., Tsang, D.C.W., 2020. Sci. Total Environ. 716, 137116.
Kumari, A., Kamaraj, N., Selvaraj, R., Nanoth, R., 2025. Environ. Monit. Assess. 197, 547.
Lakshmi, D., Akhil, D., Kartik, A., Gopinath, K.P., Arun, J., Bhatnagar, A., Rinklebe, J., Kim, W., Muthusamy, G., 2021. Sci. Total Environ. 801, 149623.
Lee, X.J., Ong, H.C., Ooi, J., Yu, K.L., Tham, T.C., Chen, W.-H., Ok, Y.S., 2022. J. Hazard. Mater. 423, 126921.
Lessa, E.F., Medina, A.L., Ribeiro, A.S., Fajardo, A.R., 2020. Arab. J. Chem. 13, 709–720.
Li, B., Pan, Y., Zhang, Q., Huang, Z., Liu, J., Xiao, H., 2019. Cellulose 26, 9163–9178.
Li, D., Li, J., Dong, H., Li, X., Zhang, J., Ramaswamy, S., Xu, F., 2021. Int. J. Biol. Macromol. 185, 49–65.
Li, J., Yang, Z., Ding, T., Song, Y.-J., Li, H.-C., Li, D., Chen, S., Xu, F., 2022. Carbohydr. Polym. 276, 118789.
Li, M., Zhang, S., Cui, S., Qin, K., Zhang, Y., Li, P., Cao, Q., Xiao, H., Zeng, Q., 2021. Sep. Purif. Technol. 277, 119493.
Liu, P., Rao, D., Zou, L., Teng, Y., Yu, H., 2021. Sci. Total Environ. 767, 145447.
Liu, Y., Wang, W., Wang, A., 2010. Desalination 259, 258–264.
Lu, J., Jin, R.-N., Liu, C., Wang, Y.-F., Ouyang, X., 2016. Int. J. Biol. Macromol. 93, 547–556.
Lunardi, V.B., Cheng, K.-C., Lin, S.-P., Angkawijaya, A.E., Go, A.W., Soetaredjo, F.E., Ismadji, S., Hsu, H.-Y., Hsieh, C.-W., Santoso, S.P., 2024. J. Hazard. Mater. 464, 132973.
Makoś-Chełstowska, P., Słupek, E., Gębicki, J., 2024. Sci. Total Environ. 945, 173910.
Manzoor, K., Ahmad, M., Ahmad, S., Ikram, S., 2019. Synthesis, ACS Omega 4, 17425–17437.
Martins, R.O., Cardoso, A.T., Borsatto, J.V., Lanças, F.M., 2025. Talanta 283, 127171.
Min, S.H., Han, J.S., Shin, E.W., Park, J.K., 2004. Water Res. 38, 1289–1295.
Miretzky, P., Cirelli, A.F., 2010. J. Hazard. Mater. 180, 1–19.
Mohammadabadi, S.I., Javanbakht, V., 2020. Int. J. Biol. Macromol. 164, 1133–1148.
Naeimi, A., Amini, M., 2018. Iran. Polym. J. 27, 461–468.
Nejadshafiee, V., Islami, M.R., 2019. Mater. Sci. Eng. C 101, 42–52.
Nieminen, J., Anugwom, I., Pihlajamäki, A., Mänttäri, M., 2022. J. Memb. Sci. 659, 120786.
Njimou, J.R., Fai, V., Sieugaing, M.T., Nguenamadje, D., Godwin, J., Genola, O.B., Noumi, G.B., Tripathy, B.C., 2025. Hybrid Adv. 9, 100392.
O'Connell, D.W., Birkinshaw, C., O'Dwyer, T.F., 2006. J. Chem. Technol. Biotechnol. 81, 1820–1828.
O'Connell, D.W., Birkinshaw, C., O'Dwyer, T.F., 2008. Bioresour. Technol. 99, 6709–6724.

Pandey, A., Kalamdhad, A., Chandra Sharma, Y., 2023. Environ. Nanotechnology, Monit. Manag. 20, 100791.
Persano, F., Malitesta, C., Mazzotta, E., 2024. Polymers. 16(9), 1292. https://doi.org/10.3390/polym16091292
Rahaman, M.H., Islam, M.A., Islam, M.M., Rahman, M.A., Alam, S.M.N., 2021. Curr. Res. Green Sustain. Chem. 4, 100119.
Rahman, M.L., Shamrih, S.A., Azlyzan, N.A., Sarjadi, M.S., Arsad, S.E., Sarkar, S.M., Kumar, S., 2025. Carbohydr. Polym. Technol. Appl. 9, 100633.
Ramesh, B., Saravanan, A., Senthil Kumar, P., Yaashikaa, P.R., Thamarai, P., Shaji, A., Rangasamy, G., 2023. Environ. Pollut. 327, 121572.
Rocky, M.M.H., Rahman, I.M.M., Biswas, F.B., Rahman, S., Endo, M., Wong, K.H., Mashio, A.S., Hasegawa, H., 2023. Chem. Eng. J. 472, 144677.
Rodrigues, F.H.A., de C. Magalhães, C.E., Medina, A.L., Fajardo, A.R., 2019. Cellulose 26, 9119–9133.
Ruthran B, V., Barman, P., Kadam, R., Kumar, A., 2022. Emergent Mater. 5, 923–943.
Saito, T., Isogai, A., 2005. Carbohydr. Polym. 61, 183–190.
Saning, A., Thanachayanont, C., Suksai, L., Watcharin, W., Techasakul, S., Chuenchom, L., Dechtrirat, D., 2024. Int. J. Biol. Macromol. 261, 129765.
Saravanan, A., Karishma, S., Kumar, P.S., Thamarai, P., Yaashikaa, P.R., 2023. Environ. Pollut. 339, 122720.
Sciban, Marina, Klasnja, Mile, 2004. Adsorpt. Sci. Technol. 22, 195–206.
Shao, Z., Lu, J., Ding, J., Fan, F., Sun, X., Li, P., Fang, Y., Hu, Q., 2021. Int. J. Biol. Macromol. 176, 217–225.
Sharma, G., Verma, Y., Lai, C.W., Naushad, M., Iqbal, J., Kumar, A., Dhiman, P., 2024. Heliyon 10, e36288.
Sharma, J., Sharma, M., Nigam, S., Joshi, M., 2023. Chem. Phys. Impact 6, 100169.
Sharma, Y.C., Srivastava, V., Singh, V.K., Kaul, S.N., Weng, C.H., 2009. Environ. Technol. 30(6), 583–609.
Sharma, Y.C., Uma, Srivastava, V., Srivastava, J., Mahto, M., 2007. Chem. Eng. J. 127, 151–156.
Shen, Y., Li, H., Zhu, W., Ho, S.-H., Yuan, W., Chen, J., Xie, Y., 2017. Bioresour. Technol. 244, 1031–1038.
Sheraz, N., Shah, A., Haleem, A., Iftikhar, F.J., 2024. RSC Adv. 14, 11284–11310.
Shorey, R., Salaghi, A., Fatehi, P., Mekonnen, T.H., 2024. RSC Sustain. 2, 804–831.
Shweta, K., Jha, H., 2015. Biotechnol. Reports 7, 95–106.
Singh, K., Arora, J.K., Sinha, T.J.M., Srivastava, S., 2014. Clean Technol. Environ. Policy 16, 1179–1191.
Singh, S., Kumar, V., Datta, S., Dhanjal, D.S., Sharma, K., Samuel, J., Singh, J., 2020. Sci. Total Environ. 709, 135895.
Song, M., Hu, X., Gu, T., Zhang, W., Deng, Z., 2022. J. Environ. Chem. Eng. 10, 107466.
Srivastava, S., Abhishek, K., and Raj, K.R., 2012. Int. J. Green Nanotechnol. 4, 46–53.
Srivastava, V., Heponiemi, A., Tuomikoski, S., Kupila, R., Bergna, D., Lassi, U., 2024. Chapter 23—in: Sarkar, A., Lassi, U.B.T.-P. of B.W. (Eds.),. Elsevier, pp. 361–383.
Srivastava, V., Lappalainen, K., Rusanen, A., Morales, G., Lassi, U., 2023. Chempluschem n/a, e202300309.
Srivastava, V., Lassi, U., 2024. Curr. Opin. Green Sustain. Chem. 46, 100890.
Srivastava, V., Sharma, Y.C., 2013. Water, Air, Soil Pollut. 225, 1776.
Suhas, Carrott, P.J.M., Ribeiro Carrott, M.M.L., 2007. Bioresour. Technol. 98, 2301–2312.
Supanchaiyamat, N., Jetsrisuparb, K., Knijnenburg, J.T.N., Tsang, D.C.W., Hunt, A.J., 2019. Bioresour. Technol. 272, 570–581.

Surendran, G., Sherje, A.P., 2022. J. Drug Deliv. Sci. Technol. 75, 103601.
Tamjidi, S., Moghadas, B.K., Esmaeili, H., Shakerian Khoo, F., Gholami, G., Ghasemi, M., 2021. Process Saf. Environ. Prot. 148, 775–795.
Tan, X., Fengfa, Z., Huiwen, W., and Ho, S.-H., 2023. Bioengineered 14, 2252157.
Tee, G.T., Gok, X.Y., Yong, W.F., 2022. Environ. Res. 212, 113248.
Thakur, S., Chaudhary, J., Kumar, V., Thakur, V.K., 2019. J. Environ. Manage. 238, 210–223.
Thibault, J.-F., Ralet, M.-C., 2003. in: Voragen, F., Schols, H., Visser, R. (Eds.),. Springer Netherlands, Dordrecht, pp. 91–105.
Ton-That, L., Nguyen, T.-P.-T., Duong, B.-N., Nguyen, D.-K., Nguyen, N.-A., Ho, T., Dinh, V.-P., 2024. Biochem. Eng. J. 212, 109525.
Tosh, B., Routray, C.R., 2014. Chem. Sci. Rev. Lett. 3, 74–92.
Touihri, M., Gouveia, S., Guesmi, F., Hannachi, C., Hamrouni, B., Cameselle, C., 2021. Desalin. Water Treat. 225, 430–442.
Varamesh, A., Abraham, B.D., Wang, H., Berton, P., Zhao, H., Gourlay, K., Minhas, G., Lu, Q., Bryant, S.L., Hu, J., 2023. J. Hazard. Mater. 457, 131824.
Voisin, H., Bergström, L., Liu, P., Mathew, A.P., 2017. Nanomaterials, 7(3), 57.
Wang, J., Liu, M., Duan, C., Sun, J., Xu, Y., 2019. Carbohydr. Polym. 206, 837–843.
Wang, R., Liang, R., Dai, T., Chen, J., Shuai, X., Liu, C., 2019. Trends Food Sci. Technol. 91, 319–329.
Wang, T., Jiang, M., Yu, X., Niu, N., Chen, L., 2022. Sep. Purif. Technol. 302, 122116.
Wang, X., Li, X., Peng, L., Han, S., Hao, C., Jiang, C., Wang, H., Fan, X., 2021. Chemosphere 279, 130504.
Wang, X.-D., Li, Y., Dai, T., He, X., Chen, M., Liu, C., Liang, R., Chen, J., 2021. Carbohydr. Polym. 260, 117811.
Wang, Y., Ma, C., Kong, D., Lian, L., Liu, Y., 2023. J. Environ. Chem. Eng. 11, 111263.
Xiao, M., Hu, J., 2017. Cellulose 24, 2545–2557.
Xue, F., He, H., Zhu, H., Huang, H., Wu, Q., Wang, S., 2019. Langmuir 35, 12636–12646.
Yao, C., Wang, F., Cai, Z., Wang, X., 2016. RSC Adv. 6, 92648–92654.
Younas, F., Mustafa, A., Farooqi, Z.U., Wang, X., Younas, S., Mohy-Ud-Din, W., Ashir Hameed, M., Mohsin Abrar, M., Maitlo, A.A., Noreen, S., Hussain, M.M., 2021. Water, 13(2), 215.
Yousif, A.M., Zaid, O.F., El-Said, W.A., Elshehy, E.A., Ibrahim, I.A., 2019. Ind. Eng. Chem. Res. 58, 4828–4837.
Yu, D., Wang, Y., Wu, M., Zhang, L., Wang, L., Ni, H., 2019. J. Clean. Prod. 232, 774–783.
Yu, S., Sun, C., Fang, S., Wang, C., Alharbi, N.S., Chen, C., 2023. Appl. Surf. Sci. 637, 157883.
Yu, X., Tong, S., Ge, M., Wu, L., Zuo, J., Cao, C., Song, W., 2013. J. Environ. Sci. 25, 933–943.
Yue, Y., Liu, Y., Zhang, W., Guo, J., Gong, Y., Yu, Y., 2022. J. Environ. Chem. Eng. 10, 107846.
Yulia, R., Husin, H., Zaki, M., Razali, N., Hisbullah, Nasution, F., Ahmadi, Nurhazanah, Yazil, M.L., Sy, Y., Syafiie, S., 2025. Energy Nexus 17, 100398.
Zakzeski, J., Bruijnincx, P.C.A., Jongerius, A.L., Weckhuysen, B.M., 2010. Chem. Rev. 110(6), 3552–3599.
Zare, E.N., Iftekhar, S., Park, Y., Joseph, J., Srivastava, V., Khan, M.A., Makvandi, P., Sillanpaa, M., Varma, R.S., 2021. Chemosphere 280, 130907.
Zeng, H., Hu, Z., Peng, C., Deng, L., Liu, S., 2021. Polymers 13(21), 3788.
Zhang, W., Song, J., He, Q., Wang, H., Lyu, W., Feng, H., Xiong, W., Guo, W., Wu, J., Chen, L., 2020. J. Hazard. Mater. 384, 121445.
Zhang, X., Wang, Y., Lu, C., Zhang, W., 2014. Carbohydr. Polym. 114, 166–169.
Zhang, Y., Ni, S., Wang, X., Zhang, W., Lagerquist, L., Qin, M., Willför, S., Xu, C., Fatehi, P., 2019. Chem. Eng. J. 372, 82–91.

Zhang, Z., Wu, C., Ding, Q., Yu, D., Li, R., 2021. Ind. Crops Prod. 173, 114100.
Zhao, B., Jiang, H., Lin, Z., Xu, S., Xie, J., Zhang, A., 2019. Carbohydr. Polym. 224, 115022.
Zhao, D., Chen, H., Zhao, Q., Qi, W., Liu, Y., Tian, Y., 2025. Environ. Technol. Innov. 38, 104140.
Zheng, G., Kang, X., Ye, H., Fan, W., Sonne, C., Lam, S.S., Liew, R.K., Xia, C., Shi, Y., Ge, S., 2024. Chinese Chem. Lett. 35, 108817.
Zhu, W., Yang, J., Hu, D., Wang, Z., 2021. Food Funct. 12, 2418–2427.

6 Ion-Exchange Resins and Polymers in Heavy Metal Ion Removal

Jesús García Serrano and
Ana María Herrera González

6.1 INTRODUCTION

Heavy metal ions are hazardous environmental pollutants that are non-biodegradable and can be toxic or carcinogenic even at very low concentrations (Briffa et al. 2020), posing a serious environmental problem. Water pollution with metallic ions has significantly increased, capturing worldwide attention (Waheed et al. 2021). Among all the methods used to remove heavy metal ions from aqueous solutions, ion-exchange is considered one of the best because it is efficient and straightforward.

Ion-exchange is a reversible chemical reaction that replaces undesirable metal ions with harmless ones (Dąbrowski et al. 2004). In the ion-exchange process, the free mobile ions of a solid, called an ion-exchanger, are exchanged for ions of similar charge dissolved in the liquid media (aqueous or nonaqueous) with which the exchanger is in contact, which means that in this process, the ions can be removed from the liquid medium when the ion-exchanger is withdrawn from the solution. The ion-exchange process offers several advantages for the removal of dissolved heavy metal ions. For example, because most ion-exchange materials possess functional groups with a strong affinity for specific metal ions, the ion-exchange process allows the selective removal of these ions (Dąbrowski et al. 2004; Hubicki and Koodynsk 2012). This process is highly effective at removing even trace amounts of heavy metal ions from solutions (Smara et al. 2007). During the ion-exchange process, the exchanger's structure practically does not change, allowing it to be regenerated and reused. Also, ion-exchange methods have high capacity for wastewater treatment and can be integrated with other technologies for improved performance (Mahmoud and Hoadley 2012). These characteristics make ion-exchange a valuable tool for addressing heavy metal contamination in water and wastewater treatment.

Ion-exchange materials can be organic or inorganic. These materials contain functional groups that, when dissociated, form two ions of opposite charge; one of the ions remains fixed to the polymer chain or crystalline structure of the material, whereas the other ion is soluble and can participate in ion-exchange processes via the exchange with free ions of a solution. Examples of inorganic ion-exchangers are zeolites (Calvo et al. 2009), montmorillonite (Liu and Yang 2022), clay (Kim 2013), and

DOI: 10.1201/9781003599098-6

soil humus (Shoba and Chudnenko 2014). Organic ion-exchange materials are essentially polymers, known as ion-exchange polymers or polyelectrolytes, which can be defined as any polymeric material capable of exchanging ions with their environment. Ion-exchange polymers contain polymer chains with ionic or ionizable groups covalently bonded to the polymer backbone. A characteristic of these polymer materials is that, when they are immersed in water or a solvent with a high dielectric constant, their ionic groups dissociate, leaving positively or negatively charged groups along the polymer chain. Also, the charge of these groups is neutralized by oppositely charged counter-ions in solution, maintaining the electroneutrality of the polymeric material. Due to their polymeric nature and their ability to ionize in aqueous environments, polyelectrolytes are highly effective for wastewater treatment. Also, polyelectrolytes are effective chemical agents for modifying the surfaces of other materials, imparting surface charge and improving their performance in removing metals from water (Chen et al. 2024). The polyelectrolytes can be used directly for heavy metal ion removal from water or indirectly as components of other ion-exchange materials, such as ion-exchange resins (IERs), polyelectrolyte complexes, polyelectrolyte-surfactant aggregates, polyelectrolyte gels, or membranes functionalized with polyelectrolytes. Among them, IERs are synthetic organic ion-exchange materials widely used to remove heavy metal ions from wastewater.

6.2 ION-EXCHANGE RESINS

IERs are a class of crosslinked polymers containing functional groups with mobile ions; these polymers are insoluble in aqueous solutions. IERs are produced by copolymerization reactions and crosslinking of monomers containing ionic functional groups or non-ionic monomers. When non-ionic monomers are used, one of the repeating units in the resulting crosslinked polymers is functionalized with positively or negatively charged groups, providing the resin with ion-exchange capacity. Different resins have also been studied for the removal and separation of heavy metals (Pramanik 2023; Silva et al. 2018).

6.2.1 Classification of Ionic Exchange Resins

The main classification of IERs is based on the electrical charge that the ionic groups attached to the resin can exchange. IERs are classified in three categories, which are cationic, anionic, and amphoteric, as shown in Figure 6.1a. Cationic resins are those that contain negatively charged functional groups, which exchange positively charged ions (cations). This class of resins can be further classified into strong acid type (if the functional groups in the resins were obtained from a strong acid, commonly sulfonic acid) or weak acid type (with functional groups obtained from a weak acid, such as carboxylic acid or phosphonic acid). Cationic resins are used to capture heavy metal ions from wastewater via ion-exchange. Figure 6.1b illustrates the ion-exchange process of a cationic exchange resin having sulfonic acid groups. On the other hand, the anionic resins have positively charged functional groups that can exchange negatively charged ions (anions), and depending on the type of

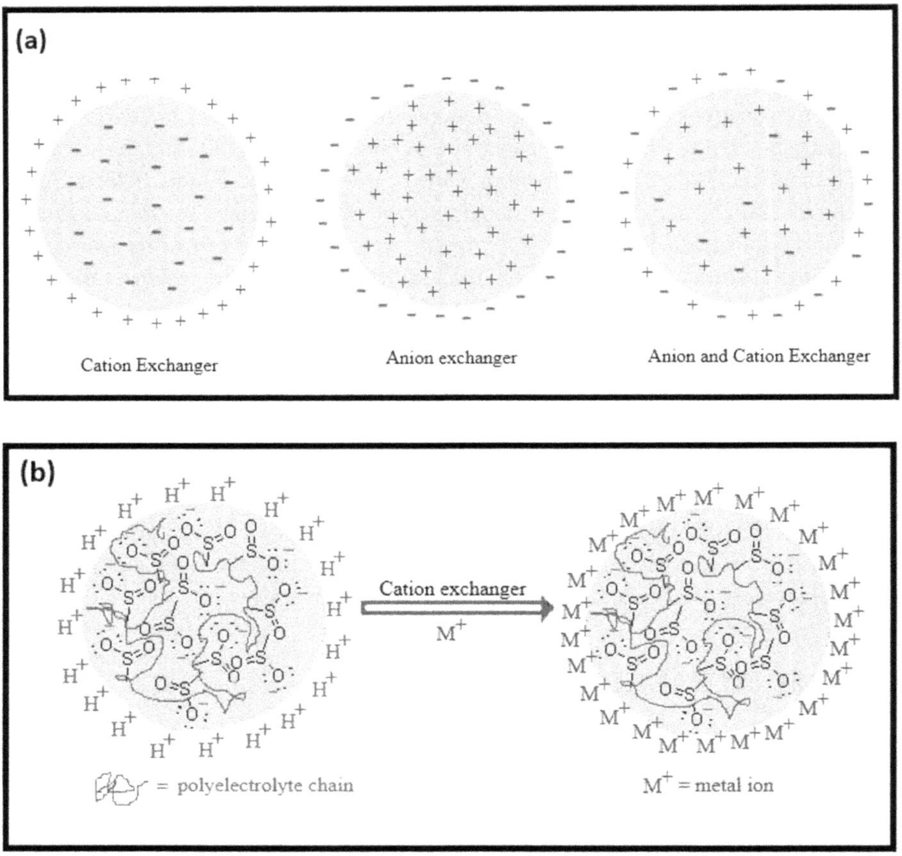

FIGURE 6.1 Types of ion-exchange resins (a) and the Ion-exchange process of a cationic exchange resin based on a polymer matrix and fixed sulfonic acid groups (b).

group exchange, these can be further classified as strong base anionic and weak base anionic, which contain functional groups commonly obtained from quaternary ammonium compounds and amine compounds, respectively. The classification of strong and weak indicates the resins' ability to dissociate at different pH values. While the strong cationic resins and strong anionic resins can dissociate over the entire pH range, the weak cationic resins and weak anionic resins can only dissociate in a small pH range (Wachinski 2016).

Amphoteric IERs contain both negatively and positively charged functional groups and, depending on solution pH, can exchange both cations and anions. Amphoteric resins are also called polyampholytes or polyzwitterions. These differ in that polyampholytes contain charged groups in different repeating units and in varying amounts along the polymeric chain. Meanwhile, polyzwitterions contain charge groups within the same repetitive unit in equal amounts. Finally, there is another special class of resins known as chelating IERs, which are essentially weak cation-exchange resins with functional groups capable of donating free electron pairs, forming complexes with metal ions. These resins contain functional groups with electron-donating elements such as sulfur, nitrogen, or oxygen. Some examples of functional groups with electron-donating elements in chelating resins are: iminodiacetic, carboxylic, hydroxamic, diphosphonic, aminophosphonic, hydroxyquinoline, isothiourea, thiol, and thiourea, among others. Resins are classified as monodentate, bidentate, or polydentate based on the number of coordination bonds with metal ions. Chelating resins are the most commonly used for the selective removal of toxic metal ions from aqueous solutions due to their significant affinity for heavy metal ions. Due to their selectivity, various synthetic, semisynthetic, and natural chelating polyelectrolytes and resins have been reported for the removal of heavy metal ions from wastewater (Chen et al. 2024; López-Maldonado et al. 2024). If only the charge of the groups responsible for ion-exchange is considered, other ion-exchange polymers, such as polyelectrolytes, polyelectrolyte complexes, polyelectrolyte-surfactant aggregates, polyelectrolyte gels, and polyelectrolyte-functionalized membranes, can be classified in the same way as resins.

6.3 PHYSICAL AND CHEMICAL PROPERTIES OF POLYMERIC ION-EXCHANGERS

The efficiency and effectiveness of a polymeric ion-exchanger depend on its structural characteristics and specific properties. Various ion-exchange polymers exhibit a wide range of properties due to their specific chemical structures, molecular weights, ionic functional groups (anionic, cationic, or both), charge densities, hydrophilicity, and other related factors. These differences result in variations in physical, chemical, and mechanical properties, impacting the durability and performance of the polymers under different conditions. Understanding key properties such as degree of crosslinking, ion-exchange capacity, solubility, swelling, mechanical strength, chemical stability, and ion-exchange selectivity, among others, is crucial for designing efficient polymeric ion-exchangers for the capture, separation, and removal of dissolved heavy metal ions from water and wastewater.

6.3.1 Chemical Structure

The properties of ion-exchange polymers are strongly influenced by their microscopic structure, specifically by the arrangement and characteristics of their components. In general, the polymers are formed with a large number of monomeric units or monomers linked by covalent bonds. The order in which the monomers are located along the polymeric chains determines the molecular structure of the polymers. Different monomer arrangements give rise to various structures, including linear, branched, cross-linked, and network structures (Figure 6.2). The structure of an ion-exchange polymer used in the removal of dissolved heavy metal ions in water consists of three essential components: a polymer matrix, ionic functional groups, and mobile ions. The polymer matrix forms the main structure of the material (resin, gel, membrane, etc.) and largely determines the ion-exchanger's mechanical, chemical, and

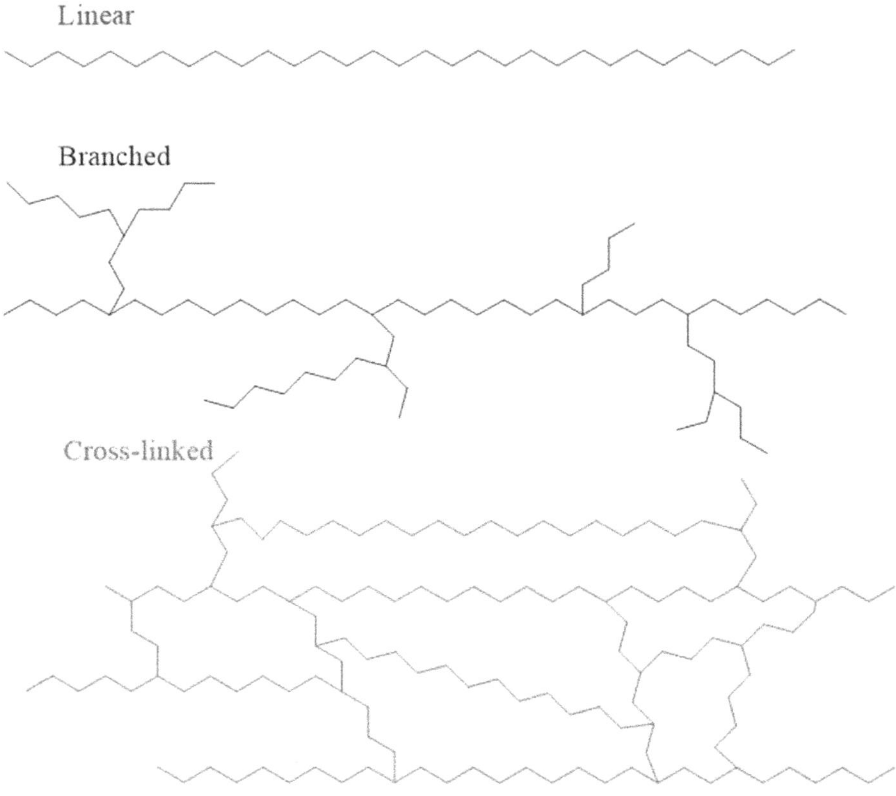

FIGURE 6.2 Polymers with linear, branched, cross-linked, or network structure.

thermal stability. Polymer matrix generally consists of a three-dimensional network of covalently crosslinked hydrophobic polymers, such as polystyrene, polyethylene, or polyacrylic acid crosslinked with divinylbenzene. Although the polymer matrix is insoluble in water, the presence of hydrophilic functional groups permanently attached to the polymer backbone allows limited hydration and swelling, enabling selective ion transport.

In cationic exchange resins, the hydrophilic groups responsible for ion-exchange, such as sulfonic, phosphonic, and carboxylic acid groups, carry free mobile cations, called counterions (typically H^+ or Na^+), which are exchanged for other metal ions in solution during the ion-exchange process. In the case of chelating ion-exchange polymers, the density and relative position of the functional groups (chelating groups), such as carboxylic acids, iminodiacetic acid, aminophosphonic acid, and thiol, as well as the distance from the matrix of chelating groups, are structural aspects that have a significant influence on the properties of these polymers. Table 6.1 lists the types of polymeric matrices, acid groups, and chelating groups commonly found in synthetic ion-exchange polymers.

Effect of crosslinking degree on the structure. Chemical or physical crosslinking is the process by which adjacent chains of the polymer matrix are linked together,

TABLE 6.1
Examples of Polymeric Matrices and Functional Groups Commonly Found in Synthetic Ion-Exchange Polymers

Polymer matrix:	Polystyrene-divinylbenzene, polyacrylic-divinylbenzene, polystyrene, polysulfone, polyacrylamide, polyacrylic acid, and polymethacrylic acid
Functional groups in cationic exchangers:	Sulfonic acid ($-SO_2(OH)$), carboxylic acid ($-COOH$), phosphonic acid ($-PO(OH)_2$), phosphinic acid ($-PO(OH)$), arsonic acid ($-AsO(OH)_2$), and selenonic acid ($-SeO_2(OH)$)
Functional groups in chelating ion-exchangers:	Carboxylic acid ($-COOH$), hydroxamic acid ($-CONHOH$), iminodiacetic acid ($-N(CH_2COOH)_2$), aminophosphonic acid ($-NH-CH_2{\_}P(=O)(OH)_2$), diphosphonic acid ($-PO(OH)_2)_2$, thiol ($-RSH$), thiourea ($-CH_2NHC=SNH_2$), amidoxime ($-C(=NOH)NH_2$), dithiocarbamate ($>N-C(=S)-S-$), and sulfonic acid ($-SO_2(OH)$)

usually through covalent bonds. This process involves the reaction of polymer-specific functional groups with bifunctional molecules (called crosslinkers) to form a three-dimensional structure interconnected with covalent bonds for chemical cross-linking or non-covalent interactions, such as hydrogen bonding and ionic interactions, for physical crosslinking. The structural characteristics of ion-exchange polymers, such as porosity, density, and distribution of ion-exchange sites within the polymer matrix, are determined for the crosslinking degree (Pérez-Maciá et al. 2015). For example, sulfonated cross-linked copolymers functionalized with carboxylic groups and hydroxyl groups with a low degree of crosslinking lead to a gel-type ion-exchange polymer, while a high degree of crosslinking results in a macroreticular structure. These two types of porous structures differ in their swelling behavior: the gel-type resins are glassy, nonporous materials when dry, but in a suitable solvent, the polymer matrix can completely swell. Instead, the macroreticular resins have permanent mesopores and exhibit less swelling (Lavarda et al. 2019). In general terms, higher levels of crosslinking in polymers result in a more rigid and less swelling resin. In contrast, lower crosslinking produces microporous polymer structures with smaller pores, greater flexibility, and increased swelling capacity. Therefore, it is essential to consider that the ion-exchange polymer must have a structure that is sufficiently rigid to avoid collapse in solution, yet not so rigid as to impede metal-ion diffusion. The relationship between the degree of crosslinking and chemical structure is critical for optimizing mechanical stability, porosity, and the arrangement of ionic functional groups, thereby facilitating the design of polymeric ion-exchangers for specific applications.

6.3.2 Ion-Exchange Capacity

The ion-exchange capacity (IEC) of an ion-exchange polymer is the number of ionic groups that participate in ion-exchange with the environment. The units used to express IEC are milliequivalents (or millimoles) of ionic groups per gram of dry polymer, or milliequivalents (or millimoles) of ionic groups per milliliter of wet polymer. The IEC of cationic exchangers is often determined using acid-base titration and back-titration (Shi et al. 2019; Shabani et al. 2020). In a back-titration procedure, the ion-exchanger is immersed in a solution of known HCl concentration for a specific time. The resulting solution is then titrated with NaOH to determine the amount of H^+ ions exchanged. The following equation calculates the IEC of the sample:

$$IEC = (C_{HCl}\, V_{HCl} — C_{NaOH}\, V_{NaOH}) \,/\, m_{dry}$$

where C is concentration, V is volume, and m is the mass of the dried sample following titration.

Other methods and techniques used for the quantification of IEC include ion chromatography (Lendero et al. 2005), thermogravimetric analysis, elemental analysis, and spectroscopic techniques, such as Fourier-transform infrared spectroscopy, Nuclear Magnetic Resonance, and energy-dispersive X-ray (Moukheiber et al. 2012).

The IEC of polymeric ion-exchangers is affected by the type and density of ionic groups, the degree of crosslinking of the polymer matrix, and the physical characteristics of the interchanger material, such as particle size and porosity. In cationic exchange polymers, the type of acid group (e.g., sulfonic acid, phosphonic acid, carboxylic acid) on the polymer directly determines its capacity to bind specific heavy metal ions. Stronger acid groups generally lead to higher exchange capacity. For example, resins containing sulfonic acid show a greater IEC than carboxylic resins due to the greater acidic character of the sulfonic acid (Shah et al. 2023). Regarding crosslinking, a high degree of crosslinking reduces the polymer's pore size, thereby restricting the diffusion of large ions. When ions in solution cannot reach the exchange centers due to steric factors, the ion-exchanger exhibits a low IEC (Marin et al. 2024). Various commercial polystyrene matrix-based strong-acid cation-exchange resins with crosslinking degrees of 2% to 8% exhibit IEC values of 1.8 to 4.8 meq/g.

6.3.3 Hydrophilicity

Cation-exchange polymers used to remove positively charged ions from water are usually insoluble in aqueous solution because they are crosslinked. Still, they exhibit a hydrophilic character due to the presence of negatively charged functional groups, such as sulfonic or carboxylic acid groups, which attract water molecules. The hydrophilicity of cation-exchange polymers is key to optimizing cation-exchange processes in solution. A high degree of hydrophilicity enhances cation mobility and selectivity, promoting the formation of water channels that directly influence metal ion capture.

6.3.4 Swelling

Ion-exchange polymers contain ionic functional groups attached to their structure. As these groups are polar, when polymers are exposed to water or other organic solvents, the ionic groups are solvated. The penetration of solvent molecules into polymeric chains leads to polymer swelling. That is, the swelling of an ion-exchange polymer is primarily due to the solvation of the ionic groups covalently bonded in the polymer matrix. However, the extent of swelling that a polymeric exchanger can achieve depends on multiple factors. For instance, the polarity of the solvent is decisive in the interaction between the solvent and the ion-exchanger; accordingly, polar solvents cause greater swelling than nonpolar ones, as they interact more strongly with the ionic groups. Other factors such as the concentration of ionic groups and the degree of crosslinking affect the polymer's interactions with different solvents. In general, the swelling of an ion-exchange polymer is directly proportional to the number of ionic groups attached to the polymer chain and is inversely proportional to the crosslinking degree (Melekaslan and Okay 2000). The ionic form of the ionic groups also influences the degree of swelling in an ion-exchange polymer. For cationic exchange resins with H^+ ions attached to each sulfonic group, the polymer matrix swells more compared with the same resin in the Na^+ form (Marin et al. 2024). This is because the water sorption depends on the type and concentration of the counterions, which decreases with the size of the hydrated counterion. In the case of a cation-exchanger, the water sorption follows the series $Li^+ > Na^+ > K^+ > Mg^{2+} > Ca^{2+}$ (Strathmann 2004). Generally, the degree of swelling of a polymeric ion-exchanger is vital in the ion-exchange process, as it can affect the polymer's capacity, ion mobility, and overall performance.

6.3.5 Mechanical and Chemical Stability

The mechanical and chemical stability of polymeric cation-exchange materials is essential for water treatment technologies, particularly for the removal of heavy metal ions from aqueous solutions. Mechanical stability refers to the ability of the polymer to withstand external forces and maintain its physical and chemical properties during operation, including resistance to deformation, degradation, and wear. Chemical stability, on the other hand, describes the polymer's resistance to chemical attack, heat, visible light, and elevated temperatures.

The chemical structure of the polymeric matrix, the nature of anionic functional groups, and the degree of crosslinking are key factors governing the chemical stability and mechanical strength of polymeric ion-exchangers. These parameters determine the performance of the ion-exchange polymers; thus, selecting an appropriate backbone and functional groups, together with the optimal degree of crosslinking, can significantly enhance the stability and durability of these materials. Cation-exchange polymers may exhibit strong-acid or weak-acid behavior, and their mechanical stability differs accordingly, with strong-acid polymers generally demonstrating greater resistance than their weak-acid counterparts. For example, strongly acidic cation-exchange polymers, such as sulfonic acid resins, exhibit high chemical and thermal stability and can be used at temperatures up to 120 °C while maintaining high efficiency over a wide pH range (Lim et al. 2024). The degree of crosslinking of the polymer matrix significantly influences mechanical and chemical stability. A higher degree of crosslinking results in denser, less flexible structures, which improves mechanical stability and resistance to oxidizing chemicals, but increases resistance to swelling. Another structural feature that enhances the chemical and mechanical stability of ion-exchange polymers is the presence of aromatic rings in their structure. Many polymers commonly used in the synthesis of ion-exchange polymers contain aromatic rings, which confer greater mechanical stability and resistance to chemical degradation.

6.3.6 Ion-Exchange Selectivity

Ion-exchange selectivity determines the relative affinity of an ion-exchanger for a specific ion in solution. The selectivity of polymeric ion-exchangers for metal ions depends on multiple factors, among the most important ones, the nature and density of functional groups, solution pH (Silva et al. 2018), the degree of cross-linking, and the relative charge and ionic radius of metal ions. Generally, the polymers and resins with sulfonic acid groups have a higher metal ion retention capacity than those containing phosphonic or carboxylic acid groups. This is due to the higher acidity and charge density of the sulfonic acid group, which allows it to exchange metal ions more readily than the carboxylic and phosphonic acid groups. However, sulfonated polymers display a lower selectivity toward a specific metal ion in comparison with the polymers with phosphonic and carboxylic acid groups, because these acid groups bind with metals through both electrostatic and covalent interactions. The pH of the solution affects the ionization of the ion-exchange functional groups, altering the strength and nature of electrostatic forces. For instance, strong acid cation exchangers maintain their charge across the entire pH range, while weak acid cation exchangers' charge is more pH-dependent. In general, at low pH, metal ion removal decreases due to the higher concentration of H^+ ions present in the solution, which compete with metal ions for negatively charged sites on the polymer (Gharbi et al. 2014). Conversely, at high pH, the concentration of H^+ ions decreases, and the interaction between negatively charged groups and metal ions increases. Another way is that high degrees of crosslinking lead to denser, less flexible polymer networks, which reduce the mobility of larger metal ions, while smaller ions are more likely to penetrate the network and interact with the polymer matrix's ion-exchange groups. The limited access of large ions results in polymeric ion-exchangers showing greater selectivity for small metal ions. Increasing the charge on a solute ion increases its affinity for an ion-exchanger in a process called *electroselectivity*. This is due to increased coulombic interactions between the ion and the oppositely charged functional groups on the exchanger (Haddad 1990). For example, it has been reported that sulfonic acid cation-exchange resins exhibit greater affinity for cations with larger ionic size and higher valency (Hubicki and Koodynsk 2012). Accordingly, the affinity series is as follows:

$$Pu^{4+} >> La^{3+} > Ce^{3+} > Pr^{3+} > Nd^{3+} > Sm^{3+} > Eu^{3+} > Gd^{3+} > Tb^{3+} > Dy^{3+}$$
$$> Ho^{3+} > Er^{3+} > Tm^{3+} > Yb^{3+} > Lu^{3+} > Y^{3+} > Sc^{3+} > Al^{3+} >> Ba^{2+} > Pb^{2+}$$
$$> Sr^{2+} > Ca^{2+} > Ni^{2+} > Cd^{2+} > Cu^{2+} > Co^{2+} > Zn^{2+} > Mg^{2+} > UO_2^{2+}$$
$$>> Tl^{+} >> Ag^{+} > Cs^{+} > Rb^{+} > K^{+} > NH^{4+} > Na^{+} > H^{+} > Li^{+}.$$

While cation-exchange resins with the carboxylic functional groups show the opposite, the affinity series for alkali and alkaline earth metal ions. The affinity is as follows:

$$H+ > Mg^{2+} > Ca^{2+} > Sr^{2+} > Ba^{2+} > Li^{+} > Na^{+} > K^{+} > Rb^{+} > Cs^{+}.$$

On the other hand, since chelating IERs adsorb metal ions by combining ionic and coordination interactions, their selectivity also depends considerably on the stability of the complexes formed between the metal ions and the chelating groups of the resin. For example, resins with iminodiacetate groups can selectively remove divalent metal ions rather than monovalent metal ions, according to the affinity series (Hubicki and Koodynsk 2012):

$$Hg^{2+} > UO_2^{2+} > Cu^{2+} > Pb^{2+} > Ni^{2+} > Cd^{2+} > Zn^{2+} > Co^{2+} > Fe^{2+} > Mn^{2+} > Ca^{2+} > Mg^{2+} > Ba^{2+} > Sr^{2+} >> Li^{+} > Na^{+} > K^{+}.$$

6.3.7 Size and Shape

The implementation of an ion-exchange polymer in a water treatment process depends on its physical state. In this respect, the synthesis and processing methods for ion-exchange polymers are highly versatile, allowing them to be obtained in the form of gels, membranes, sponges, fibers, or porous particles. Generally, cation-exchange polymer particles are small, porous, and spherical or bead-like. For applications such as water softening and ion-exchange chromatography, the typical particle size range is around 0.84–0.30 mm, corresponding to the 20–50 mesh range (Hubicki and Koodynsk 2012). The particle size of polymeric ion-exchangers is a key factor influencing their physical and chemical properties. Smaller particles provide a larger surface area, more accessible ion-exchange sites, and improved ion mobility. This allows greater interaction between the ion-exchanger and the metal ions in solution, thereby enhancing the capture of heavy metal ions. However, smaller particles can also create more resistance to the flow of the solution.

6.4 MECHANISM OF HEAVY METAL IONS CAPTURE BY ION-EXCHANGE RESINS AND POLYMERS

IERs and polymers are effective at capturing heavy metal ions dissolved in water by selectively binding and exchanging ions. Cation exchange polymers can capture heavy metal ions through the interaction between their negatively charged functional groups and the metal cations, mainly by ion-exchange and, in some cases, by coordination interactions, as shown in Figure 6.3. The capture of metal ions by ion-exchange between the ionic polymer and metal ions dissolved in a liquid phase, where the metal ions bind to the polymer via electrostatic interactions to form ion pairs, is known as the ion-pairing mechanism. In contrast, the capture of metal ions by coordination interactions with the ionic polymer, forming complexes that can be considered chelates, is often called the coordination bonding mechanism or chelating interaction.

The ion pairing mechanism is based on the electrostatic attraction between the heavy metal ions (cations) and the groups with negative charge (anions) attached to the polymer chains; this interaction forms cation-anion ion pairs. In this mechanism, positive counterions bound to the functional groups of the cation-exchange

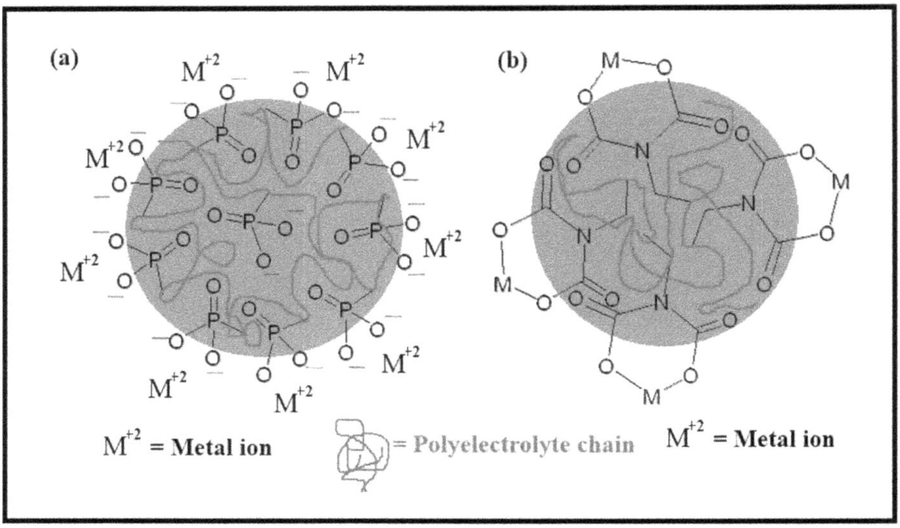

FIGURE 6.3 Types of interactions between heavy metal ions and ion-exchange polymers: (a) electrostatic association and (b) coordination complexation.

polymer are exchanged for heavy metal ions, and the replaced ions, typically sodium or hydrogen ions, remain in solution, while maintaining the charge neutrality within the polymer. This exchange process is reversible, allowing heavy metal ions to be replaced with other ions to regenerate the polymer. In contrast, the coordination bonding mechanism is a specific type of coordination where the functional groups fixed on the polymeric chains act as electron donors, bonding to the metal ions through coordinate covalent bonds, forming insoluble complexes with a structure in a ring-like shape, called a chelate ring, which can be easily separated from the aqueous phase. Functional groups containing sulfur, nitrogen, or oxygen atoms can act as electron donors, forming covalent bonds with metal atoms.

The type of interaction established between the cation-exchange polymer and the heavy metal ions is determined by the specific properties of the polymeric material, such as the nature and concentration of the functional groups and their relative positions within the polymer matrix. Other factors that influence the type of heavy-metal ion capture mechanism include the concentration and charge of the metal ions, solution pH, ionic strength, temperature, and dielectric constant. Specifically, the nature and concentration of the polymer's functional groups are key factors in the capture of metal ions. Strongly charged groups, such as sulfonic acid on the polymer, together with a high concentration, generate greater electrostatic attraction, which favors interaction with heavy metal ions through ion-exchange. Additionally, a high concentration of anionic functional groups on the polymer, combined with a high concentration of heavy metal ions in solution, increases electrostatic interactions, which leads to faster ion-exchange until the polymer is saturated with ion pairs. Another parameter that affects the strength of electrostatic interactions is the solution's pH. Changes in pH affect the dissociation of certain anionic groups. A change in pH alters the equilibrium of these groups by altering their charge, thereby influencing their ability to engage in electrostatic interactions. At low pH, sulfonic, phosphonic, and carboxylic acid groups are more protonated, leading to competition between heavy metal cations and hydrogen ions (H^+) for the polymer's ion-exchange sites. On the contrary, at high pH values, the acid groups are less protonated and therefore more available for the capture of metal ions. However, it must be considered that at high pH, heavy metal ions can form hydroxides, which decreases the concentration of metal ions available for ion-exchange with the polymer. This is why the best efficiency for capturing heavy metal ions in strong-acid cation-exchange resins is typically observed in the pH range of 4 to 8 (Demirbas et al. 2005).

6.5 METHODS OF PREPARING ION-EXCHANGE RESINS AND POLYMERS FOR SELECTIVE SEPARATION OF HEAVY METAL IONS

Because IERs contain polyelectrolytes, it is important to describe the mechanism or pathway of their synthesis. Ionic polymers, called polyelectrolytes, are macromolecules with flexible chains, cross-linked chains, conjugated polymers, random, graft, and block copolymers, as well as hyperbranched and dendritic structures. Depending on their structure and molecular weight, they are water-soluble or insoluble. Generally, high-molecular-weight, cross-linked polyelectrolytes are water-insoluble and are used in ion-exchange processes due to these structural characteristics. The most useful ion-exchange reactions in polyelectrolytes are reversible, allowing the polyelectrolyte to be reused multiple times. Typically, polymeric IERs are manufactured in spherical form, free of stress and strain, to resist physical degradation. These polymeric resins are stable at high temperatures and function effectively over a wide pH range. Polymeric IERs are completely insoluble in most aqueous and organic solvents and consist of a cross-linked polymer matrix with charged functional groups (Sherrington 1995). The ion-exchanger processes in the cationic and anionic polyelectrolytes are illustrated in Figure 6.4.

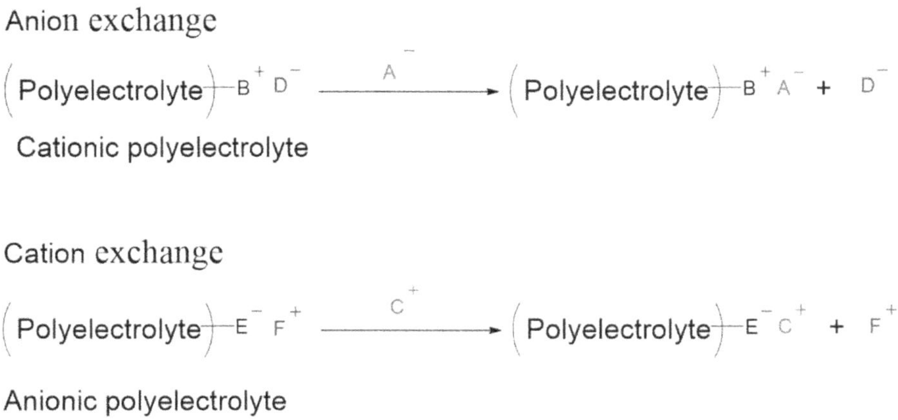

FIGURE 6.4 Exchange of ions between insoluble cationic and anionic polyelectrolytes and a solution phase with ions.

The chemical structures of polyelectrolytes are accessible using well-established polymer synthesis techniques, specifically chain or stepwise polymerization of suitable monomers, as well as chemical modification of reactive precursor polymers.

6.5.1 Chain Polymerization (Homopolymerization and Copolymerization)

Chain and step polymerization methods lead to polyelectrolytes with 100% functionality. However, characterizing these polyelectrolytes at the molecular level can be challenging because their conformation in aqueous media is highly sensitive to ionic strength and interactions with other substances.

There is a wide range of polyelectrolytes, which can be classified based on their origin into natural, synthetic, and semi-synthetic. According to the electric charge of the ionic groups in the polyelectrolytes, they can be classified as polycations, polyanions, or polyampholytes (Koetz and Kosmella 2007). Most cationic polyelectrolytes are based on amino groups or ammonium ions (Bolto and Gregory 2007). In addition to amines and their derivatives, the literature describes the synthesis of cationic polyelectrolytes containing other heteroions, such as phosphonium, sulfonium, boronium, cobaltocenium, imidazolium, and pyridinium (Laschewsky 2012). In contrast, the most common ionic groups in polyanions are carboxylates, sulfates, sulfonates, phosphates, and phosphonates (Koetz and Kosmella 2007). Finally, polyampholytes are those polyelectrolytes that contain both anionic and cationic groups within the same polymer chain. Polyampholytes can be synthesized by copolymerizing anionic and cationic monomers or by homopolymerizing zwitterionic monomers (Koetz and Kosmella 2007). Additionally, chain polymerization typically involves monomers that contain carbon-carbon double bonds and ionic groups. Monomers with carbon-carbon double bonds can react with themselves, forming polymers by converting these double bonds into single bonds.

Generally, the synthesis of electrolyte monomers is challenging because many chemical reactions are not feasible in the presence of ionogenic groups (acidic or basic). The synthesis of these monomers often requires the use of highly polar solvents, many of which are protic solvents such as water or alcohol. These solvents can compete with the desired product in some chemical reactions, reducing the yield of the desired product. However, the synthesis of electrolyte monomers is feasible, and some of them are commercially used, such as acrylic acid, methacrylic acid, vinylphosphonic acid, sodium vinylsulfonate, and sodium styrene sulfonate. Additionally, recent literature describes the synthesis of monomers containing other ionogenic groups, such as *o*- and *p*-methacryloylaminophenylarsonic acid (Percino et al. 2000), (2-((2-cyanalyl) oxy)ethyl) phosphonic acid, (2-((2-diethylcarbamoyl)allyl)oxy)ethyl)phosphonic acid, and *o*-acrylamidophenylboronic acid (Moszner et al. 2001).

In this context, polyelectrolytes capable of proton exchange, specifically polyanions, are of particular interest for metal ion recovery. Among the polyelectrolytes synthesized via the chain mechanism and used for proton exchange in metal ion recovery are those containing carboxylic, sulfonic, or phosphoric acid groups, which can be obtained as homopolymers or copolymers. Figure 6.5 illustrates some of the monomers commonly used to produce IERs through the chain polymerization mechanism.

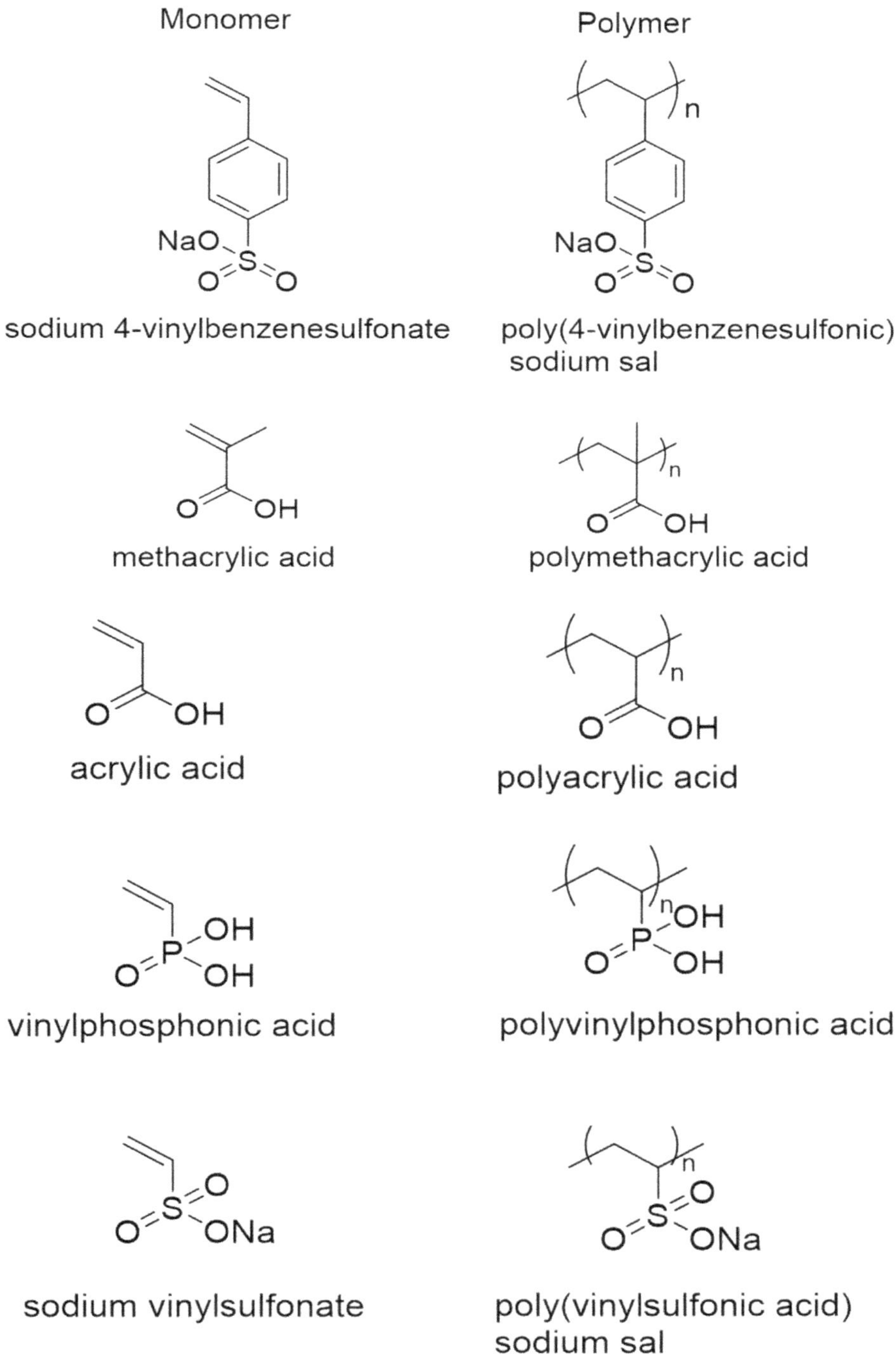

FIGURE 6.5 Polyanionic polyelectrolytes and their monomers.

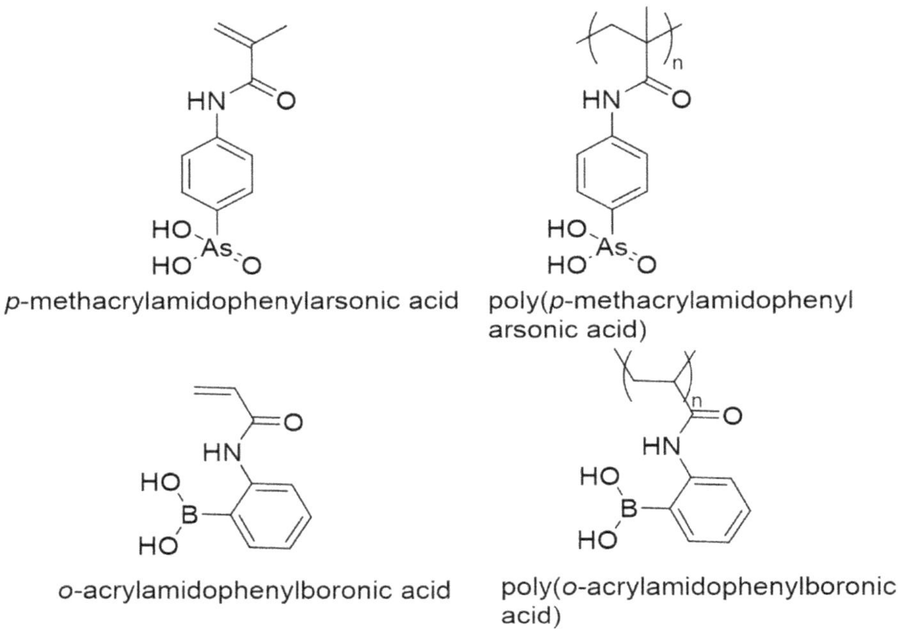

FIGURE 6.5 (Continued)

The most common polyelectrolyte is sodium polystyrene sulfonate, crosslinked with 3–8% divinylbenzene to ensure structural stability (Kun and Kunin 1967; Liberti and Helfferich 1983). Polymer resins with a higher degree of crosslinking (>8% divinylbenzene) are more expensive and difficult to produce, but they exhibit greater mechanical and chemical resistance.

6.5.2 Step Polymerization

The polymers obtained in steps are often prepared by condensation reactions of polyfunctional monomers, using common organic reactions that, depending on the functional groups in the monomers, eliminate or condense small molecules such as water, ethanol, methanol, or hydrochloric acid. The repeating constitutional units in the polymer are linked together by ester (RCOOR'), amide ($RCONH_2$), urethane (RNHCOOR'), sulfide (RSR'), ether (R-O-R'), imine ($R_2C=NR'$), carbonate (ROCOOR') groups, among others, which are formed during the polymerization reaction. However, polymers can also be obtained via the step mechanism without the formation of organic functional groups, as in the case of Bakelite.

The phenol-formaldehyde polymers were the first synthetic IERs, and (Baekeland 1909) reported their synthesis. The resin was prepared under basic conditions and referred to as Bakelite. Bakelite is obtained from phenol and formaldehyde via a step-growth mechanism and has been used in cation-exchange processes (Figure 6.6a). The synthesis of cation-exchange polymers via a stepwise mechanism has also been reported.

One study described the preparation of polymers by reacting aniline, *m*-toluidine, *m*-phenylenediamine, or urea with formaldehyde, furfural, or furfuryl alcohol. The polymer that exhibited the best combination of properties was the one prepared with *m*-phenylenediamine and formaldehyde (Schwartz et al. 1940) (Figure 6.6b).

In this sense, using the step-by-step polymerization mechanism, one of the most widely used proton exchange resins, known commercially as Dowex 30 (Bauman 1946), was synthesized (Figure 6.6c).

6.5.3 Chemical Modification of Polymers

The limitations encountered in the synthesis of polyelectrolytes from electrolyte monomers can be overcome by chemical modification of the polymers, in which ionic groups are introduced as a final step in the synthetic route (Cölfen 2001). The literature describes the use of various chemical reactions to modify polymers and produce polyelectrolytes. For example, in polymers containing epoxides as pendant groups, ring opening can be performed with sodium sulfite or phosphoric acid in the presence of a potassium salt to obtain the respective polyelectrolytes with sulfonate and phosphonate groups, which are useful for cationic exchange (Choi and Nho 1999), Figure 6.7.

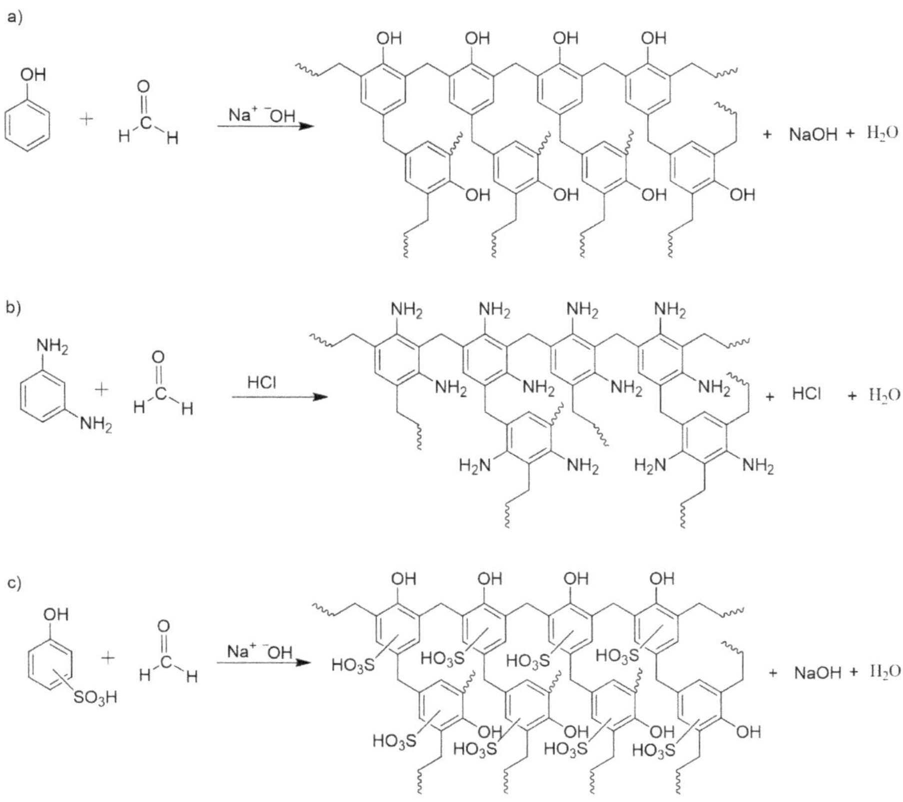

FIGURE 6.6 Synthesis of polyelectrolytes via a step mechanism.

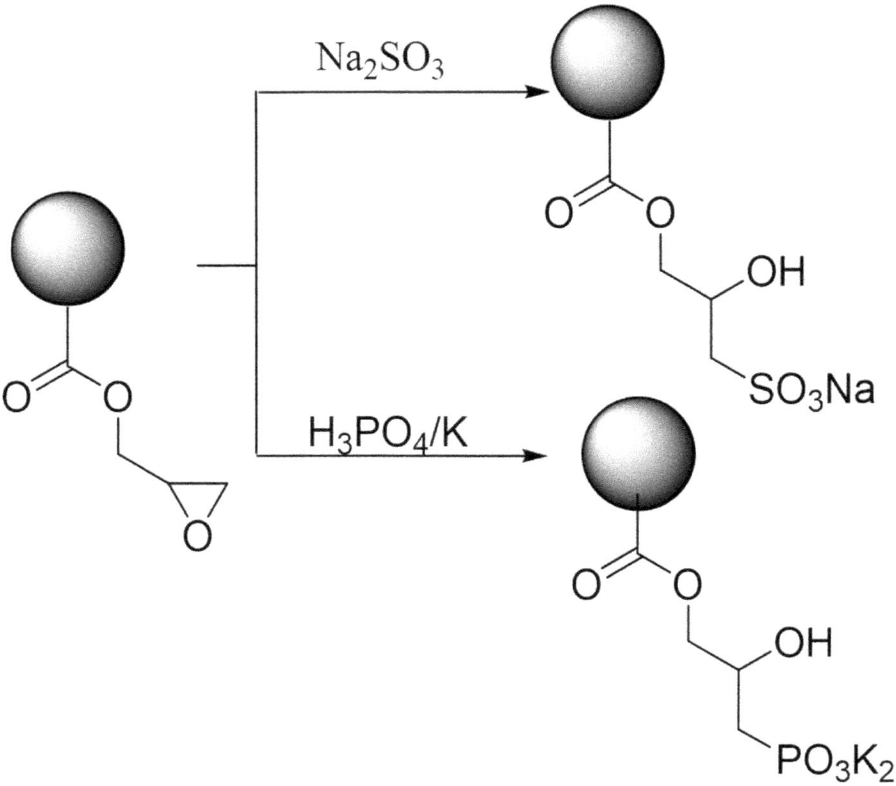

FIGURE 6.7 Synthesis of polyelectrolytes with sulfonate and phosphonate by opening epoxides.

6.6 SYNTHETIC POLYMER-BASED ION-EXCHANGERS USED FOR THE REMOVAL OF HEAVY METAL IONS

Ion-exchange polymers are highly effective at capturing and removing heavy metal ions from water and wastewater. The versatility of these organic ion-exchangers lies in the variety of possible functional groups that can be attached to the polymer chains during or after synthesis. Ion-exchange polymers can be designed with specific functional groups that interact with metallic cations via ion-exchange or complexation, making them a versatile tool for water purification. The selectivity of an ion-exchange polymer toward specific metal ions depends on the nature of its functional groups, as well as the physicochemical properties of the metal ions. This section presents some of the synthetic ion-exchange polymers reported in the literature for the selective removal of heavy metal ions from water and wastewater (see Table 6.2). Special attention was paid to selectivity and removal efficiency, depending on the nature of the functional groups on the polymer chains and the polymer's characteristics, such as the matrix type and degree of crosslinking. It is essential to note that ion-exchange polymers of natural origin were not considered; for this topic, see Chapter 5 of this book.

TABLE 6.2
Ion-Exchange Polymers Used for Heavy Metal Ions Removal from Water

Ion-exchange polymer	Functional group	Target cations	pH	Adsorption capacity	Reference
Resin Purolite C100	Sulfonic acid	Ni^{2+}	5–6	71 mg/g	Reda Hamed and Abdallah (2024)
		Cu^{2+}		67 mg/g	
		Zn^{2+}		66 mg/g	
		Pb^{2+}		62 mg/g	
Resin Purolite C100	Sulfonic acid	Cu^{2+}	11	1.3664 mg/g	Jasim and Ajjam (2024)
		Pb^{2+}		1.3823 mg/g	
Polyacrylic acid–polyurethane	Carboxylic acid	Pb^{2+}	6	257 mg/g	Singh et al. (2024)
Resin Dowex G-26	Sulfonic acid	Pb^{2+}	5	45.45 mg/g	Elfeghe, Sheng et al. (2022)
		Cd^{2+}		78.12 mg/g	
Resin Puromet™ MTS9570	Sulfonic and phosphonic acids	Pb^{2+}	3.5	38.46 mg/g	Elfeghe, Anwar et al. (2022)
		Cd^{2+}		51.02 mg/g	
Divinylbenzene-maleic anhydride	Carboxylic and sulfonic acids	Pb^{2+}	2–3.6	439.29 mg/g	Feng et al. (2025)
		Cu^{2+}		294.75 mg/g	
		Cd^{2+}		358.19 mg/g	
Copolymer based on glycidyl methacrylate and methylene bisacrylamide	Iminodiacetic acid	Cd^{2+}	5	368.7 mg/g	Jiang et al. (2015)
		Cu^{2+}		151 mg/g	
Polymer based on dimethyl 2,2'-([1,1'-biphenyl]-4-ylazanediyl)diacetate	Iminodiacetic acid	Pb^{2+}	7	1138 mg/g	Anito et al. (2020)
Cross-linked dianionic polyelectrolyte	Aminopropyl phosphonic acid	Pb^{2+}	4	10.1 mmol/g	Ali et al. (2015)
		Cu^{2+}	5.3	4.1 mmol/g	
Resin based on poly(methyl acrylate)	Aminophosphonic acid	Pb^{2+}	5	1.01 mmol/g	Lv et al. (2022)

6.6.1 Cationic Exchange Resins and Polymers Containing Acid Functional Groups

Sulfonic ($-SO_3H$), carboxylic (–COOH), and phosphonic ($-PO(OH)_2$) are the most commonly used functional groups in the development of cation exchange polymers, because they allow interactions through ion-exchange or complexation with the heavy metal ions, depending on the solution pH.

Ion-exchange polymers with sulfonic acid groups. Among the cation exchange polymers with acid groups, those containing sulfonic acid are the most studied for the removal of heavy metal ions, since the sulfonic group confers a high ion-exchange capacity and increases the hydrophilic character of the polymers, making them very effective for the capture and removal of metal cations. Also, in aqueous solution, the sulfonic acid group dissociates over the entire pH range, carrying a negative charge ($-SO_3^-$), thus establishing only electrostatic interactions with the metal cations, which gives rise to an ion-exchange capture mechanism. A wide variety of synthetic cation-exchange resins and polymers with sulfonic acid groups have been studied for heavy metal ions removal from aqueous solutions. For instance, Reda Hamed and Abdallah (Reda Hamed and Abdallah 2024) studied the removal of divalent heavy metal ions (Ni^{2+}, Cu^{2+}, Zn^{2+}, and Pb^{2+}) from wastewater by an ion-exchange process using the commercial resin Purolite C100 (Na^+), which is a strong acid cation-exchange resin with microporous structure (gel type) formed of a polymeric matrix of polystyrene crosslinked with divinylbenzene and sulfonic acid groups that act as ion-exchange sites. In its sodium form, this resin exhibits an ion-exchange capacity of 2 eq/L, a moisture retention of 44–48%, and a relative density of 1.29. In this study, the influence of various treatment parameters, including solution pH (2–8), contact time (10–300 minutes), and initial heavy metal concentration (0.2–2.0 mg/L), on removal efficiency was investigated. The authors observed that pH significantly affects the heavy metal adsorption capacity and identified that the optimum pH for metal ion removal is between 5 and 6. The study found that the affinity of metal ions toward the Purolite C100 resin follows the sequence $Ni^{2+} > Cu^{2+} > Zn^{2+} > Pb^{2+}$. The authors verified that under optimized removal conditions (one hour of contact time, pH 5 to 6, and initial heavy metal concentrations of 1 mg/L), the resin reached maximum adsorption capacities of 71 mg/g for Ni^{2+}, 67 mg/g for Cu^{2+}, 66 mg/g for Zn^{2+}, and 62 mg/g for Pb^{2+}. For the same purpose, heavy metals removal, Jasim and Ajjam used the same IER to remove Pb^{2+} and Cu^{2+} from aqueous solutions containing different metal ion concentrations (Jasim and Ajjam 2024) and found that an initial pH greater than 6 improves the deprotonation of the sulfonic acid groups, increasing the ion-exchange capacity of the resin. Optimal recovery of Pb^{2+} and Cu^{2+} was achieved at pH 11, with a metal ion concentration of 50 ppm and a contact time of 60 minutes. The maximum adsorption capacities for Cu^{2+} and Pb^{2+} were 1.3664 and 1.3823 mg/g, respectively.

In a previous study, the water-soluble anionic polyelectrolyte poly(sodium 4-styrenesulfonate) was used to remove heavy metal ions from water using ultrafiltration (Chen et al. 2018). The authors studied how competition from cations commonly found in natural water (H^+, Li^+, Na^+, K^+, Mg^{2+}, Ca^{2+}, Sr^{2+}, and Ba^{2+}) affects the removal of heavy metal cations (Pb^{2+}, Cu^{2+}, Co^{2+}, and Ni^{2+}) in binary (M^{2+}/Na^+) and multicationic systems. Cation selectivity coefficients were determined from experimental results in the binary system and an ion-exchange model treating this polyelectrolyte as a strong cation exchanger. The affinity sequence of cations for poly(sodium 4-styrensulfonate) was as follows $Ba^{2+} > Pb^{2+} > Sr^{2+} > Ca^{2+} > Cu^{2+} > Co^{2+} > Ni^{2+} > Mg^{2+} > H^+ > K^+ > Na^+ > Li^+$. The higher selectivity for divalent cations compared to monovalent cations was associated with their greater charge, which generates greater electrostatic attraction between the cations and the sulfonate groups of poly(sodium 4-styrenesulfonate). In the solutions containing multiple divalent and monovalent cations, the divalent cations associate more strongly with the polymer, and the affinity sequence follows the same order as the binary system $Pb^{2+} > Cu^{2+} > Co^{2+} > Ni^{2+}$.

Ion-exchange polymers with carboxylic acid groups. The carboxylic acid groups have a $pKa \approx 3.2–4.5$ (Franz 2001); consequently, they do not dissociate over the entire pH range. At pH values above the pKa, the carboxylic acid dissociates, and its interaction with metal cations occurs through ion-exchange. However, at pH values below the pKa, the carboxylic acid groups do not dissociate but can interact with metal ions by forming coordination bonds, thereby capturing them through complexation. Thus, in ion-exchange polymers containing carboxylic acid groups, the mechanism of interaction with metal cations strongly depends on the pH of the aqueous solution. Recently, Singh and coworkers synthesized polyurethane foam copolymers functionalized with polyacrylic acid via radiation grafting (Singh et al. 2024). They studied the removal of Pb^{2+} from wastewater using a 67% grafted foam with a highly porous lattice structure and a pore size of 100–200 μm. According to the authors, the initial pH of the solution affects the adsorption capacity of the metal ions. The maximum adsorption capacity was reached at pH 5, suggesting that the mechanism responsible for the capture of Pb^{2+} ions in the functionalized copolymer is through electrostatic interactions, since at pH values above 4.25 (pKa value of acrylic acid), the dissociation of the acrylic acid group occurs, producing acrylate anions that interact with Pb^{2+} cations in an ion-exchange process. The adsorption capacity of Pb^{2+} was approximately 257 mg/g at the experimental conditions of ~25 °C, 10 min contact time, and pH of 5.

Ion-exchange polymers with phosphonic acid groups. The phosphonic acid has two acidic protons, so it has two dissociation constants. For example, aromatic phosphonic acids exhibit pKa_1 values ranging from 1.09 to 1.7 and pKa_2 values ranging from 6.1 to 7.1 (Franz 2001). In aqueous solutions with pH values lower than the pKa1, phosphonic acid groups do not dissociate, and ion-exchange polymers containing such groups can interact with metal cations by forming coordination bonds. When the solution pH is between pKa_1 and pKa_2, the phosphonic acid groups exhibit a negative charge ($-PO_3H^-$), which favors electrostatic interaction with metal ions; however, complexation with metal ions is still possible. At pH values greater than the pka_2, the functional group dissociates completely ($-PO_3^{2-}$), increasing the electrostatic attraction of the metal cations, so that the capture mechanism is established by ion-exchange. Several authors have reported the selectivity of phosphonic acid ion-exchange polymers for specific metal cations. For example, Ihara et al. (Ihara et al. 2001) investigated the metal ion selectivity of a macroreticular poly(styrene-co-divinylbenzene) cation exchange resin containing phosphonic acid groups attached to phenyl groups of the copolymer. In this study, 20 metal ions were analyzed in nitric, hydrochloric, sulfuric, and phosphoric acid media and the selectivity sequences observed were very similar, following the order: $Mo^{6+} > Fe^{3+} > U^{6+} > Bi^{3+} > Lu^{3+} > Al^{3+} > Gd^{3+} > La^{3+} > Cr^{3+} > Pb^{2+} > Mn^{2+} > Cd^{2+} > Cu^{2+} > Ca^{2+} > Co^{2+} > Zn^{2+} > Ba^{2+} > Sr^{2+} > Ni^{2+} > Mg^{2+}$. These results demonstrate that the resin shows a preference for trivalent cations over divalent ones and exhibits the highest selectivity for Pb^{2+} among the divalent metal ions studied. According to the authors, the high selectivity of the resins for Pb^{2+} over Ba^{2+} is the opposite of that of conventional cation-exchange resins with sulfonic acid groups, which prefer Ba^{2+} to Pb^{2+}. Regarding the interaction of the phosphonic acid resin with the trivalent metal ions and with Mo^{6+} and U^{6+} ions in nitric acid medium, two groups of metal ions with different adsorption behaviors were identified, grouping the ions into two categories: Mo^{6+}, U^{6+}, Fe^{3+} and Lu^{3+}, Bi^{3+}, Al^{3+}, Gd^{3+}, La^{3+}, Cr^{3+}, which was related to two interaction mechanisms, one through the formation of nitrate complexes and the other by cation exchange.

Ion-exchange polymers with a mixture of acid groups. Recently, several researchers have studied the removal of heavy metal ions from aqueous solution using ion-exchange polymers containing mixtures of two acid groups (sulfonic/carboxylic or sulfonic/phosphonic), which is essential for understanding the effect of these acid groups on the removal mechanism. This topic is vital, as combining the ability of carboxylic and phosphonic groups to interact with metal cations through complexation reactions with the high ion-exchange capacity of the sulfonic group can improve the efficiency and selectivity of ion-exchange polymers. In this respect, Elfeghe and coworkers investigated the use of the commercial IERs Dowex G-26 and Puromet MTS9570 to remove Pb^{2+} (Elfeghe et al. 2022) and Cd^{2+} (Elfeghe, Anwar et al. 2022) ions from aqueous solutions, in a pH range of 1.5 to 5.5. The Dowex G-26 resin is a strongly acidic gel-type cation exchange resin with a styrene-divinylbenzene matrix, functionalized with sulfonic acid groups as the ion-exchange sites. In contrast, Puromet MTS9570 is a macroporous cation-exchange resin with a polystyrene matrix crosslinked with divinylbenzene, difunctionalized with sulfonic and phosphonic acid groups. The authors studied the influence of operating parameters (initial metal concentration, contact time, temperature, resin dosage, and solution pH) on the adsorption of Pb^{2+} and Cd^{2+} ions onto the resins via ion-exchange. They found that, with an initial metal concentration of 1000 mg/L, an adsorption time of 30 min, room temperature, and a resin dosage of 25 g/L, both resins removed 100% of Pb^{2+} ions in an initial solution of pH 3.5; similarly, the Puromet MTS9570 resin removed 98.9% of Cd^{2+} ions at an initial pH of 5.0. Instead, the Dowex G-26 resin removed 99.2% of Cd^{2+} ions at an initial pH of 5, but with a lower resin dosage (10 g/L). The maximum adsorption capacities to Pb^{2+} and Cd^{2+} ions were 45.45 mg/g and 78.12 mg/g, respectively, for Dowex G-26 resin, which contains only sulfonic acid as an interchange site. For the Puromet MTS9570 resin difunctionalized with sulfonic and phosphonic acid groups, the maximum adsorption capacities for Pb^{2+} and Cd^{2+} were 38.46 mg/g and 51.02 mg/g, respectively. This indicates that the sulphonic acid group is more efficient than the phosphonic acid group for bonding and adsorbing both ion types.

In a further study, Feng et al. (2025) used *p*-divinylbenzene and maleic anhydride to synthesize crosslinked mesoporous polymeric hollow spheres, which were functionalized with carboxylic acid groups and difunctionalized with carboxylic and sulfonic acid groups, where the sulfonic groups were estimated within the range of 4.13 at%–7.98 at%. Both ion-exchange polymers were used for the adsorption of bivalent heavy metal ions (Pb^{2+}, Cu^{2+}, and Cd^{2+}), and they showed a higher affinity for Cu^{2+} and a lower affinity for Pb^{2+}. However, the dual-functional polymer showed superior adsorption capacity at both high and low metal ion concentrations compared to the sulfonate-free polymer, indicating the crucial role of sulfonate groups in the adsorption process. According to the authors, carboxyl and sulfonate groups play an essential role in the removal of heavy metal ions. Carboxyl groups are the primary contributors to adsorption due to their strong complexation with metals. In contrast, sulfonate groups enhance mass transfer by increasing interchain spacing, negative surface charge, and coordination with metal ions, suggesting that both ion-pairing and coordination bonding are key mechanisms in the adsorption process. Furthermore, pH sensitivity enabled the desorption of the metal ions and the regeneration of the polymer by adjusting the solution's acidity. The maximum adsorption capacity for Pb^{2+} was 439.29 mg/g, for Cu^{2+} it was 294.75 mg/g, and for Cd^{2+} it was 358.19 mg/g.

6.6.2 Ion-Exchange Resins and Polymers Containing Chelating Functional Groups

Chelating-type ion-exchange polymers, or chelating resins, have functional groups capable of forming strong complexes with specific heavy-metal ions. The functional groups of these polymers usually contain nitrogen, oxygen, and/or sulfur atoms, which are capable of donating electron pairs to bind to metal ions through two or more coordination bonds. For example, the thiol groups on the resins act as ligands because sulfur atoms can form complexes with metal ions, especially with divalent metal ions, as is illustrated in Figure 6.8. Some functional groups that have proven effective for the selective removal of heavy metal cations include iminodiacetic and aminophosphonic acids.

Ion-exchange polymers with iminodiacetic acid groups. Iminodiacetic acid is a chelating agent that exhibits a high affinity for binding heavy metals by forming covalent coordination bonds. This group typically acts as a bidentate or tridentate ligand, using two oxygen atoms from the carboxylate groups and the nitrogen atom of the amino group (Figure 6.9). The iminodiacetic acid groups have been exploited in the design of chelating-type ion-exchange polymers due to their strong binding capacity. The literature reports that several commercial chelating resins with an iminodiacetate functional group are highly selective for copper, nickel, and cobalt (Botelho Junior et al. 2019). For example, Jiang et al. (2015) synthesized a chelating resin by free-radical copolymerization of iminodiacetic acid modified glycidyl methacrylate with N,N′-methylenebisacrylamide. They synthesized a porous resin and used it to remove numerous heavy metal ions, including Cu^{2+}, Pb^{2+}, Cd^{2+}, Zn^{2+}, Co^{2+},

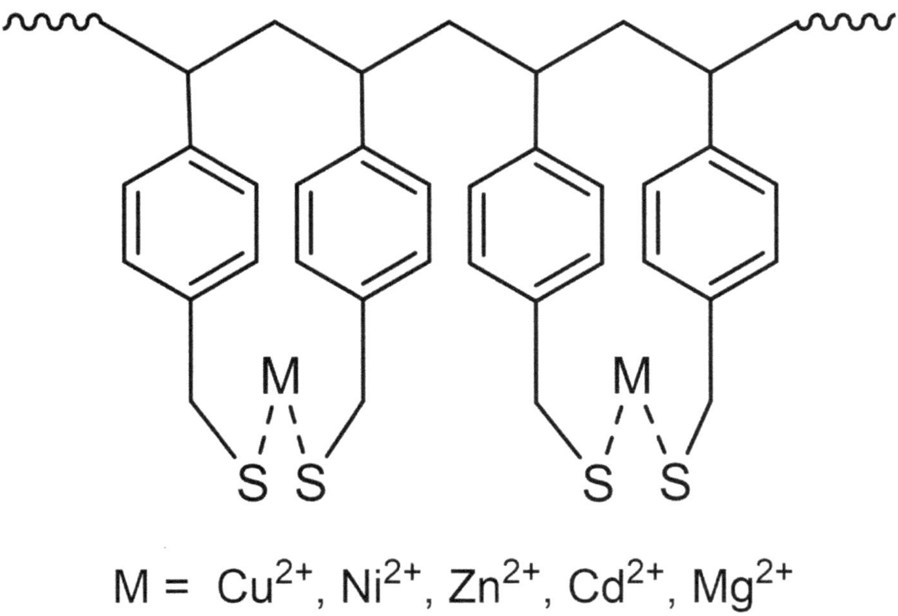

FIGURE 6.8 Formation of chelate complexes between thiol groups and divalent metal ions.

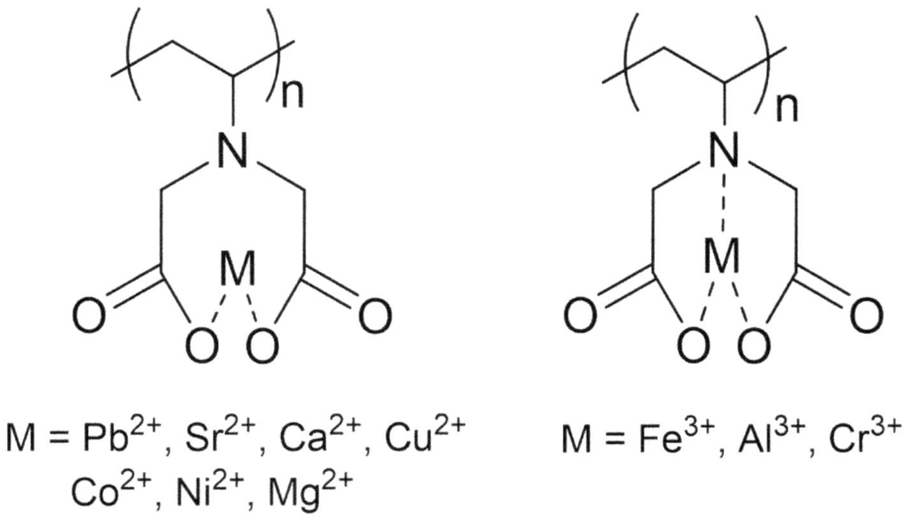

FIGURE 6.9 Interaction of the iminodiacetic acid group with divalent and trivalent metals.

Ni^{2+}, Cr^{3+}, and Mn^{2+}, from aqueous solutions. The authors conclude that the adsorption capacities of all metal ions were low at pH 1; however, as pH increased, they increased. The optimal pH values at which the resin showed the highest adsorption capacity were 5 for Cu^{2+} and Cd^{2+}, and 6 for Mn^{2+}, Co^{2+}, Ni^{2+}, Cr^{3+}, Pb^{2+}, and Zn^{2+}. According to the results, the resin exhibited a higher adsorption capacity for Cd^{2+} (368.7 mg/g), indicating the strongest affinity of the iminodiacetic functional group for the Cd^{2+} cation via covalent bonds. The descending orders of affinity were $Cd^{2+} > Cu^{2+} > Mn^{2+} > Ni^{2+} > Zn^{2+} > Co^{2+} > Cr^{3+} > Pb^{2+}$.

In another work, Anito et al. (Anito et al. 2020) reported the synthesis of an iminodiacetic acid-functionalized hypercrosslinked polymer and its application for the purification of water contaminated with toxic metal ions. The polymer was synthesized in two steps. First, the dimethyl 2,2'-([1,1'-biphenyl]-4-ylazanediyl)diacetate monomer was crosslinked with 1,4-dimethoxybenzene. After, the dimethyl iminodiacetate groups were hydrolyzed to obtain the iminodiacetic acid groups The analysis of the selectivity of polymer for metal ion capture in a solution with different metal ions, such as Pb^{2+}, Hg^{2+}, Cd^{2+}, Co^{2+}, Fe^{3+}, Zn^{2+}, Mg^{2+}, and Na^{+} revealed that the polymer was highly efficiency for toxic heavy metal ions, such as Pb^{2+}, Hg^{2+}, and Cd^{2+}, and its ability to capture competitive alkali metals, such as Mg^{2+} and Na^{+}, was negligible. The polymer with iminodiacetic groups exhibited a Pb^{2+} removal capacity of 1138 mg/g and good recyclability. Furthermore, the polymer displayed high affinity for Pb^{2+} in the pH range of 1 to 12, with a removal efficiency exceeding 99.6%. The unique properties of the hypercrosslinked polymer for heavy metal chelation are due to the iminodiacetic acid group, which can bind metal ions.

Recently, Czupryński et al. (2022) investigated the efficiency of an IER system for removing Ni, Hg, and Cr from flue gas desulphurization wastewater in the presence of competing metals such as Ca, Mg, Al, Fe, and Mn. One of the IERs studied was

Purolite S930 (Na^+ form), which is a macroporous resin of polystyrene crosslinked with divinylbenzene containing iminodiacetic functional groups, and the results of the study showed that the selectivity sequence of the resin was Ni > Cr > Hg > Fe > Al > Mn> Ca, Mg. The resin reduced the Ni content from 89.3 mg/dm³ to less than 0.1 mg/dm³. According to the authors, the higher affinity of the iminodiacetic group for Ni was related to its form in wastewater, where the dominant species in solution was the divalent cation Ni^{2+}. Therefore, Ni was mainly removed in the ion-exchange process through the formation of complexes between the Ni^{2+} cation and the carboxyl groups and the nitrogen atom of the iminodiacetic functional group.

Ion-exchange polymers with aminophosphonic acid groups. The aminophosphonic acid group can act as a multidentate chelating ligand, since the nitrogen atom of the amino group and the oxygen atoms of the phosphonic acid group allow the creation of chelate rings through the formation of multiple coordination bonds. The incorporation of aminophosphonic functional groups into an ion-exchange polymer can increase its affinity for heavy metal ions. In this respect, Ali's research group (Ali et al. 2015) reported the synthesis of a new cross-linked dianionic polyelectrolyte (CDAPE) containing (aminopropyl)phosphonic acid groups, which was obtained by the cyclotermopolymerization of [(diallylamino)propyl]phosphonic acid hydrochloride (90 mol%) with the cross-linker 1,1,4,4-tetraallylpiperazinium dichloride (10 mol%). The CDAPE was used to remove toxic Pb^{2+} and Cu^{2+} ions from aqueous solutions and industrial wastewater. The polyelectrolyte exhibited a high ion-exchange capacity (IEC) of 5.02 mmol/g. The adsorption studies of metal ions in the pH range of 2.8–5.3 revealed that pH has a strong effect on the adsorption capacity for Cu, with adsorption at pH 2.8 being almost zero, whereas at pH 5.3 it was 4.1 mmol/g. The optimal pH for Pb^{2+} uptake was 4. The polyelectrolyte exhibited maximum adsorption capacities of 3.83 mmol/g for Pb^{2+} and 10.1 mmol/g for Cu^{2+}. Also, the authors of this study evaluated the effectiveness of the (aminopropyl)phosphonic-functionalized polyelectrolyte in industrial wastewater samples containing Cu, Ni, Cr, Cd, and Pb. They found that heavy metal concentrations in the solutions, after treatment with the polyelectrolyte, were reduced to levels below the detection limit. The researchers considered that the presence of amine and phosphate groups favored ionic exchange and the formation of complexes via coordination bonds with metal ions, making these functional groups responsible for the capture and removal of the metal ions.

In another study, Lv et al. (2022) synthesized a chelating resin based on poly(methyl acrylate) functionalized with aminophosphonic acid, which was evaluated in the selective separation of Pb^{2+} ions in strong acid media. In the static selective adsorption experiments of bivalent heavy metal ions onto the resin, carried out in the hexa-solute system Pb^{2+}-Cu^{2+}-Cd^{2+}-Zn^{2+}-Co^{2+}-Ni^{2+}, at pH 5 and metal cation concentrations of 1.0 mmol/L, the researchers found that the resin with aminophosphonic acid showed the selectivity order of $Pb^{2+} >> Cu^{2+} > Cd^{2+} > Zn^{2+} > Co^{2+} > Ni^{2+}$ and the adsorption amount of Pb^{2+} was 1.010 mmol/g. However, in dynamic selective separation experiments with a simulated hexa-solute system (Pb^{2+}-Cu^{2+}-Cd^{2+}-Zn^{2+}-Co^{2+}-Ni^{2+}), ~0.2 mmol/L, pH 2, the resin with aminophosphonic groups maintained high affinity and excellent selectivity for Pb^{2+}. Interestingly, in this case, the sequence of the replacement effect was $Ni^{2+} > Co^{2+} > Zn^{2+} > Cu^{2+} > Cd^{2+} > Pb^{2+}$, approximately opposite to the order of static selective adsorption. According to the results of FTIR and XPS analyses, the authors proposed a possible adsorption mechanism, which considers that the amide, phosphonic, and amino groups of the resin act as adsorption sites for Pb^{2+} and that the joint action of these adsorption sites favored the formation of stable complexes between Pb^{2+} and the resin, macroscopically showing a highly selective separation of lead ions.

6.7 ADVANTAGES AND DISADVANTAGES OF USING ION-EXCHANGE RESINS AND POLYMERS FOR THE REMOVAL OF HEAVY METAL IONS

IERs and polymers are highly effective at removing heavy metal ions from water, due to their ability to interact with metal cations via ion-exchange or chelation. These materials are designed with functional groups that selectively attract and bind metal ions, providing several benefits for the capture and elimination of heavy metals from water and wastewater, including:

- **High removal efficiency**. IERs and polymers often achieve heavy-metal ion removal efficiencies exceeding 90% under optimal conditions. For example, different cationic exchange resins can achieve removal efficiencies of 94.4% for Cu^{2+}, 92.9% for Pb^{2+}, and almost 100% for Ni^{2+} when operated under optimal pH and resin dose conditions (Jasim and Ajjam 2024; Araucz et al. 2020).
- **Selectivity**. Many IERs and polymers, especially those containing chelating groups, can make a selective removal of specific heavy metal ions from water, even when other ions are present. For instance, several chelating resins and polymers with iminodiacetate functional groups have been reported to be highly selective for Cu^{2+}, Ni^{2+}, Co^{2+} (Botelho Junior et al. 2019), Cd^{2+} (Jiang et al. 2015), Pb^{2+} (Anito et al. 2020), and other resins with aminophosphonic acid are selective for Ni^{2+} (Czupryński et al. 2022) and Pb^{2+} (Lv et al. 2022).
- **Regeneration and reusability**. During the ion-exchange process, the resin or polymer structure does not change, allowing it to be regenerated and reused.
- **Variety of physical forms**. Ion-exchange polymers can be processed in multiple physical forms, including gels, beads, fibers, sponges, and membranes, each tailored for specific performance needs.
- **Versatility**. Ion-exchange polymers can be integrated into advanced systems, such as membranes and adsorbents, providing them with new functionalities, such as surface charge, ion transport, and selectivity, to improve their performance.

The application of the IERs and polymers presents certain disadvantages, such as:

- **Limited removal capacity**. IERs and polymers have limited metal-ion removal capacity. This is because, when their ion-exchange sites are saturated, the removal efficiency decreases dramatically, especially at high metal ion concentrations.

- **pH-dependent removal efficiency**. The removal capacity and selectivity of many IERs and polymers depend on the pH of the metal ion solution. Because pH affects the dissociation of specific functional groups, it influences the interactions between metal ions and the polymer's functional groups.
- **Factors Affecting Synthesis**. To manufacture efficient ion-exchange polymers for the capture and removal of heavy metal ions, several factors must be controlled, including the composition of the polymer matrix, the degree of crosslinking, the nature and concentration of ionic functional groups, and the reaction conditions, among others.

6.8 ADVANCEMENTS AND CHALLENGES

Ion-exchange polymers offer significant advantages for heavy metal removal from water, including efficiency, selectivity, regeneration capabilities, and integration with other water treatment systems. However, continued innovation in polymer design and the emergence of materials with outstanding properties have opened new avenues for the development of more efficient and sustainable IERs and polymers.

Recently, polymeric ion-exchange materials have seen significant advancements, including:

- **Hybrid ion-exchange resins**, which combine the ion-exchange properties of conventional resins with the enhanced adsorption capacity of inorganic nanoparticles or metal oxides. This combination produces resins with greater adsorption selectivity, adsorption capacity, and faster kinetics than those without the inorganic component, enabling more efficient removal of heavy metal ions (Sodzidzi et al. 2024; Maletin et al. 2024).
- **Next-generation membranes**. Recently, the use of ion-exchange polymers or polyelectrolytes to modify or synthesize membranes has received considerable attention, leading to the development of filtration and adsorption membranes with advanced functionalities (Chen et al. 2024). Among these, cation exchange membranes fabricated by the layer-by-layer method with the capacity to adsorb heavy metal ions and with high removal efficiency of metals like Cu^{2+}, Ni^{2+}, and Pb^{2+} (Hosseini et al. 2020) or ion-exchange -nanofiltration membranes that can achieve highly efficient indiscriminate heavy metal ion removal (Fu et al. 2022).
- **Molecularly imprinted polymers (MIPs)**. Recent advancements in MIPs for the removal of heavy metal ions focus on enhanced selectivity and efficiency through the design of cavity-based polymeric templates that serve as molecular targeting binding sites (Khan et al. 2023). Most MIPs are synthesized using the free radical copolymerization technique employing vinyl functional monomers and divinyl crosslinking agents, in the presence of a solvent or a porogen (pore-forming substance) (Sharma and Kandasubramanian 2020). Some of the most commonly used functional monomers include: vinylpyridine, *p*-vinylbenzoic acid, methacrylic acid, methyl methacrylic acid, itaconic acid, and methacrylamide. Resins with high capacity and selectivity for heavy metal ions have been prepared using molecular imprinting polymerization

(Wu et al. 2022). For example, molecularly imprinted resins composed of a random copolymer of methacrylate and methacrylamide were developed for the removal of hydrophilic metal complexes, such as $CdCl_2$ and $ZnCl_2$. The imprinted IERs were found to exhibit a significant increase in metal retention capacity compared to non-imprinted controls (Ashraf et al. 2011).

6.9 KEY CHALLENGES

While IERs and polymers offer several advantages for capturing and removing heavy metal ions, significant challenges must be addressed, including optimizing synthetic strategies to control their structure and properties, and incorporating functional groups to remove specific heavy metal ions. Additional challenges include developing efficient and non-toxic regeneration methods, increasing the adsorption capacity and lifetime of resins, reducing production costs, and enhancing stability in complex wastewaters.

6.10 CONCLUSION

The removal of heavy metal ions from aqueous solutions using IERs and polymers has proven to be a favorable approach for wastewater treatment. Several parameters, such as the composition of the polymer matrix, the degree of crosslinking, and the nature and concentration of functional groups, must be controlled during synthesis and preparation to obtain polymers with high ion-exchange capacity and selectivity, as well as mechanical and chemical stability. The adsorption capacity and selectivity of ion-exchange polymers are key factors in the removal of heavy metal ions, as wastewater may contain multiple metal cations at varying concentrations. Cation-exchange polymers capture heavy metal ions via electrostatic attraction and ion-exchange. In some cases, coordination chemistry between the metal ions and the functional groups also contributes to the capture process. Polymer chemistry plays a crucial role in the design and synthesis of IERs. It enables the development of cross-linked polymers that are insoluble in aqueous media and can exchange ions. The reaction mechanisms involved in their synthesis can vary significantly depending on the ionic functional groups present in their structures. Different ion-exchange polymers used for the removal of heavy metal ions from water were reviewed in the current chapter. Ion-exchange polymers offer several advantages, such as high efficiency, selectivity, and regeneration, and can be integrated into advanced systems, such as membranes, adsorbents, and nanomaterials.

REFERENCES

Ali, Shaikh A., Izzat W. Kazi, and Nisar Ullah. 2015. "New Chelating Ion-Exchange Resin Synthesized via the Cyclopolymerization Protocol and Its Uptake Performance for Metal Ion Removal." *Industrial & Engineering Chemistry Research* 54 (40): 9689–98. https://doi.org/10.1021/acs.iecr.5b02267.

Anito, Dejene Assefa, Tian-Xiong Wang, Zhi-Wei Liu, Xuesong Ding, and Bao-Hang Han. 2020. "Iminodiacetic Acid-Functionalized Porous Polymer for Removal of Toxic Metal Ions from Water." *Journal of Hazardous Materials* 400 (December): 123188. https://doi.org/10.1016/j.jhazmat.2020.123188.

Araucz, Katarzyna, Andreas Aurich, and Dorota Kołodyńska. 2020. "Novel Multifunctional Ion Exchangers for Metal Ions Removal in the Presence of Citric Acid." *Chemosphere* 251 (July): 126331. https://doi.org/10.1016/j.chemosphere.2020.126331.

Ashraf, Syed, Angela Cluley, Ckarlos Mercado, and Anja Mueller. 2011. "Imprinted Polymers for the Removal of Heavy Metal Ions from Water." *Water Science and Technology* 64 (6): 1325–32. https://doi.org/10.2166/wst.2011.423.

Bauman, W. C. 1946. "Improved Synthetic Ion Exchange Resin." *Industrial & Engineering Chemistry* 38 (1): 46–50. https://doi.org/10.1021/ie50433a026.

Bolto, B., Gregory, J. 2007. "Organic Polyelectrolytes in Water Treatment." *Water Research* 41: 2301–2324. https://doi.org/10.1016/j.watres.2007.03.012.

Botelho Junior, Amilton Barbosa, David B. Dreisinger, and Denise C. R. Espinosa. 2019. "A Review of Nickel, Copper, and Cobalt Recovery by Chelating Ion Exchange Resins from Mining Processes and Mining Tailings." *Mining, Metallurgy & Exploration* 36 (1): 199–213. https://doi.org/10.1007/s42461-018-0016-8.

Briffa, Jessica, Emmanuel Sinagra, and Renald Blundell. 2020. "Heavy Metal Pollution in the Environment and Their Toxicological Effects on Humans." *Heliyon* 6 (9): e04691. https://doi.org/10.1016/j.heliyon.2020.e04691.

Calvo, Benjamin, Laureano Canoira, Fernando Morante, et al. 2009. "Continuous Elimination of Pb2+, Cu2+, Zn2+, H+ and NH4+ from Acidic Waters by Ionic Exchange on Natural Zeolites." *Journal of Hazardous Materials* 166 (2–3): 619–27. https://doi.org/10.1016/j.jhazmat.2008.11.087.

Chen, Ming, Karen Shafer-Peltier, Stephen J. Randtke, and Edward Peltier. 2018. "Competitive Association of Cations with Poly(Sodium 4-Styrenesulfonate) (PSS) and Heavy Metal Removal from Water by PSS-Assisted Ultrafiltration." *Chemical Engineering Journal* 344 (July): 155–64. https://doi.org/10.1016/j.cej.2018.03.054.

Chen, Ming, Ruibo Xu, Yichen Wu, Jianglei Xiong, Seda Zeynep Keleş, and Nicholas P. Hankins. 2024. "Application of Polyelectrolytes for Contaminant Removal and Recovery during Water and Wastewater Treatment: A Critical Review." *Journal of Water Process Engineering* 64 (July): 105528. https://doi.org/10.1016/j.jwpe.2024.105528.

Choi, Seong-Ho, and Young Chang Nho. 1999. "Electrochemical Properties of Polyethylene Membrane Modified with Sulfonic and Phosphonic Acid Groups." *Korean Journal of Chemical Engineering* 16 (6): 725–30. https://doi.org/10.1007/BF02698343.

Cölfen, Helmut. 2001. "Double-Hydrophilic Block Copolymers: Synthesis and Application as Novel Surfactants and Crystal Growth Modifiers." *Macromolecular Rapid Communications* 22 (4): 219–52. https://doi.org/10.1002/1521-3927(20010201)22:4%253C219::AID-MARC219%253E3.0.CO;2-G.

Czupryński, Piotr, Maciej Płotka, Piotr Glamowski, Witold Żukowski, and Tomasz Bajda. 2022. "An Assessment of an Ion Exchange Resin System for the Removal and Recovery of Ni, Hg, and Cr from Wet Flue Gas Desulphurization Wastewater—a Pilot Study." *RSC Advances* 12 (9): 5145–56. https://doi.org/10.1039/D1RA09426B.

Dąbrowski, A., Z. Hubicki, P. Podkościelny, and E. Robens. 2004. "Selective Removal of the Heavy Metal Ions from Waters and Industrial Wastewaters by Ion-Exchange Method." *Chemosphere* 56 (2): 91–106. https://doi.org/10.1016/j.chemosphere.2004.03.006.

Demirbas, Ayhan, Erol Pehlivan, Fethiye Gode, Turkan Altun, and Gulsin Arslan. 2005. "Adsorption of Cu(II), Zn(II), Ni(II), Pb(II), and Cd(II) from Aqueous Solution on Amberlite IR-120 Synthetic Resin." *Journal of Colloid and Interface Science* 282 (1): 20–25. https://doi.org/10.1016/j.jcis.2004.08.147.

Elfeghe, Salem, Shams Anwar, and Yahui Zhang. 2022. "Adsorption and Removal Studies of Cadmium Ion onto Sulphonic/Phosphonic Acid Functionalization Resins." *The Canadian Journal of Chemical Engineering* 100 (10): 3006–14. https://doi.org/10.1002/cjce.24400.

Elfeghe, Salem, Qiuyue Sheng, Abbas Mamudu, Lesley Anne James, and Yahui Zhang. 2022. "Recovery of Lead (II) Ions from Aqueous Solutions Using G-26 and MTS9570 Resins with Sulfonic/Phosphonic Functional Groups." *Minerals* 12 (10): 1312. https://doi.org/10.3390/min12101312.

Feng, Han, Yen Nan Liang, Penghua Wang, Chun Po Hu, Wantai Yang, and Xiao Hu. 2025. "Highly Regenerative Dual-Functionalized Mesoporous Polymeric Hollow Spheres for Efficient Heavy Metal Removal." *Separation and Purification Technology* 363 (August): 132120. https://doi.org/10.1016/j.seppur.2025.132120.

Franz, Robert G. 2001. "Comparisons of pKa and Log P Values of Some Carboxylic and Phosphonic Acids: Synthesis and Measurement." *AAPS PharmSci* 3 (2): 1–13. https://doi.org/10.1208/ps030210.

Fu, Zheng-Jun, Shang-Kun Jiang, Xin-Yi Chao, et al. 2022. "Removing Miscellaneous Heavy Metals by All-in-One Ion Exchange -Nanofiltration Membrane." *Water Research* 222 (August): 118888. https://doi.org/10.1016/j.watres.2022.118888.

Gharbi, Soumaya, Jameleddine Khiari, and Bassem Jamoussi. 2014. "Studies, Synthesis and Characterization of Chelation Ion-Exchange Properties of Copolymer Resin Derived from 8-Hydroxyquinoline-Formaldehyde-Pyrogallol." *American Chemical Science Journal* 4 (6): 874–89. https://doi.org/10.9734/ACSJ/2014/10877.

Haddad, P.R. 1990. "Chapter 2 An Introduction to Ion-Exchange Methods." In *Ion Chromatography: Principles and Applications*, edited by P.R. Haddad, P.E. Jackson. Journal of Chromatography Library, ISSN: 0301-4770, 46 (C): 15-27. Elsevier. https://doi.org/10.1016/s0301-4770(08)61134-3.

Hosseini, S.M., H. Alibakhshi, E. Jashni, et al. 2020. "A Novel Layer-by-Layer Heterogeneous Cation Exchange Membrane for Heavy Metal Ions Removal from Water." *Journal of Hazardous Materials* 381 (January): 120884. https://doi.org/10.1016/j.jhazmat.2019.120884.

Hubicki, Zbigniew, and Dorota Koodynsk. 2012. "Selective Removal of Heavy Metal Ions from Waters and Waste Waters Using Ion Exchange Methods." In *Ion Exchange Technologies*, edited by Ayben Kilislioglu. InTech. https://doi.org/10.5772/51040.

Ihara, Toshihiro, Akinori Jyo, and Kazunori Yamabe. 2001. "Metal Ion Selectivity of Nacroreticular Chelating Cation Exchange Resins with Phosphonic Acid Groups Attached to Phenyl Groups of a Styren-Divinylbenzene Copolymer Matrix." *Separation Science and Technology* 36 (15): 3511–28. https://doi.org/10.1081/SS-100107917.

Jasim, Ansam Qays, and Sata Kathum Ajjam. 2024. "Removal of Heavy Metal Ions from Wastewater Using Ion Exchange Resin in a Batch Process with Kinetic Isotherm." *South African Journal of Chemical Engineering* 49 (July): 43–54. https://doi.org/10.1016/j.sajce.2024.04.002.

Jiang, Jun, Xiao-Shuang Ma, Ling-Yun Xu, et al. 2015. "Applications of Chelating Resin for Heavy Metal Removal from Wastewater." *E-Polymers* 15 (3): 161–67. https://doi.org/10.1515/epoly-2014-0192.

Khan, Muhammad Shahzeb, Shan E. Zehra Syeda, and Anna Maria Skwierawska. 2023. "Recent Advancements in Molecularly Imprinted Polymers for the Removal of Heavy Metal Ions and Dyes." *Desalination and Water Treatment* 289 (March): 123–44. https://doi.org/10.5004/dwt.2023.29458.

Kim, Cheol Gyu. 2013. "Adsorption/Ion Exchange of Metal Ions on Clay Mineral Surfaces." *Asian Journal of Chemistry* 25 (10): 5884–86. https://doi.org/10.14233/ajchem.2013.OH118.

Koetz, Joachim, and Sabine Kosmella. 2007. *Polyelectrolytes and Nanoparticles*. Springer Laboratory Manuals in Polymer Science. Springer.

Kun, Kenneth A., and Robert Kunin. 1967. "The Pore Structure of Macroreticular Ion Exchange Resins." *Journal of Polymer Science Part C: Polymer Symposia* 16 (3): 1457–69. https://doi.org/10.1002/polc.5070160323.

Laschewsky, André. 2012. "Recent Trends in the Synthesis of Polyelectrolytes." *Current Opinion in Colloid & Interface Science* 17 (2): 56–63. https://doi.org/10.1016/j.cocis.2011.08.001.

Lavarda, Giulia, Silvia Morales-delaRosa, Paolo Centomo, Jose M. Campos-Martin, Marco Zecca, and Jose L. G. Fierro. 2019. "Gel-Type and Macroporous Cross-Linked Copolymers Functionalized with Acid Groups for the Hydrolysis of Wheat Straw Pretreated with an Ionic Liquid." *Catalysts* 9 (8): 675. https://doi.org/10.3390/catal9080675.

Lendero, Nika, Jana Vidič, Peter Brne, Aleš Podgornik, and Aleš Štrancar. 2005. "Simple Method for Determining the Amount of Ion-Exchange Groups on Chromatographic Supports." *Journal of Chromatography A* 1065 (1): 29–38. https://doi.org/10.1016/j.chroma.2004.10.072.

Liberti, Lorenzo, and Friedrich G. Helfferich. 1983. *Mass Transfer and Kinetics of Ion Exchange*. NATO ASI Series, Series E: Applied Sciences 71. Springer. https://doi.org/10.1007/978-94-009-6899-8.

Lim, Seung Joo, Wang Kyu Choi, and Mansoo Choi. 2024. "Changes in Physicochemical Characteristics of Cation Exchange Resins by High Dose Gamma Irradiation." *Nuclear Engineering and Technology* 56 (5): 1777–80. https://doi.org/10.1016/j.net.2023.12.033.

Liu, Xiantang, and Gang Yang. 2022. "Mechanisms for Cation Exchange at the Interfaces of Montmorillonite Nanoparticles: Insights for Pb2+ Control." *Colloids and Surfaces A: Physicochemical and Engineering Aspects* 641 (May): 128556. https://doi.org/10.1016/j.colsurfa.2022.128556.

López-Maldonado, Eduardo Alberto, Youness Abdellaoui, Mahmoud H. Abu Elella, et al. 2024. "Innovative Biopolyelectrolytes-Based Technologies for Wastewater Treatment." *International Journal of Biological Macromolecules* 273 (July): 132895. https://doi.org/10.1016/j.ijbiomac.2024.132895.

Lv, Yingzhi, Tanshang Li, Yanni Jiang, et al. 2022. "Highly Selective Separation of Pb(II) with a Novel Aminophosphonic Acid Chelating Resin from Strong-Acidic Hexa-Solute Media." *Separation and Purification Technology* 300 (November): 121818. https://doi.org/10.1016/j.seppur.2022.121818.

Mahmoud, Akrama, and Andrew F.A. Hoadley. 2012. "An Evaluation of a Hybrid Ion Exchange Electrodialysis Process in the Recovery of Heavy Metals from Simulated Dilute Industrial Wastewater." *Water Research* 46 (10): 3364–76. https://doi.org/10.1016/j.watres.2012.03.039.

Maletin, Marija, Jasmina Nikić, Vesna Gvoić, et al. 2024. "Optimization and Efficiency of Novel Magnetic-Resin-Based Approaches for Enhanced Nickel Removal from Water." *Processes* 12 (10): 2287. https://doi.org/10.3390/pr12102287.

Marin, Nicoleta Mirela, Mihai Nita Lazar, Marcela Popa, Toma Galaon, and Luoana Florentina Pascu. 2024. "Current Trends in Development and Use of Polymeric Ion-Exchange Resins in Wastewater Treatment." *Materials* 17 (23): 5994. https://doi.org/10.3390/ma17235994.

Melekaslan, D., and O. Okay. 2000. "Swelling of Strong Polyelectrolyte Hydrogels in Polymer Solutions: Effect of Ion Pair Formation on the Polymer Collapse." *Polymer* 41 (15): 5737–47. https://doi.org/10.1016/S0032-3861(99)00808-3.

Moszner, Norbert, Frank Zeuner, Steffen Pfeiffer, Ivan Schurte, Volker Rheinberger, and Marco Drache. 2001. "Monomers for Adhesive Polymers, 3. Synthesis, Radical Polymerization and Adhesive Properties of Hydrolytically Stable Phosphonic Acid Monomers." *Macromolecular Materials and Engineering* 286 (4): 225–31. https://doi.org/10.1002/1439-2054(20010401)286:4%253C225::AID-MAME225%253E3.0.CO;2-T.

Moukheiber, E., G. De Moor, L. Flandin, and C. Bas. 2012. "Investigation of Ionomer Structure through Its Dependence on Ion Exchange Capacity (IEC)." *Journal of Membrane Science* 389 (February): 294–304. https://doi.org/10.1016/j.memsci.2011.10.041.

Percino, M. Judith, Víctor M. Chapela, René Gutiérrez-Pérez, and Ana M. Herrera. 2000. "New O- and p-Methacryloylaminophenylarsonic Monomers. Useful Building Units for Water-Soluble Polymeric and Nonlinear Optical Materials." *Designed Monomers and Polymers* 3 (2): 155–60. https://doi.org/10.1163/156855500300142843.

Pérez-Maciá, María A., David Curcó, Roger Bringué, Montserrat Iborra, and Carlos Alemán. 2015. "Atomistic Simulations of the Structure of Highly Crosslinked Sulfonated Poly(Styrene-Co-Divinylbenzene) Ion Exchange Resins." *Soft Matter* 11 (11): 2251–67. https://doi.org/10.1039/C4SM02417F.

Pramanik, A.K. 2023. "Ion Exchange Resins for Selective Separation of Toxic Metals." In *Materials Research Foundations*, 1st ed., edited by Inamuddin, vol. 137. Materials Research Forum LLC. https://doi.org/10.21741/9781644902219-4.

Reda Hamed, Mohamed Ahmed, and Marwa Abdelfattah Abdallah. 2024. "Applications of Ion Exchange in Environmental Remediation: Removal of Heavy Metals From Wastewater." *Remediation Journal* 34 (4): e21788. https://doi.org/10.1002/rem.21788.

Schwartz, M. C., W. R. Edwards, and Grace Boundreaux. 1940. "Removal of Chlorides and Sulfates by Synthetic Resins." *Industrial & Engineering Chemistry* 32 (11): 1462–66. https://doi.org/10.1021/ie50371a013.

Shabani, Mehri, Habibollah Younesi, Maxime Pontié, Ahmad Rahimpour, Mostafa Rahimnejad, and Ali Akbar Zinatizadeh. 2020. "A Critical Review on Recent Proton Exchange Membranes Applied in Microbial Fuel Cells for Renewable Energy Recovery." *Journal of Cleaner Production* 264 (August): 121446. https://doi.org/10.1016/j.jclepro.2020.121446.

Shah, Khizar Hussain, Noor S. Shah, Gul Afshan Khan, et al. 2023. "The Cr(III) Exchange Mechanism of Macroporous Resins: The Effect of Functionality and Chemical Matrix, and the Statistical Verification of Ion Exchange Data." *Water* 15 (20): 3655. https://doi.org/10.3390/w15203655.

Sharma, Gaurav, and Balasubramanian Kandasubramanian. 2020. "Molecularly Imprinted Polymers for Selective Recognition and Extraction of Heavy Metal Ions and Toxic Dyes." *Journal of Chemical & Engineering Data* 65 (2): 396–418. https://doi.org/10.1021/acs.jced.9b00953.

Sherrington, D. C. 1995. "Polymer-Supported Synthesis." In *Chemistry of Waste Minimization*, edited by J. H. Clark. Springer Netherlands. https://doi.org/10.1007/978-94-011-0623-8_6.

Shi, Yu, Chika Eze, Binyu Xiong, et al. 2019. "Recent Development of Membrane for Vanadium Redox Flow Battery Applications: A Review." *Applied Energy* 238 (March): 202–24. https://doi.org/10.1016/j.apenergy.2018.12.087.

Shoba, V. N., and K. V. Chudnenko. 2014. "Ion Exchange Properties of Humus Acids." *Eurasian Soil Science* 47 (8): 761–71. https://doi.org/10.1134/S1064229314080110.

Silva, Rene A., Kelly Hawboldt, and Yahui Zhang. 2018. "Application of Resins with Functional Groups in the Separation of Metal Ions/Species—a Review." *Mineral Processing and Extractive Metallurgy Review* 39 (6): 395–413. https://doi.org/10.1080/08827508.2018.1459619.

Singh, Krishan Kant, Narendra K. Goel, Amit Kanjilal, et al. 2024. "Radiation Grafted Polyacrylic Acid–Polyurethane Foam Copolymer for Efficient Toxic Metal Removal from Aqueous Waste: A Sustainable Approach to Waste Management." *Polymer Bulletin* 81 (15): 13435–60. https://doi.org/10.1007/s00289-024-05340-8.

Smara, Abdelaziz, Rachid Delimi, Eric Chainet, and Jacqueline Sandeaux. 2007. "Removal of Heavy Metals from Diluted Mixtures by a Hybrid Ion-Exchange/Electrodialysis Process." *Separation and Purification Technology* 57 (1): 103–10. https://doi.org/10.1016/j.seppur.2007.03.012.

Sodzidzi, Zizikazi, Zebron Phiri, Jemal Fito Nure, Titus A. M. Msagati, and Lueta-Ann De Kock. 2024. "Adsorption of Toxic Metals Using Hydrous Ferric Oxide Nanoparticles Embedded in Hybrid Ion-Exchange Resins." *Materials* 17 (5): 1168. https://doi.org/10.3390/ma17051168.

Strathmann, 2004. "H. Preparation and Characterization of Ion-Exchange Membranes." In *Membrane Science and Technology Ion-Exchange Membrane Separation Processes,* edited by H. Strathmann. Elsevier. https://doi.org/10.1016/S0927-5193(04)80034-2.

Wachinski, Anthony M. 2016. *Environmental Ion Exchange: Principles and Design*, 2nd ed. CRC Press. https://doi.org/10.1201/9781315368542.

Waheed, Abdul, Nadeem Baig, Nisar Ullah, and Wail Falath. 2021. "Removal of Hazardous Dyes, Toxic Metal Ions and Organic Pollutants from Wastewater by Using Porous Hyper-Cross-Linked Polymeric Materials: A Review of Recent Advances." *Journal of Environmental Management* 287 (June): 112360. https://doi.org/10.1016/j.jenvman.2021.112360.

Wu, Hanrong, Guo Lin, Chenchen Liu, Shiyu Chu, Chao Mo, and Xiaobo Liu. 2022. "Progress and Challenges in Molecularly Imprinted Polymers for Adsorption of Heavy Metal Ions from Wastewater." *Trends in Environmental Analytical Chemistry* 36 (December): e00178. https://doi.org/10.1016/j.teac.2022.e00178.

7 Technologies for Disposal/Extraction of Adsorbed Heavy Metal Ions

Joyeeta Bose, Arpita Jash, Sudip Chakraborty, and Jaya Sikder

7.1 INTRODUCTION

The presence of heavy metals in natural environments has become a significant global concern, posing substantial risks to human health, ecological systems, and well-being. Elements such as cadmium (Cd^{2+}), lead (Pb^{2+}), arsenic (As^{3+}/As^{5+}), nickel (Ni^{2+}), chromium (Cr^{3+}/Cr^{6+}), copper (Cu^{2+}), and mercury (Hg^{2+}) are commonly introduced into ecosystems through several anthropogenic activities, including leather tanning, mining, battery production, electroplating, and the manufacture of pesticides and fertilizers (Yenduri et al., 2025; Agarwal & Chaudhry, 2015; Raji et al., 2023; Razzak et al., 2022; Shen et al., 2024; Shih et al., 2019; Song et al., 2024; B. Wang et al., 2023; Yesil & Tugtas, 2019; Ziheng Zhang et al., 2025). Unlike organic pollutants, heavy metals are non-biodegradable. They are often undergoing chemical transformations that exist in soils, sediments, and aquatic systems, which impact their mobility and bioavailability. These toxic metals entered the food chain through water consumption or plant uptake, resulting in chronic exposure. For example, kidney function and bone structure are affected by cadmium; arsenic is associated with cancer and cardiovascular diseases present in groundwater across South Asia and Latin America; neurological development in children may be observed in the presence of lead; mercury is a potent neurotoxin in its methylated form. The European Union (EU), the World Health Organization (WHO), and the US Environmental Protection Agency (EPA), among others, have established stringent limits on the permitted concentrations of heavy metals in water. Nevertheless, it remains a significant technological and economic challenge to manage such pollution. Traditional treatment methods like chemical precipitation generate toxic sludge as a by-product, whereas membrane filtration, reverse osmosis, and ion-exchange are often cost-intensive (Du et al., 2020; Fu & Wang, 2011; Kazakis, 2025; Kim et al., 2022; Raji et al., 2023; Shen et al., 2024; Ze Zhang et al., 2024). Chemical leaching, adsorption, and electrochemical regeneration techniques have gained dominance due to their environmental compatibility and operational flexibility, which is more sustainable. Figure 7.1 highlights both disposal

DOI: 10.1201/9781003599098-7

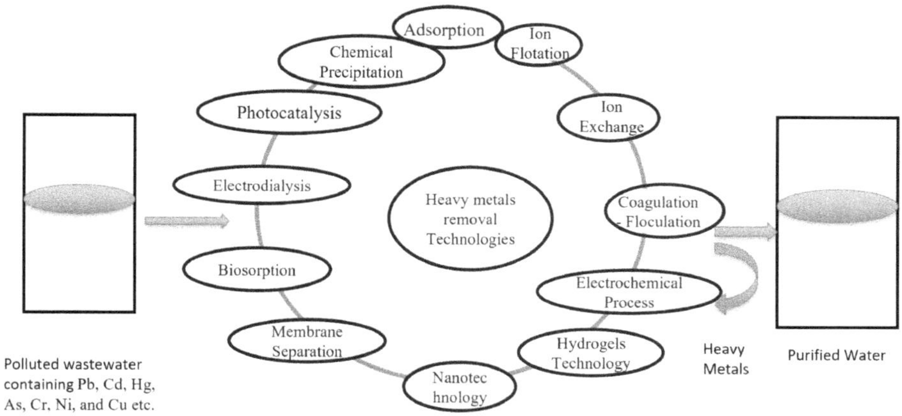

FIGURE 7.1 Overview of technologies for heavy metal removal, including landfilling, incineration, electrochemical regeneration, chemical leaching, and others, categorized by stage of treatment and mechanism (Shrestha et al., 2021).

and recovery approaches by mechanism and stage of treatment, providing an overview of key technologies used for heavy metal removal, along with a comparison (Shrestha et al., 2021).

In adsorption, ions or molecules from a fluid phase adhere to the surface of a solid material, known as an adsorbent, and it is a surface phenomenon. Adsorption is widely regarded for its low operational cost, simplicity, and adaptability to varying wastewater compositions, making it one of the most preferred treatment methods. Clays and zeolite minerals, which occur naturally, and coconut shells, rice husks, and orange peels, which are biobased materials, are used as adsorbents (Jain et al., 2021; Akhtar et al., 2025; Shabbirahmed et al., 2025). To enhance surface properties and improve selectivity and adsorption capacity, bio-adsorbents are usually chemically or thermally modified. For example, biomass-derived activated carbon becomes a viable option for large-scale applications, which offers a high surface area and a porous network (B. Wang et al., 2023). Removal efficiencies of more than 90% have been reported at optimum conditions for metal ions such as Pb(II) and Cu (II) (Twizerimana & Wu, 2024). Likewise, up to 46% metal removal from sewage sludge leachates has been demonstrated using supported liquid membrane systems (Yesil & Tugtas, 2019). Despite these advantages, conventional adsorbents face limitations, including surface saturation, reduced performance in multi-metal systems, and the potential for secondary pollution due to improper disposal of spent adsorbents. These drawbacks necessitate the development of integrated solutions that involve the regeneration or reuse of metal-laden materials, which are likely to serve effectively in catalysis or energy applications.

Metal-organic frameworks (MOFs), ion-exchange resins, and nanocomposite adsorbents have shown considerable promise, being synthetic and engineered materials that can be tailored for specific contaminants by differing surface area, pore structure, and functional groups (Renu et al., 2017). Physisorption by Van der Waals forces, chemisorption through covalent or ionic bonding, ion-exchange, and surface complexation with functional moieties such as –OH, –COOH, and $-NH_2$ are adsorption mechanisms involved in metal adsorption. System conditions, like pH, temperature, competing ions, solution chemistry, and the characteristics of the adsorbent, on which the adsorption efficiency depends. As such, Kim et al. (2022) demonstrated enhanced heavy metal extraction from paper incineration ash using a combined EDTA and oxalic acid treatment system. Similarly, efficient removal and regeneration performance of granular activated carbon (GAC F-400) in nickel-laden wastewater was reported by Kinhikar (2021). Improving removal performance and facilitating material recovery requires condition-specific optimization and hybrid approaches, which is a significant finding here (Kim et al., 2022; Binbin Zhou et al., 2025; Velempini et al., 2023). In adsorption, metal-laden adsorbents inherently generate a secondary waste stream, despite being highly effective in capturing metal ions from aqueous media. The safe disposal or reuse of these materials gives new challenges, including environmental, economic, and logistical issues (Fouda-Mbanga et al., 2024).

The capacity of an adsorbent is significantly reduced once it reaches saturation, making it unable to bind additional metal ions. These metal-loaded adsorbents pose a serious risk of secondary contamination if they are discarded without adequate treatment. The release of previously adsorbed heavy metals may occur due to changes in pH, microbial activity, or rainwater infiltration, which undermine the efficacy of the original remediation efforts (Fu & Wang, 2011). Extraction and disposal are two primary post-treatment strategies to eliminate this risk. Disposal methods can prevent future leaching by stabilizing or immobilizing the adsorbed metals, often through techniques such as vitrification, encapsulation, or solidification. Instead, extraction techniques emphasize recovering the adsorbed metals by electrochemical regeneration, chemical leaching, or acid/base desorption, allowing for both metal recovery and the potential reuse of the regenerated adsorbent (Lata et al., 2015). The fate of spent adsorbents is a serious decision point in the treatment cycle, as shown in Figure 7.2. Appropriate treatment unlocks pathways for resource recovery, reduced reliance on raw materials, and sustainable reuse. Improper disposal resulted in long-term environmental hazards. Extensively used high-value metals such as Ni and Cu in electrical, metallurgical, and manufacturing sectors are especially relevant economically (Iyer et al., 2024; Syeda et al., 2022). Supports circular economy principles and sustainable industrial practices by recovering metals from wastewater, thereby reducing dependence on primary mining. Moreover, especially in large-scale industrial applications, recurring treatment costs associated with adsorbent regeneration are reduced. Therefore, post-treatment technologies that enable safe disposal or effective metal extraction serve the purpose of both preventing environmental pollution and contributing to resource recovery, aligning with sustainability goals and long-standing economic viability (Razzak et al., 2022; B. Wang et al., 2023).

This chapter covers the basics of outside adsorption, with an emphasis on the critical post-treatment stage. The choices regarding the fate of metal-laden adsorbents have a significant impact on both the potential for resource recovery and environmental safety. Assessing their operational requirements, ecological insinuations, and suitability under different regulatory contexts, it first analyzes established disposal practices, such as incineration and landfilling. To evaluate their mechanisms,

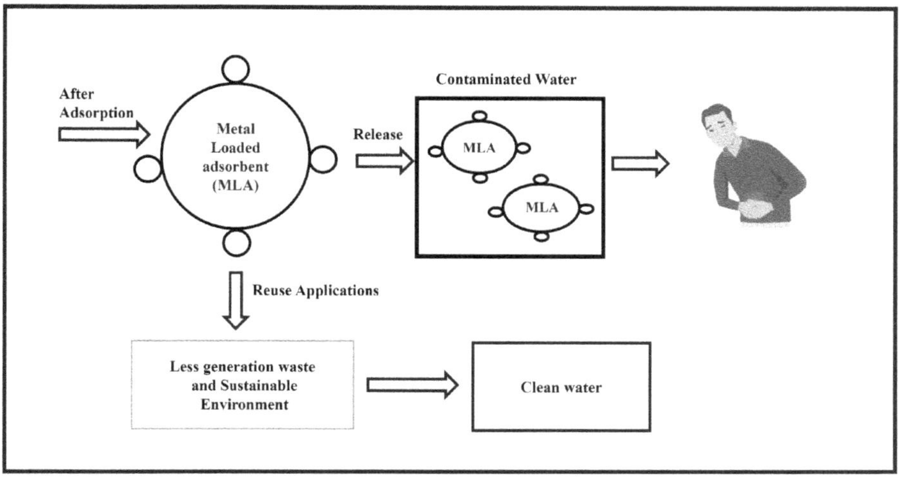

FIGURE 7.2 Outlines both pathways: metal disposal routes aimed at environmental safety and recovery strategies aligned with economic and circular economy benefits. Fouda-Mbanga et al. (2024).

efficiency, and scalability for industrial use, the following methods are examined: chemical leaching, electrochemical regeneration, extraction-based methods, thermal desorption, solvent extraction, ion-exchange, and bioremediation. How these approaches can be integrated within circular economy frameworks is deliberated, including opportunities for recovering high-value metals for waste-to-energy conversion. The rationale for selecting between disposal and recovery is explicated in terms of economic profit, regulatory compliance, and environmental risk. Although an analysis of current challenges and emerging research directions provides insight into the future of sustainable management of heavy metal-laden adsorbents, real-world case studies validate the practical applications of this approach. Table 7.1 shows the typical concentrations of heavy metal ions in industrial and natural waters. The chapter's purpose is to support informed decision-making for researchers, engineers, industry policymakers, and practitioners by combining economic and environmental outlooks with technical assessment.

7.2 METAL DISPOSAL TECHNIQUES

At the end of the service life of adsorbents after use, removal must be managed responsibly to prevent secondary environmental pollution and minimize solid residues containing heavy metals. The careless disposal of these adsorbents can pollute soil and groundwater, posing serious ecological risks. These adsorbed materials often contain high levels of toxic metals, such as lead (Pb), zinc (Zn), cadmium (Cd), or arsenic (As), which can leach out over time (Hesong Wang et al., 2024). Disposal methods can be broadly classified into two types. Containment-oriented techniques' objectives are to immobilize contaminants and prevent their migration. While destructive or transformative approaches either thermally degrade hazardous

TABLE 7.1
Typical Levels of Metal Concentrations Found in the Environment and Wastewater

Metal	Typical concentration in natural water (mg/L)	Typical concentration in industrial wastewater (mg/L)	Major sources	Reference
Arsenic (As)	0.001–0.05	0.1–5.0	Mining, smelting, pesticide manufacturing	Alidokht et al., (2021)
Cadmium (Cd)	0.0001–0.005	0.1–10.0	Battery manufacturing, electroplating, pigments	Bhatnagar & Minocha (2009)
Chromium (Cr)(Total)	0.001–0.1	1–50	Tannery effluents, metal plating, dyeing	Misganaw et al., (2024)
Copper (Cu)	0.001–0.05	0.1–20	Mining, smelting, plumbing, and electroplating	W. Zhou et al., (2022a)
Lead (Pb)	0.001–0.01	0.1–50	Battery industry, paints, plumbing	Ali et al., (2025)
Manganese (Mn)	0.001–0.2	0.1–10	Mining, alloy industry, fertilizers	Van Wendel De Joode et al., (2016)
Mercury (Hg)	0.00001–0.001	0.01–2.0	Chlor-alkali, electronics, mining	Kurniawan et al., (2023)
Nickel (Ni)	0.001–0.05	0.1–20	Electroplating, alloy manufacturing, fertilizers	Peng et al., (2014)
Zinc (Zn)	0.005–0.1	0.2–50	Galvanization, paints, dyes	Lim et al., (2021)
Cobalt (Co)	0.0001–0.01	0.1–5	Metal refining, pigments	Hassan & Abbar (2025)

constituents or reduce them into less toxic forms (Gusiatin & Rouhani, 2023). Due to their large-scale viability and cost-effectiveness, among all techniques, landfilling and incineration are the most widely used. Nevertheless, evolving substitutes are being explored to reduce the environmental burden and align with the goals of the circular economy. By embedding materials and processes that actively stabilize heavy metals, recent developments have improved the effectiveness of disposal strategies and ecological safety. For instance, organic chelating agents such as TS-300 and DDP have demonstrated stabilization efficiencies exceeding 95% for Pb, Cd, and Zn in various ash and adsorbent matrices. Even at extreme pH conditions, these additives diminish the leachability of metals and broaden the environmental stability of treated wastes (Ze Zhang et al., 2024; Xu et al., 2013).

Parallel to this, metallurgical slag-based cementitious materials (MSCM), such as low-carbon binders, offer promising alternatives to conventional Portland cement. While increasing mechanical strength and reducing CO_2 emissions, these materials show high immobilization capacities (up to 99.9%) for heavy metals, notably Pb, Zn, Cr, and Hg (Deng et al., 2022). Such consolidative methods decrease long-term environmental risk and ensure regulatory obedience. This section examines the primary disposal technologies employed for metal-laden solid wastes following adsorption-based treatment, focusing on their stabilization efficiency, operational mechanisms, and environmental effects. The following subsections explore the technical details of incineration, landfilling, ash stabilization, and other advanced disposal systems.

7.2.1 Landfilling

Heavy metals such as cadmium (Cd), lead (Pb), mercury (Hg), chromium (Cr), and nickel (Ni) are among the most hazardous contaminants due to their persistence, non-biodegradability, and bioaccumulative potential. These metals are widely used in industrial and customer applications, including paints, rechargeable batteries, switches, pigments, and electronic devices (Agarwal & Chaudhry, 2015; Yenduri et al., 2025; Aucott, 2006). Worldwide mercury production in 2004 was approximately 1,260 metric tons, with a significant portion ending up in the waste stream through products such as fluorescent lamps and switches. Cadmium, used widely in electroplating and later in battery manufacturing, reached its peak utilization in the United States at 6,000 tons per year in 1969. Cadmium-bearing products continue to enter municipal waste streams, despite recycling efforts (Aucott, 2006).

Landfilling remains one of the principal techniques for managing solid waste containing heavy metals, predominantly for spent adsorbents. Engineered landfills feature leachate collection systems, composite liners, and gas control measures to separate trash and minimize environmental impact. However, landfill environments progress, undergoing discrete geochemical phases: initial adjustment, acid formation, transition, stabilization, and methane generation (Fouda-Mbanga et al., 2024). During the acid formation phase, microbial disintegration leads to the generation of organic acids and a drop in pH, often below 5.0, which increases the solubility of Zn and Cd. pH rises toward 7.5–8.5 as the system progresses into the methanogenic phase, encouraging the precipitation of metals as sulfides or hydroxides (Velempini et al., 2023).

Studies of older landfill sites have revealed that leachate often contains increased concentrations of toxic metals. Cadmium levels in landfill leachate have been demonstrated to exceed 2.0 mg/L, exceeding the recommended drinking water limits of 0.005 mg/L by more than 400 times (Aucott, 2006). Arsenic concentrations are substantially above the groundwater quality thresholds in some leachates, which have ranged from 0.1 to 1.2 mg/L. If containment systems degrade or fail, the chance of leachate seepage exists even in lined landfills. Moreover, remobilization of previously stabilized metals may happen due to aerobic oxidation, air intrusion, or fluctuating moisture and temperature conditions (Velempini et al., 2023).

Their chemical form stimulates the stability of adsorbed or immobilized metals in landfills. Metals embedded in ceramic matrices or vitrified slag show reduced mobility, while those in more reactive oxide or salt forms are more prone to leaching (Shen et al., 2024). A study using aluminum nitride developed from secondary aluminum dross to treat municipal solid waste incineration fly ash (MSWI FA) documented reduction efficiencies of 100% for Zn and 67% for Fe. The resulting vitrified transparent glass exhibited leaching concentrations for Ni, Cu, Zn, and Pb below 0.1 mg/L, meeting Toxicity Characteristic Leaching Procedure (TCLP) regulatory thresholds (Shen et al., 2024). Thermal and structural analyses confirmed that the vitrified product exhibited high chemical stability and mechanical strength (Figure 7.3).

While engineered containment strategies such as leachate management and liner systems improve environmental safety, long-term projections remain uncertain. Modeling suggests landfill behavior can deviate from expectations under arid or acidic conditions. Available data for landfills operating beyond 60 years indicate no consistent increase in heavy metal leachate concentrations over time; however, comprehensive long-term behavior assessments are still needed (Aucott, 2006).

Iron oxides and humic substances within landfill matrices can improve metal binding, but their effectiveness over multi-decadal timeframes remains a topic of scientific interest (Velempini et al., 2023). The Resource Conservation and Recovery Act (RCRA) in the United States requires double-liner systems and post-closure monitoring for harmful pollutants in landfills. Simultaneously, policies like Extended Producer Responsibility (EPR) aim to reduce the influx of metal-containing products into the waste stream through recycling and material substitution strategies (Aucott, 2006).

Minimizing the deposition of heavy metals in landfills is imperative, given their toxicity and long-term persistence. The method is increasingly viewed as a last alternative, although landfilling propositions controlled sequestration in the short term (Adu & Aneke, 2025). Approaches that endorse metal recovery, adsorbent regeneration, and resource circularity offer more sustainable long-term solutions to managing metal-laden wastes.

7.2.2 Incineration

Municipal solid waste incineration (MSWI) plays a vital role in integrated waste management systems, particularly in densely populated areas such as China and the European Union, where land availability is inadequate. Nevertheless, the effective

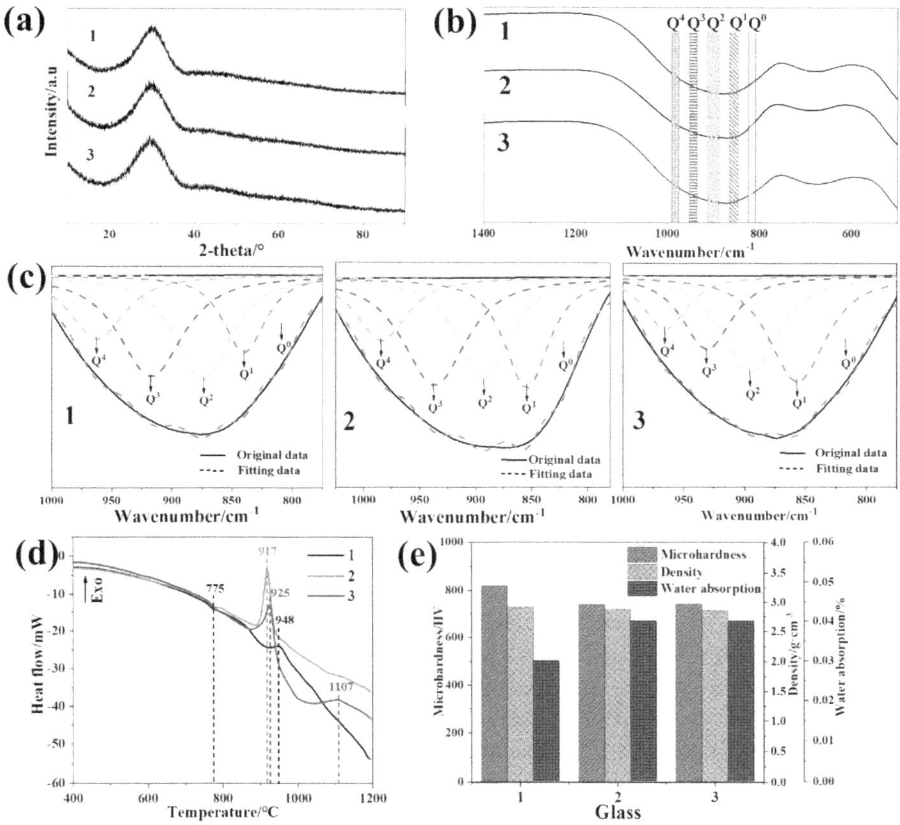

FIGURE 7.3 (a) XRD patterns of the transparent glass; (b) FT-IR spectra of the transparent glass; (c) Gaussian deconvolution of the central absorption peak; (d) DSC heating curve from 400 °C to 1200 °C at a rate of 10 °C/min; (e) Microhardness, density and water absorption values of the transparent glass (from Shen et al. (2024)).

treatment of MSWI residues, primarily fly ash, remains crucial to guarantee environmentally safe disposal. Despite these advantages, the incineration process generates fly ash, a by-product that is considered acceptable particulate matter and constitutes approximately 3–5% of the incinerated mass. In China alone, this corresponds to over three million tons annually, much of which contains high concentrations of leachable heavy metals, primarily Pb, Zn, and Cd. The classification of MSWI fly ash as hazardous waste (e.g., HW18 under Chinese regulations) necessitates appropriate stabilization before landfill disposal to minimize metal leaching and associated environmental risks(Ze Zhang et al., 2024). Various methods like incineration and landfilling are employed to handle this waste, with incineration favored for its ability to reduce waste volume by 10–15% and recover energy. However, incineration produces fly ash containing heavy metals and chloride ions, posing environmental hazards if untreated. Traditional treatment methods include vitrification, chemical washing, and solidification/stabilization (S/S), with cement-based S/S being popular for its cost-effectiveness and efficiency in stabilizing hazardous metals. Despite this, S/S typically leads to landfilling, which wastes materials like fly ash that contain minerals (SiO_2, Al_2O_3, CaO) similar to cement additives, thereby presenting opportunities for reuse in construction. Research shows that using fly ash in cement systems, particularly Limestone Calcined Clay Cement (LC3), can reduce environmental impact and improve durability, thanks to LC3's lower carbon footprint, resistance to chloride-induced corrosion, and capacity to immobilize heavy metals effectively (Kim et al., 2022). Challenges remain in optimizing fly ash substitution rates and ensuring mechanical performance without pretreatment, but ongoing studies aim to address these issues to advance sustainable waste valorization and construction practices. Advanced characterization methods such as XRD (X-ray diffraction), FT-IR (Fourier Transform Infrared Spectroscopy), and XPS (X-ray photoelectron spectroscopy) confirmed the conversion of heavy metals to more stable, low-valence forms and the integration of sulfur-based functional groups. AFM (Atomic force microscopy) analysis revealed the formation of robust, chain-like macromolecular matrices that effectively immobilized the metals within a three-dimensional network (Shen et al., 2024).

Environmental risk metrics further validated the effectiveness of Dithiocarbamate (DTC) chelation. The Risk Assessment Code (RAC) and Nemero Index demonstrated significant reductions in ecological risk after chelator treatment. As illustrated in Figure 7.4 (Placeholder1) 4(a) and 4(b), the RAC indices for Pb, Zn, and Cd were markedly reduced post-chelation, shifting their classification from 'high' or 'very high risk' to 'low' or 'negligible risk' levels. Similarly, the Nemero Index showed a transition of MSWI FA from 'seriously polluted' to 'safe' status, emphasizing the environmental merit of chelator stabilization (Ze Zhang et al., 2024).

Diisopropyl dithiophosphate (DDP) has also shown strong stabilization performance. At a dosage of just 1%, DDP achieved over 95% immobilization efficiency for Pb, Cd, and Zn, with leachate concentrations remaining well below hazardous landfill limits (Pb: 5 mg/L; Cd: 0.5 mg/L; Zn: 75 mg/L) across a pH range of 1–13 (Xu et al., 2013). In terms of efficiency and pH tolerance, DDP outperformed traditional inorganic stabilizers, such as sodium sulfide and lime. Fractionation studies revealed a shift from exchangeable and carbonate-bound metal forms to more stable organic-bound fractions, lowering environmental and biological risks.

Physical stabilization via encapsulation is another viable option. Metallurgical slag-based cementitious material (MSCM), produced from blast furnace slag, steel slag, desulfurization ash, and phosphoric acid sludge, has been used to immobilize MSWI FA at contents up to 60 wt.% (Deng et al., 2022). Treated blocks exhibited compressive strengths ranging from 8.9 to 13.0 MPa and immobilization rates above 98.7% for Pb, Zn, Cr, and Hg. Leachate concentrations of Zn and Pb were reduced to <0.11 mg/L, far below Chinese regulatory thresholds. XRD and SEM (scanning electron microscope) analyses identified the formation of calcium silicate hydrate (C-S-H), Friedel's salt, ettringite (AFt), and C-A-S-H phases, which immobilize metals through precipitation, ion-exchange, and physical encapsulation, ensuring both chemical stability and mechanical integrity (Ye et al., 2025). While incineration offers clear benefits for waste volume reduction and energy recovery, its by-product, fly ash, presents a persistent environmental challenge due to heavy metal content (Xue & Liu, 2021). Stabilization technologies, such as chemical chelation

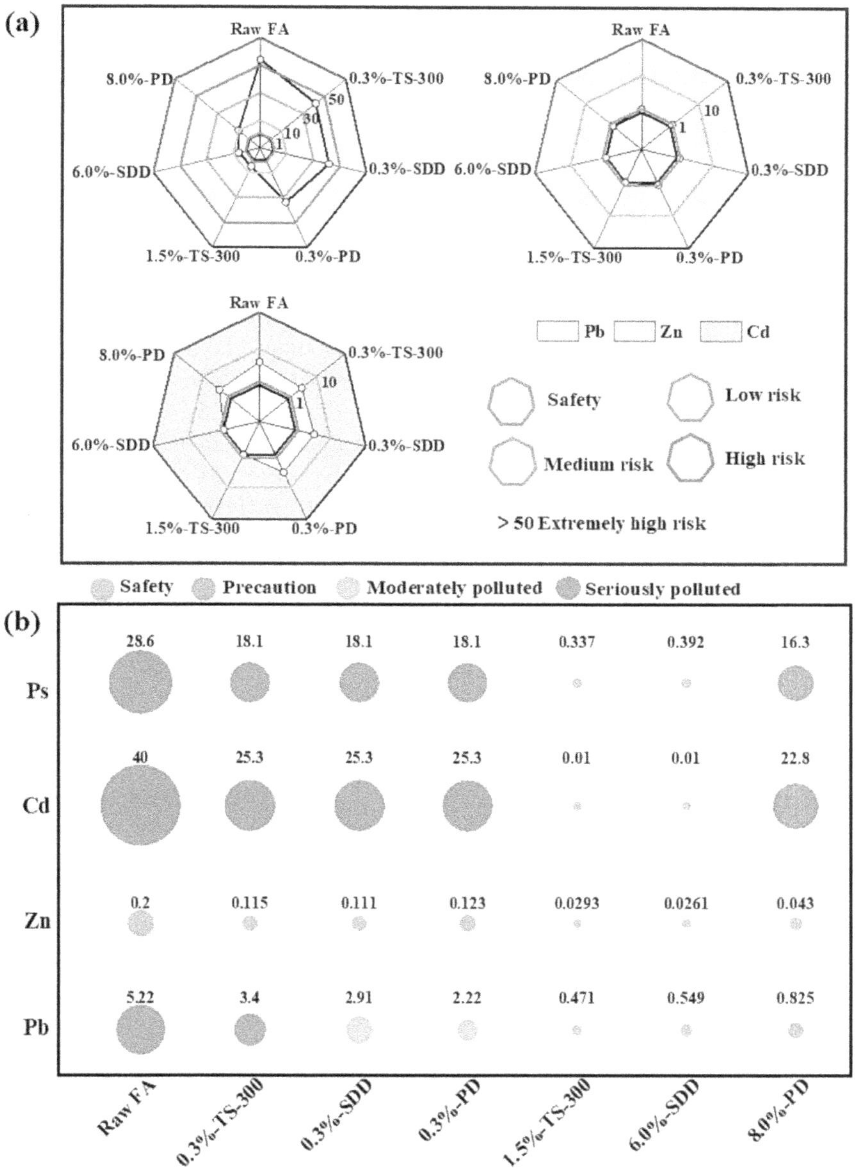

FIGURE 7.4 (a) Risk Assessment Code (RAC) values for Pb, Zn, and Cd in MSWI fly ash before and after chelation; (b) Environmental risk classification of MSWI fly ash pre- and post-chelation, evaluated using the Nemero index. Adapted from Ze Zhang et al. (2024).

and encapsulation, lower leachability to meet national and international disposal standards and enhance long-term environmental safety. Continued research on durability, cost-effectiveness, and large-scale applicability will be essential for optimizing these approaches in practice.

7.3 TECHNOLOGIES FOR METAL ION EXTRACTION

7.3.1 Chemical Leaching

Chemical leaching depends largely on the choice of elution agent, yet many of the most frequently used agents present notable drawbacks. For example, acidic agents can significantly damage soil mineral structures, alter key physicochemical characteristics, and compromise the stability of soil aggregates. Synthetic chelating compounds, such as sodium ethylenediaminetetraacetate (EDTA-2Na), may persist in the environment and contribute to secondary pollution issues. Inorganic acids and salts can negatively affect soil pH and disrupt soil structure, while natural organic acids pose a lower environmental risk but still encounter practical limitations like high remediation costs, restricted solubility, and less-than-ideal efficiency. Popular surfactants, including Triton and sodium dodecyl sulfate (SDS), demonstrate solid performance in extracting heavy metals; however, concerns about bioaccumulation and elevated costs remain (Gebeyehu et al., 2025; Iyer et al., 2024).

Inorganic acids, such as nitric acid (HNO_3), sulfuric acid (H_2SO_4), phosphoric acid (H_3PO_4), and hydrochloric acid (HCl), are commonly used due to their ability to disrupt metal-matrix bonds and dissolve a wide range of metals. For instance, using H_2SO_4 and H_2O_2, cobalt (Co) and nickel (Ni) recovery from lithium-ion battery waste has reached efficiencies of up to 98% and 96%, respectively, under optimized conditions (Binbin Zhou et al., 2025). The addition of oxidizing or reducing agents, such as H_2O_2, pyrite (FeS_2), glucose, or ethanol, often improves leaching performance by converting metals to more soluble oxidation states. In Co recovery, H_2O_2 facilitates the reduction of Co^{3+} to Co^{2+}, thereby improving dissolution rates (Van Wendel De Joode et al., 2016). Chemical chelation has emerged as a propitious tactic for immobilizing metals in MSWI FA. DTC-based chelating agents have shown significant promise in reducing the mobility of heavy metals in fly ash. In a recent study, three DTC chelators, TS-300 (thiocarbamate-based dithiocarbamate), PD (pyrrolidine dithiocarbamate), and SDD (sodium dithiocarbamate), were assessed under two landfill scenarios: mixed landfill and compartmentalized landfill conditions. Under mixed landfill conditions, even low dosages (0.3%) of these chelators resulted in leachate heavy-metal concentrations below regulatory limits (GB16889–2008). Higher dosages were necessary in more aggressive acid-rain simulated scenarios: 1.5% for TS-300, 6.0% for SDD, and 8.0% for PD to ensure compliance. For instance, TS-300 reduced Pb leaching from 3.096 mg/L to 0.066 mg/L, a stabilization efficiency of over 97%.

Recent studies have quantified the effectiveness of acid leaching in excluding heavy metals. In Figure 7.5, Du et al. showed that 7 M acetic acid (HAc) and 3 M hydrochloric acid (HCl) reduced Ni from 604.84 mg/kg to 74.52 and 59.71 Mg/Kg, and cadmium from 4.29 Mg/Kg to 0.39 and 0.43 mg/kg, respectively, well below soil quality limits. However, as leaching increased at higher HAc concentrations due to destabilization of Ca-As complexes, indicating complex behavior beyond pH effects(Du et al., 2020).

To overcome the environmental drawbacks of strong acids, such as corrosivity, high toxicity, and the need for neutralization, organic acids have gained interest. Citric acid, lactic acid, methane sulfonic acid, and biodegradable reductants (e.g., fruit peel powders or tea waste) have proven effective for extracting metals from

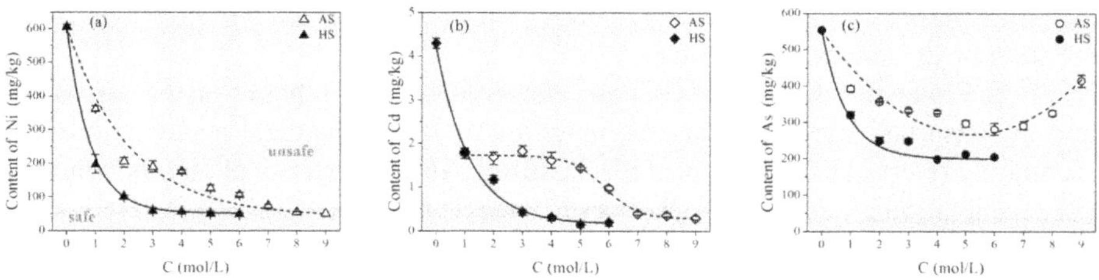

FIGURE 7.5 Variation in heavy metal concentrations in slag after treatment with different concentrations of acetic acid (AS) and hydrochloric acid (HS) compared to threshold values (Du et al., 2020).

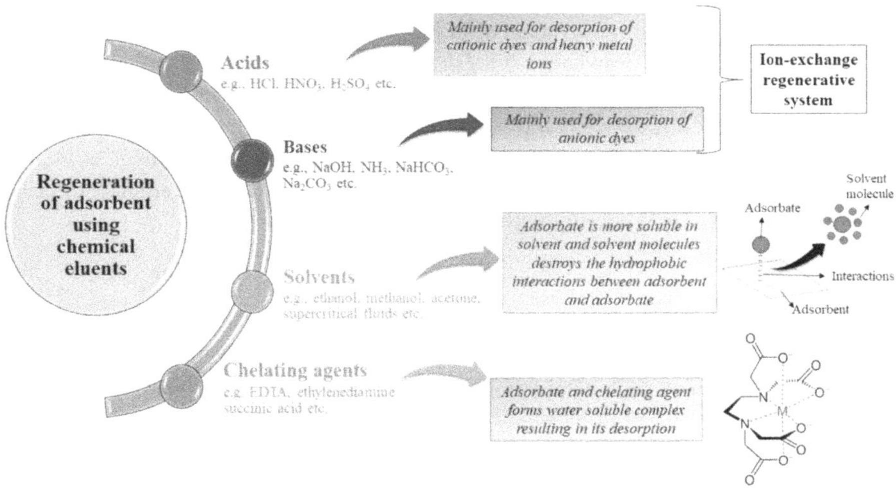

FIGURE 7.6 Schematic representation of adsorbent regeneration using different chemical eluents, showing mechanisms that facilitate desorption of bound contaminants. Adapted from Bagotia (2025).

spent adsorbents and low-grade ores. Citric acid, for example, forms stable complexes with Co^{2+} and Ni^{2+} and can achieve leaching efficiencies exceeding 70% under mild conditions (Shih et al., 2019; Syeda et al., 2022).

Hybrid systems utilizing natural reductants (e.g., bagasse pith or glucose) in combination with organic acids significantly enhance dissolution rates. In one study, the combination of tartaric acid and glucose recovered over 80% of zinc from metallurgical slag within 3 hours at 70 °C. (Ziheng Zhang et al., 2025).

The underlying leaching mechanism is governed by metal-adsorbent interactions (Figure 7.6) and chemical equilibria, which are influenced by matrix composition, redox conditions, temperature, pH, and contact time. Low pH usually favors metal solubilization but may destabilize certain matrix minerals; redox potential is critical for elements such as Cr, As, and Co; mineral phases such as gehlenite or Ca-As complexes may resist dissolution; and higher temperatures or prolonged contact times usually enhance extraction efficiency, though they may also encourage reagent degradation or secondary precipitation.

To facilitate the desorption process, the binding forces between the adsorbent and the adsorbed substances are disrupted by different chemical agents, a process commonly referred to as adsorbent regeneration. Among the critical factors influencing regeneration efficiency, pH plays a pivotal role (Lata et al., 2015). Alterations in the pH level can significantly influence the chemical characteristics and even the physical structure of the adsorbent, potentially enhancing or impairing its adsorption capacity. The regeneration mechanism often relies on ionic exchange, where changes in pH lead to the replacement of ions, either cations or anions, between the reagent solution and the contaminants previously adsorbed onto it.

Chemical leaching represents a technically robust and scalable solution for extracting heavy metals from waste solids and regenerating adsorbents (Tang et al., 2019). Although strong mineral acids offer high extraction efficiencies, their environmental and operational drawbacks have driven the development of greener alternatives (Lata et al., 2015). Organic acids and hybrid systems demonstrate considerable potential for reducing toxicity and waste generation. Future investigations should prioritize improving the use of low-cost, selective, and renewable reagents, as well as integrating leaching with downstream metal recovery processes, to support circular economy objectives and comply with environmental regulations.

7.3.2 Electrochemical Regeneration

Electrochemical remediation is emerging as a sustainable and practical technique for removing heavy metal ions from aqueous environments. Unlike traditional physicochemical methods, this approach utilizes redox reactions to convert and immobilize metal ions directly onto conductive surfaces, thereby avoiding the use of chelating agents or harmful chemicals. This method has demonstrated effective and selective removal of dissolved heavy metal ions, including lead (Pb^{2+}), nickel (Ni^{2+}), cadmium (Cd^{2+}), and copper (Cu^{2+}), among others (Choumane & Peulon, 2022). It operates through careful pH regulation and the application of an electric potential, which promotes the formation of stable, adherent oxide films on electrode surfaces, thereby transforming the metal ions into insoluble forms. For instance, lead can be oxidized to β-PbO_2 under acidic conditions, achieving removal efficiencies of up to 99.99%, in line with environmental and drinking water quality standards. The study further indicates that metals with multiple oxidation states, such as Cu^{2+} and Ni^{2+}, can be effectively removed via electrodeposition. During this process, copper is reduced to Cu_2O, while nickel forms γ-NiO(OH), with both transformations optimized through cyclic voltammetry to determine the ideal potential ranges. In contrast, cadmium, which does not exhibit multiple oxidation states in aqueous environments, cannot be electrodeposited and therefore remains in solution under similar electrochemical conditions. Nevertheless, its presence does not interfere with the concurrent removal of other metals, allowing for effective selective separation (Dhokpande et al., 2024; Du et al., 2020). The scalability and operational simplicity of this electrochemical system set it apart from conventional methods. Experimental results confirm its effectiveness not only in controlled model solutions but also in complex wastewater effluents. In multi-metal systems, a stepwise process involving sequential pH adjustments and applied potentials enables the precise separation of Pb, Ni, and Cu as distinct oxide films, which are further verified through XRD and SEM analyses. Cadmium, on the other hand, is subsequently captured via sorption onto manganese oxide-based films, providing a complementary pathway to electrochemical removal.

The method demonstrates excellent energy efficiency, with estimated consumption values ranging from 0.7 to 2.1 kWh per kilogram of metal removed—substantially lower than those of conventional electrochemical systems, which can reach up to 42.3 kWh/kg for Cu and Ni. These low energy requirements, combined with ambient-temperature operation, minimal secondary pollution, and low maintenance needs, highlight its practicality for real-world applications, particularly in decentralized or resource-limited settings. Moreover, the technique's selectivity and reproducibility make it well-suited for metal recovery and recycling, aligning with circular economy principles. The conversion of metal contaminants into functional oxide films with defined morphology and crystallinity further enhances both environmental remediation and resource valorization potential (Choumane & Peulon, 2022; Dhokpande et al., 2024; Du et al., 2020).

In summary, this electrochemical method represents a substantial and durable advancement in the remediation of adsorbed heavy metal ions. It combines the goals of environmental protection, energy efficiency, and materials recovery, thereby offering a versatile alternative to outdated treatment methods. Future research could focus on integrating renewable energy sources or combining them with biosorption and membrane technologies to enhance the removal of single-oxidation-state metals, such as cadmium.

7.3.3 Thermal Desorption

Thermal desorption uses high temperatures to remove heavy metal ions from solid matrices such as industrial sludge, adsorbents, and contaminated soils. It operates in two modes: low-temperature (100–400°C) for volatile metals like As and Hg, and high-temperature (400–600°C) for refractory metals such as Cd and Pb. Removal efficiencies can exceed 95% for Hg at 300°C and reach up to 98% for Cd and Pb around 500°C. Performance depends on residence time (20–60 min), heating rate (5–20°C/min), and carrier gas composition (Mahmoud et al., 2010). However, the method requires substantial energy (0.5–1.2 kWh/kg), and may release toxic gases and incur high equipment costs. Integrated strategies, including gas scrubbing and catalytic oxidation, are being developed to improve efficiency and sustainability. Despite its effectiveness, thermal desorption poses challenges related to high energy consumption, potential emission of toxic gases, and equipment costs (Chen et al., 2014). To mitigate these issues, integrated approaches combining thermal desorption with gas scrubbing, catalytic oxidation, and material stabilization have been explored. Future research aims to optimize energy efficiency and develop sustainable heat recovery methods for industrial-scale applications.

7.3.4 Solvent Partitioning

Solvent separation, also known as liquid–liquid extraction, has emerged as an important technology for the separation and purification of heavy and rare metals from complex aqueous systems. In this process, metal ions are transferred from an aqueous phase, usually an acidic leachate derived from thermally treated waste, into an immiscible organic phase containing an extractant with a high affinity for specific metals. The principle is based on the formation of metal extractant complexes that preferentially migrate into the organic layer, from which metals can later be stripped and recovered (Kazakis, 2025; Pandey et al., 2023; Shih et al., 2019). The most commonly employed extractants, along with their chemical properties, target metals, and optimal pH ranges, are summarized in Table 7.2.

A major challenge in solvent extraction is achieving high selectivity, particularly when separating metals with similar chemical properties, such as cobalt (Co) and nickel (Ni). Parameters such as pH, temperature, extractant concentration, metal oxidation state, and aqueous-phase composition all influence separation efficiency. For instance, the selective extraction of Co over Ni using Cynax 272 is highly pH-dependent, with optimal results typically obtained between pH 4.5 and 6.5 (Barakat, 2011). Another persistent issue is the co-extraction of impurities, such as calcium, magnesium, and aluminum, which are commonly present in thermally treated residues.

These competing ions can reduce extraction yield and purity, complicating flow recovery. To mitigate these effects, researchers have proposed methods such as pre-leaching purification, the use of complexing agents, and chemical modification of extractants (Shih et al., 2019; W. Zhou et al., 2022b).

On an industrial scale, process configuration, whether mixer-settlers, pulsed columns, or membrane-based extraction systems, significantly influences both operational efficiency and economic viability. Recently, process intensification strategies, including microfluidic solvent extraction and solvent-impregnated resins, have attracted

TABLE 7.2
Common Solvent Extractants Used for Adsorbed Heavy Metal Ions

Extractant type	Example extractants	Target metals	Separation characteristics	References
Organophosphorus Acids	DEHPA, Cyanex 272, Cyanex 301	Co^{2+}, Ni^{2+}, Mn^{2+}	High selectivity for Co over Ni (pH-dependent); widely used in battery recycling	Sadehlari et al., (2025), Rodrigues et al., (2022)
Phosphine Oxides	Cyanex 923, TOPO	U^{6+}, Th^{4+}, Fe^{3+}	Strong metal–oxygen bonding; suitable for actinides and rare earths	Tumanov et al., (2023)
Sulfoxide Derivatives	Cyanex 925	Fe^{3+}, Cu^{2+}	Strong affinity for high oxidation state ions	El-Nadi (2017)
Deep Eutectic Solvents	ChCl–Urea, ChCl–Citric Acid	Cu^{2+}, Ni^{2+}, Zn^{2+}	Biodegradable; excellent for eco-friendly leaching/extraction	Najafi et al., (2025)
Synergistic Mixtures	Cyanex 272 + TBP, D2EHPA + TOPO	Co^{2+}/Ni^{2+}, Mn^{2+}/Li^{+}	Enhanced separation by tuning the extractant ratios and the acidity of the solution	Shuai et al., (2025)

attention for their potential to minimize solvent loss, enhance phase-separation kinetics, and increase throughput (Wong et al., 2019). From a sustainability standpoint, solvent extraction offers several advantages over traditional pyrometallurgical routes, such as lower energy consumption, greater flexibility in handling diverse feedstocks, and the feasibility of closed-loop operation (Ziheng Zhang et al., 2025). However, environmental concerns regarding solvent volatility, toxicity, and recyclability persist, necessitating ongoing innovation in developing greener, biodegradable solvents and in implementing waste minimization practices.

In summary, solvent extraction is a powerful and flexible method for selectively recovering critical heavy metal ions from thermally treated waste materials. Its success hinges on careful solvent design, process control, and integration with upstream and downstream treatment steps. The following sections of this chapter will explore specific solvent systems, the extraction process, separation challenges, and case studies from recent industrial and laboratory-scale applications.

7.3.5 Ion-Exchange Processes

Ion-exchange is a well-established and widely utilized method for removing adsorbed heavy metal ions from aqueous solutions, appreciated for its high efficiency, fast reaction kinetics, and strong metal uptake capacity. The process is based on the reversible exchange of metal ions in solution with ions bound to a solid-phase material known as an ion-exchanger. Both natural and synthetic ion-exchange resins are used for this purpose; however, synthetic variants are generally preferred in industrial applications because of their superior selectivity, mechanical stability, and consistent performance across a wide range of operating conditions (Qasem et al., 2021; Razzak et al., 2022; Shih et al., 2019).

Synthetic ion-exchange resins are typically classified according to their functional groups. Strong acid cation exchangers contain sulfonic acid groups ($-SO_3H$), providing high ionization and effective performance across a broad pH range. In contrast, weak acid cation exchangers are characterized by carboxylic acid groups ($-COOH$). These functional groups host hydrogen ions, which can be displaced by multivalent heavy metal cations as wastewater passes through the resin column. The overall ion-exchange reactions can be represented as follows (Fu & Wang, 2011):

$$n\,R\text{–}SO_3H + M^{n+} \rightarrow (R\text{–}SO_3)_nM^{n+} + n\,H^+$$
$$n\,R\text{–}COOH + M^{n+} \rightarrow (R\text{–}COO)_nM^{n+} + n\,H^+$$

Several operational parameters, such as solution pH, temperature, initial metal ion concentration, and contact time, significantly influence the efficiency of the ion-exchange process (Gode & Pehlivan, 2006). The ionic charge and valence state of the metal ions also play a crucial role, as higher-valence cations generally exhibit stronger electrostatic interactions with the resin matrix. For example, studies have shown that Ce^{4+} and Fe^{3+} ions are exchanged more readily than Pb^{2+} under identical conditions, indicating a selectivity trend governed by ionic charge density and hydration energy.

In addition to synthetic resins, natural ion-exchangers such as zeolites have gained attention due to their abundance, low cost, and satisfactory performance in metal ion removal. Zeolites are crystalline aluminosilicates with a permeable structure capable of selectively exchanging cations within their lattice. Clinoptilolite is among the most extensively studied natural zeolites, exhibiting a strong capacity to remove metal ions such as Pb^{2+}, Zn^{2+}, and Cu^{2+} (Fu & Wang, 2011).

Efforts to enhance the performance of natural zeolites have focused on surface modification techniques. For instance, coating clinoptilolite with amorphous iron oxides has been shown to markedly improve its ion-exchange capacity and sorption kinetics, thereby expanding the range of metal ions it can effectively remove. In drinking water treatment applications, modified clinoptilolite Fe composites have demonstrated the ability to reduce Cu, Zn, and Mn concentrations to levels suitable for both human consumption and agricultural reuse (Shih et al., 2019). Despite these promising developments, the large-scale application of natural zeolites is still in its early stages (Coronas & Santamaria, 2004). Most investigations have been confined to laboratory-scale experiments, and challenges remain regarding material regeneration, long-term durability, and standardization for industrial deployment. In contrast, synthetic ion-exchange resins are well-established in commercial water treatment operations due to their superior exchange efficiency, reproducibility, and scalability.

In conclusion, ion-exchange remains a robust and adaptable approach for the remediation of adsorbed heavy metal ions, with potential applications in both industry and the environment. Future work should focus on scaling up natural ion-exchangers, optimizing hybrid systems, and enhancing regeneration techniques to improve the overall sustainability of ion-exchange technologies (Dhokpande et al., 2024; Qasem et al., 2021).

7.3.6 Bioremediation

Bioremediation represents an effective and environmentally sustainable approach for the elimination or transformation of hazardous contaminants, including adsorbed heavy metal ions, from polluted environments such as soil and water (Saravanan et al., 2023). This process utilizes biological systems, either living or inert, such as algae, bacteria, fungi, or plants, to immobilize or convert toxic metal ions into less detrimental forms. While organic contaminants may be completely mineralized, adsorbed heavy metal ions are not biodegradable; bioremediation focuses on sequestration or chemical transformation to less bioavailable states (Biao Zhou et al., 2023). Based on the mode of implementation, there exist two primary approaches: in situ and ex situ. In situ techniques involve treating contaminants directly at the contaminated site without excavation or removal of the affected material. Common examples include bioventing, biosparging, phytoremediation, and bioslurping, which rely on enhancing native microbial activity or plant uptake to reduce contaminant concentrations in soils and groundwater (Ojha & Tiwary, 2021; Maitra, 2018). These techniques are often more economical and less invasive, making them suitable for large-scale or difficult-to-access sites. In contrast, ex situ methods—such as bioreactors, biopiles, land farming, and windrow composting—require excavation and transfer of contaminated material to a controlled treatment system. Although ex situ approaches provide greater control over environmental parameters (such as temperature, aeration, and nutrient availability) and can achieve faster remediation rates, they generally involve higher operational and maintenance costs (Ajorloo et al., 2022; Fouda-Mbanga et al., 2024).

Bioremediation of adsorbed heavy metal ions mainly occurs through two biological mechanisms: bioaccumulation and biosorption. Bioaccumulation is a biologically active, metabolism-driven process where living organisms internalize metal ions across their cellular membranes. Once inside the cell, metal ions are sequestered by various intracellular ligands, such as proteins or peptides, reducing their toxicity. This process is often species-specific and influenced by environmental factors such as pH, temperature, and metal concentration. Representative examples of bioaccumulation capacity for numerous biological systems are summarized in Table 7.3, highlighting the diversity of organisms and their specific target metals.

Notably, *Ochrobactrum intermedium* BPS-20 achieved peak bioaccumulation of 88.5% for lead and 76.7% for Ni at alkaline conditions (pH 9), highlighting the strong influence of environmental pH on uptake efficiency. While bioaccumulation requires metabolic energy and living organisms, biosorption is a passive process in which metal ions bind to the surface of biological materials, often dead or nonliving biomass, via physicochemical interactions. These include ion-exchange, electrostatic attraction, surface complexation, and chelation.

Biosorbents can consist of bacterial biomass, fungal mycelia, algal mats, agricultural by-products, and various biopolymers. Biosorption offers several practical advantages: it is cost-effective, requires no external nutrient supply or aeration, operates effectively over a broad pH range, and generates minimal secondary pollution. This makes it particularly suitable for wastewater treatment, where the rapid removal of dissolved metal ions is critical. The presence of surface functional groups, such as carboxyl, hydroxyl, amino, and phosphate moieties, on biosorbents provides active sites for metal ion binding. Mechanistically, biosorption may proceed through multiple pathways, which can overlap with bioaccumulation when living biomass is employed. The key distinction lies in energy dependence and rate: biosorption typically occurs rapidly via surface interactions and is often metabolism-independent, whereas bioaccumulation is a slower, metabolism-dependent process involving intracellular sequestration regulated by membrane transport systems (Pandey et al., 2023; Qomariyah et al., 2022; Razzak et al., 2022; Syeda et al., 2022).

TABLE 7.3
Examples of Bioaccumulation Capacity in Various Biological Systems

Species/Organism	Target Metals	Accumulation	Reference
Nasturtium officinale	Cu, Zn	Cu 1.0 mg/l, Zn 10.0 mg/l	Klimek-Szczykutowicz et al., (2019)
Hydrilla verticillata	Cd, Pb	Cd (0–20 μM), Pb (40 μM)	H. Zhang et al., (2020)
Typha latifolia	Zn, Cu	271 ± 17 (Zn), 47 ± 4 (Cu)	Hejna et al., (2020)
Ochrobactrum intermedium BPS-20	Ni, Pb	2400 (Ni), 850 (Pb) mg/L	Sharma & Shukla (2021)

TABLE 7.4
Common Biosorbents and Their Metal Binding Capacities

Biosorbent	Target metal(s)	Binding capacity (mg/g)	Reference
Aspergillus niger biomass	Pb(II), Cu(II)	13.40 ± 0.95 mg/g Pb(II) and 10.80 ± 1.45 mg/g Cu(II)	Akar & Tunali (2006)
Banana peel powder	Pb, Cu	89.3 Pb and 5.8 mg/g Cu	Mohd Salim et al., (2016)
Sawdust	Cu(II), Ni(II), Zn(ii)	8.1 mg/g Cu(II), 9.9 mg/g NI(II) and 17.1 mg/g Zn(II)	Shukla & Pai (2005)
Algal biomass (*Sargassum*)	Cd(II), Pb(II)	240 mg/g Cd(II) and 350 mg/g Pb(II)	González Fernández et al., (2023)

Representative examples of biosorbent performance for various heavy metals are presented in Table 7.4, highlighting common biosorbents, their target metals, and binding capacities.

Unlike bioaccumulation, biosorption is a passive process in which metal ions adhere to the surface of biological material through physicochemical interactions. It does not rely on cellular metabolism and can be performed by living and non-living biomass. Biosorption occurs through ion-exchange, electrostatic attraction, chelation, and surface complexation. Materials used in biosorption include microbial biomass (e.g., bacteria, yeast, and fungi), agricultural residues, algae, and industrial by-products. Functional groups such as hydroxyl, carboxyl, amine, and phosphate present on the biomass surface play a central role in metal binding (Syeda et al., 2022).

The effectiveness of bioremediation depends on a complex mix of biological and environmental factors. Bioaccumulation is limited by microbial health and nutrient availability; biosorption is influenced more by surface chemistry and sorbent structure. Advanced imaging and spectroscopic techniques such as FTIR, SEM, XPS, and AFM are increasingly used to investigate biomass–metal interactions at the molecular level, revealing the structural and chemical transformations that govern removal efficiency(Oliveira et al., 2014). A classification of bioremediation techniques is presented in Figure 7.7, which summarizes the key in situ and ex situ approaches relevant to heavy metal remediation.

Bioremediation, encompassing bioaccumulation and biosorption, is a robust and environmentally sound strategy for treating environments contaminated with heavy metal ions (Khoshyomn et al., 2024). While bioaccumulation relies on metabolic uptake by live organisms, biosorption exploits passive binding on biomass surfaces. Integrating these technologies into real-world wastewater and soil treatment offers a viable pathway toward sustainable remediation practices. A comparative evaluation of bioremediation against other regeneration and extraction methods is essential for selecting the most suitable approach in various scenarios. Table 7.5 compares various regeneration and extraction techniques for adsorbed heavy metal ions, highlighting their underlying principles, advantages, limitations, and typical applications. Continued research is needed to enhance the selectivity, recovery, and reusability

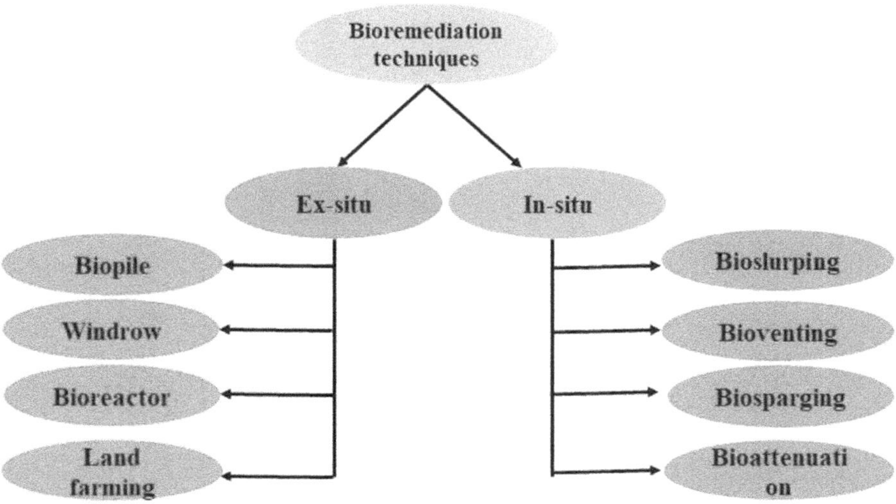

FIGURE 7.7 Classification of bioremediation techniques. Adapted from Razzak et al. (2022).

of biosorbents and scale up promising laboratory systems for industrial applications. Innovations in microbial engineering, nanobiotechnology, and reactor design are anticipated to further enhance the efficacy and versatility of bioremediation in tackling adsorbed heavy metal ion pollution (Firincă et al., 2025).

7.4 INTEGRATION WITH CIRCULAR ECONOMY PRINCIPLES

The management of heavy metal-laden waste must transition from traditional disposal methods to sustainable recovery and reuse strategies that align with circular economy principles (Figure 7.8). The circular economy goals to minimize waste, optimize resource efficiency, and promote material recovery rather than disposal. In heavy metal extraction and disposal, this approach involves recovering valuable metals, repurposing waste into functional materials, and utilizing energy-efficient conversion methods to reduce environmental impact (Rai et al., 2021). Integrating these principles can enhance industrial sustainability, reduce dependency on virgin metal sources, and mitigate environmental hazards associated with heavy metal pollution.

7.4.1 Waste-to-Energy Conversion

Waste-to-energy (WTE) conversion is a progressive strategy for transforming a variety of waste materials, including biomass contaminated with heavy metals, into valuable energy resources. This approach not only helps to reduce waste accumulation but also supports sustainable energy generation, aligning with the principles of the circular economy and industrial symbiosis. WTE systems typically offer a significant reduction in the volume (by 85–90%) and weight (by 65–80%) of waste,

TABLE 7.5
Comparison of Different Regeneration/Extraction Techniques for Adsorbed Heavy Metal Ions

Technique	Principle	Advantages	Limitations	Applications	Reference
Chemical Desorption	Uses acids, alkalis, or chelators to release adsorbed metal ions	Cost-effective, rapid, and high regeneration efficiency	pH-dependent, potential adsorbent damage, secondary pollution from leachates	Regeneration of adsorbents with Pb^{2+}, Cd^{2+}, Cr(VI), etc.	Iyer et al., (2024)
Electrochemical Regeneration	Uses redox reactions and electrolysis to desorb and convert metal ions	Selective, energy-efficient, no chemical additives	High setup cost, may not work for single-oxidation-state metals	Pb^{2+}, Cu^{2+}, Ni^{2+}, Cd^{2+} systems; decentralized water treatment	Choumane & Peulon (2022)
Biological Regeneration	Uses microbes to metabolize or immobilize adsorbed metals	Eco-friendly, low-cost	Slow kinetics, limited to specific conditions, and few reports for metals	Inorganic pollutants: potential future for Cr, Pb adsorbents	Fouda-Mbanga et al., (2024)
Ion-Exchange	Reversible replacement of metal ions on resins/zeolites	High efficiency, selective, reusable resins	Regeneration is needed, affected by the solution chemistry	Drinking water treatment, industrial wastewater	Dhokpande et al., (2024)

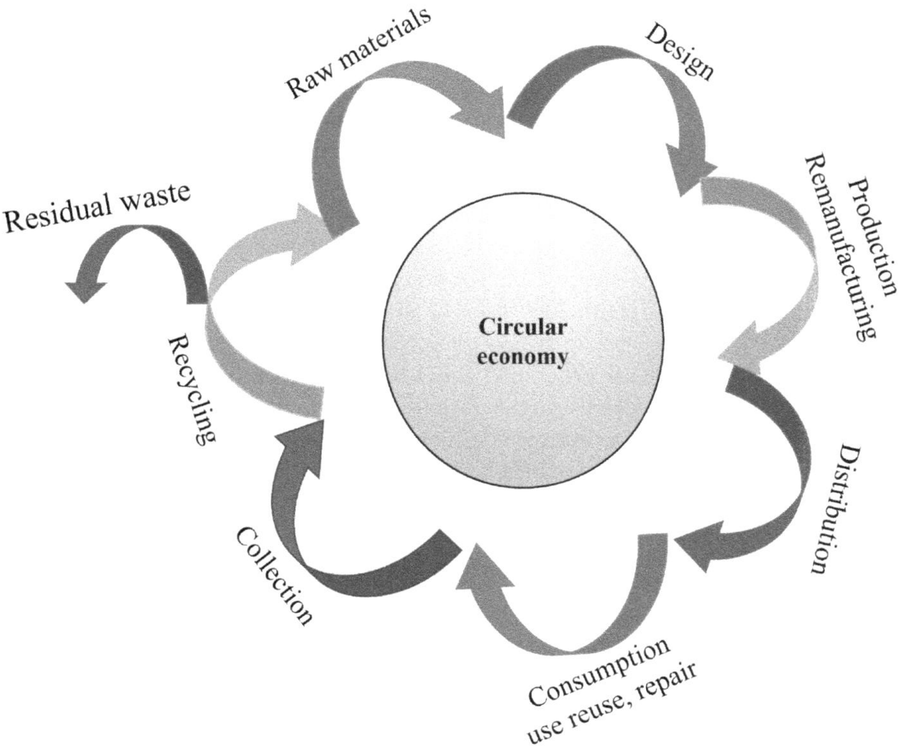

FIGURE 7.8 Circular economy approach for heavy metal recovery. Adapted from Hagelüken and Goldmann (2022).

while enabling energy utilization (Huawei Wang et al., 2019). In 2016, approximately 2.04×10^8 tons of municipal solid waste (MSW) were collected in China, with about 36.24% (approximately 7.4×10^7 tons) processed through WTE facilities for energy recovery. WTE systems commonly employ thermal methods, including gasification, pyrolysis, and incineration. Gasification involves the partial oxidation of carbon-rich waste to produce syngas, a combination of carbon monoxide, hydrogen, and methane, while facilitating the recovery of heavy metals from the solid residues generated after treatment. Pyrolysis decomposes organic matter in the absence of oxygen, yielding biochar, bio-oil, and syngas. In this process, heavy metals are primarily concentrated in the biochar, which can subsequently be processed for further metal extraction (Islam et al., 2024). Incineration with energy recovery combusts waste to generate heat that is converted into electricity, requiring advanced air pollution control systems to capture heavy metals from both flue gas and bottom ash, preventing atmospheric emissions (Ziheng Zhang et al., 2025).

Antimony trioxide (Sb_2O_3), commonly found in plastics and textiles in MSW, is transformed during incineration, with approximately 2–3% ending up in fly ash or air pollution control residues (APCs). These by-products, including both WTE fly ash and APCs, can contain high concentrations of toxic constituents such as dioxins and heavy metals, resulting in their classification as hazardous waste by Chinese authorities and multilateral environmental organizations. To mitigate the risks associated with these residues before final disposal, chemical stabilization and solidification treatments have been recognized as among the most effective management strategies (Verma et al., 2016; Appels et al., 2008). Beyond thermal conversion, biochemical methods such as anaerobic digestion and microbial fuel cells are also becoming increasingly important in WTE. Anaerobic digestion utilizes microorganisms to break down organic waste, producing biogas; however, heavy metals may accumulate in the digestate, necessitating stabilization prior to disposal. These are an emerging bioelectrochemical technology capable of simultaneously treating wastewater and recovering metals while producing electricity, making them compelling solutions for sustainable energy generation and wastewater management (Huawei Wang et al., 2019). Figure 7.9 illustrates the various pathways in waste-to-energy conversion, highlighting the fate of heavy metals at each stage of the process.

Applying WTE technologies offers multiple benefits, including reduced landfill dependency, energy recovery, and improved resource efficiency. However, challenges such as emission control, heavy metal stabilization in combustion residues, and economic feasibility must be addressed to facilitate wider industrial adoption.

7.5 RATIONALE FOR DISPOSAL OR RECOVERY OF ADSORBED IONS

Managing metal-laden adsorbents following wastewater treatment is a significant issue that requires consideration of economic feasibility, environmental stewardship, and technological sustainability. Although adsorption is widely recognized for its simplicity, efficiency, and adaptability in heavy metal ion removal, the handling of spent adsorbents presents both environmental challenges and opportunities for resource recovery and material reuse (Sonone et al., 2021). This section discusses the multifaceted rationale for either disposing of or recovering adsorbed ions. Environmental regulations increasingly emphasize the minimization of secondary pollution risks associated with the disposal of spent adsorbents. Heavy metals like As, Cd, Pb, Hg, and so forth are toxic and persistent, posing serious ecological and health risks if re-released into the environment (Vardhan et al., 2019; Qasem et al., 2021). Improper disposal of metal-laden adsorbents without adequate containment measures can lead to the leaching of toxic metals into soil and groundwater, potentially violating environmental protection standards and compromising water quality regulations (Ahmaruzzaman, 2011). To mitigate such outcomes, many national and international frameworks, including those recommended by the US Environmental Protection Agency (USEPA) and the European Union's Waste Framework Directive, mandate strict containment or recycling of hazardous wastes. Regulatory bodies advocate for controlled landfilling only after stabilization or solidification of the waste or, preferentially, the recovery of valuable metals to reduce the pollutant load and material demand on mining operations (Wagner & Raymond, 2015). Additionally, in processes such as photovoltaic (PV) module recycling, adsorption-based water treatment is complemented by subsequent physicochemical and tertiary recovery steps to extract and reclaim metals, thereby fulfilling the dual objectives of pollution mitigation and resource valorization.

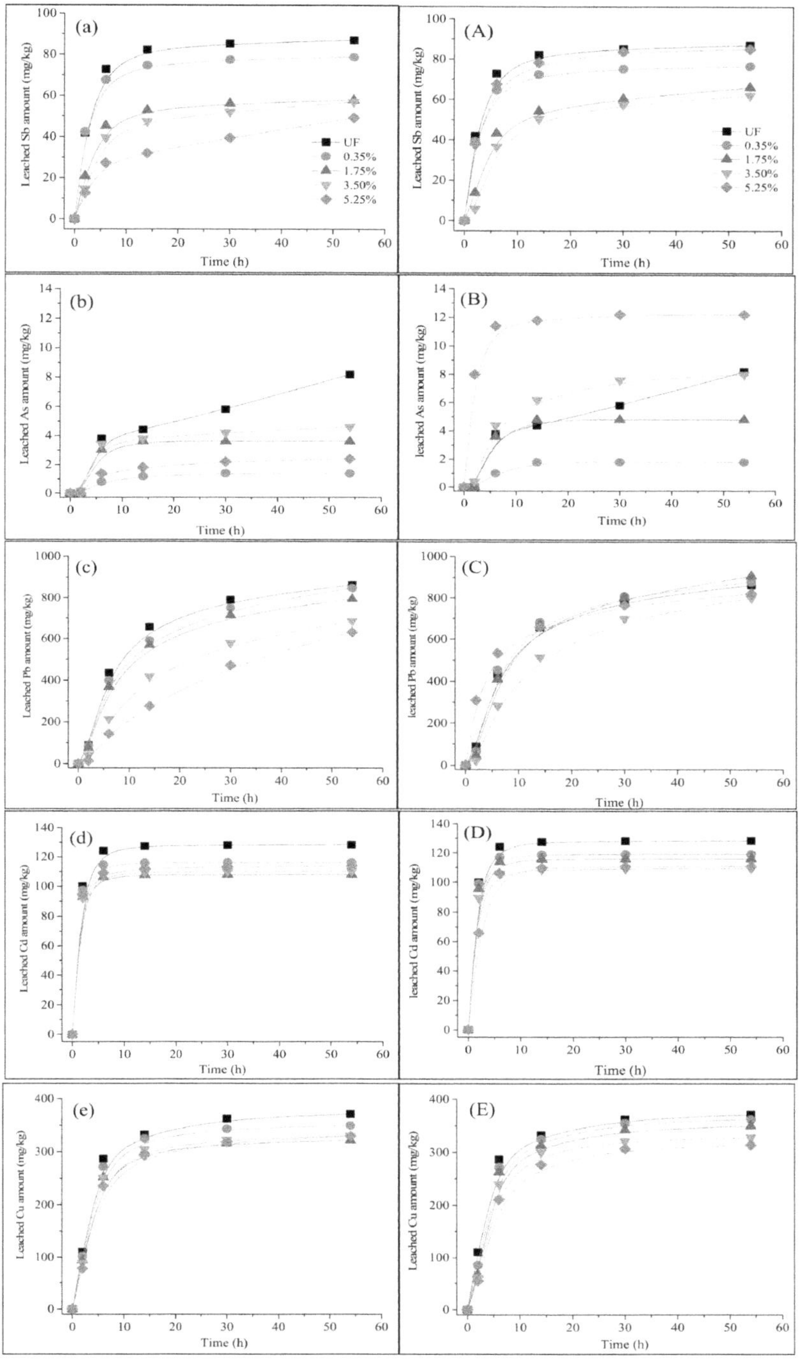

FIGURE 7.9 Waste-to-energy conversion and heavy metal management. Adapted from Huawei Wang et al. (2019).

From an economic standpoint, metal recovery from spent adsorbents offers substantial benefits. Adsorbed ions often include valuable or scarce metals, such as Cu, Ni, and Zn, as well as rare elements like indium (In) or platinum (Pt), in industrial wastewater. Recovery and reuse of these metals can offset wastewater treatment costs and contribute to a circular economy. Mining, electroplating, battery manufacturing, and semiconductor production produce effluents rich in recoverable metals. Recovery processes, including acid desorption, thermal treatment, and electrochemical regeneration, can reclaim metals for resale or reintegrate them into industrial processes (Florek et al., 2025; Gulliani et al., 2023). Integrating recovery systems helps reduce dependence on virgin raw materials while creating additional revenue opportunities. Moreover, when regeneration is efficiently achieved, recovered adsorbents can be reused multiple times, leading to lower operating costs and a reduced environmental footprint.

The ability to reuse adsorbents plays a crucial role in assessing the sustainability and cost-effectiveness of the adsorption process. Various materials, such as activated carbon, zeolites, biochar, and nanomaterials, can be regenerated using chemical, thermal, or microbial techniques. Chemical regeneration using acids or bases has proven particularly effective for desorbing metal ions and restoring the adsorption capacity of spent materials. For example, magnetic nanocomposite adsorbents, such as Fe_2O_3-loaded activated carbon, can be easily separated and regenerated through magnetic fields combined with mild chemical treatments, minimizing material loss. Similarly, biosorbents derived from agricultural residues or biomass can often be regenerated under low-energy conditions, exhibiting strong durability and consistent performance over multiple reuse cycles (Fouda-Mbanga et al., 2024; Renu et al., 2017).

However, the regeneration potential is strongly influenced by factors such as the nature of the metal ions, the type of adsorbent and functional groups, binding strength, pH sensitivity, and process conditions during adsorption and desorption. The feasibility of reusing adsorbents also depends on the degradation of structural integrity or sorption capacity over cycles. Therefore, life-cycle analysis and cost-benefit evaluations are crucial in determining whether regeneration or safe disposal is more suitable in each context.

7.6 CASE STUDIES

Depending on the process scale, nature of metal contamination, and regulatory requirements, the application of heavy metal disposal and extraction technologies varies significantly across industries (Babel & del Mundo Dacera, 2006). While integrating sustainability principles, several sectors, including water treatment, metallurgy, and mining, have adopted advanced techniques for handling waste contaminated with heavy metals. The following case studies highlight industrial applications of disposal and recovery strategies, emphasizing technological advancements and the environmental benefits they provide.

The metallurgical and mining industries are significant sources of heavy metal contamination, primarily due to smelting, ore processing, and extraction operations. Their waste streams from these industries comprise considerable concentrations of metals, notably Hg, As, and Pb, necessitating effective management strategies to avoid environmental contamination. One noteworthy example is the implementation of bioremediation, where sulfate-reducing bacteria have been used to precipitate metal sulfides, thereby significantly reducing metal toxicity in tailing ponds at the Puchuncaví copper smelting site in Chile (Hwang & Jho, 2018). To recover precious metals while promoting cyanide detoxification and heavy metal contamination in effluents, hydrometallurgical methods have been widely implemented in South African gold mines (Kuyucak & Akcil, 2013). Wastewater undergoes a multi-stage treatment process, including evaporation and crystallization, to recover valuable metals while minimizing the discharge of liquid waste. The introduction of zero-liquid discharge (ZLD) schemes in metallurgical plants in China has further demonstrated the sustainable management of heavy metals (Yao et al., 2024). Smelting and high-temperature metal extraction, also known as pyrometallurgical approaches, have been successfully integrated with lead recycling plants in the United States, enabling over 85% metal recovery from spent lead-acid batteries (Verma et al., 2016). The economic and environmental benefits of such recovery processes demonstrate the relevance of circular economy principles in the metallurgical industry. The mining and metallurgical industries, with a focus on process optimization, waste valorization, and ensuring regulatory compliance, are steadily improving their methods for recovering heavy metals. To enhance efficiency and sustainability, future advancements are anticipated to involve artificial intelligence (AI)-driven process monitoring and automation.

7.7 CHALLENGES AND FUTURE DIRECTIONS

The management and extraction of heavy metals face challenges due to socioeconomic constraints and technological limitations (Pet et al., 2024). These obstacles differ widely between developed and developing countries, shaped by industrial policies, economic capacity, environmental concerns, climate impact, industrial demand, ethical principles, and population pressures. Investment in suitable cleaner technologies, adoption of decentralized systems, and a combination of policy reforms are required to address these hurdles. The principles of Industry 4.0 and the United Nations' Sustainable Development Goals (SDGs) can also further enhance efficiency, sustainability, and resilience in the management of adsorbed heavy metal ions. Significant gaps persist in terms of scalability, efficiency, and affordability, although technological advancements in adsorbed heavy metal ion removal and recovery have shown potential. Developed nations can adopt high-tech electrochemical treatment, membrane filtration, and nanotechnology-driven adsorption methods to address these challenges. As outdated and inefficient technologies dominate the management of adsorbed heavy metal ions, these methods are often cost-prohibitive for developing countries (Hariyani et al., 2025). The limited technical expertise and high operational costs of advanced treatment systems play a critical role as bottlenecks to the widespread acceptance of these systems. One of the necessary technological challenges is the advancement of cost-effective and energy-efficient processes. Many prevailing adsorbed heavy metal ion treatment practices are energy-intensive, which limits their feasibility in regions with unbalanced infrastructure. For example, membrane-based filtration and electrochemical recovery need substantial power input, which is often unavailable or too expensive in low-income economies. The prerequisite for low-energy, alternative, and renewable-powered treatment technologies is imperative. Industry 4.0 solutions, namely IoT-enabled real-time monitoring and AI-driven process optimization, can enhance the efficiency of existing systems and reduce operational costs (Okuyelu & Adaji, 2024). Additionally, the foremost hurdle is the generation of secondary waste. Many existing adsorbed heavy metal ions require additional treatment before disposal, as removal processes can produce toxic sludge or concentrated brine. The lack of efficient waste valorization practices intensifies environmental risks. The use of biobased adsorbents and the integration of circular economy principles into waste treatment are emerging areas of research. Blockchain-based traceability systems and digital twins can enhance transparency in waste management, ensuring obedience with universal environmental standards (Hariyani et al., 2025).

The management of heavy metal disposal and extraction encounters substantial implementation barriers in developing countries, where socio-economic constraints often offset technological boundaries. Inefficiently skilled personnel, limited financial resources, inadequate laboratory and monitoring infrastructure, and insufficient public awareness or regulatory enforcement are the key challenges. In zones facing acute contamination risks, these limitations frequently result in a complete absence or inadequate implementation of advanced heavy metal removal technologies (Fouda-Mbanga et al., 2024). High-efficiency techniques, such as membrane filtration and nanocomposite adsorbents, and supported liquid membranes, are available. Their adoption in low-income areas remains minimal due to unaffordable costs and operational complications (Dawam et al., 2025). On the contrary, low-cost alternatives, such as biochar derived from agricultural waste, modified natural materials, and clay minerals, have shown promise for large-scale implementation. Biochar is highly suitable for decentralized applications in rural and peri-urban areas due to its cost-effectiveness, ease of local production, and effectiveness in removing metals such as cadmium (Cd), lead (Pb), and arsenic (As) from wastewater (Liang et al., 2021; Hegazy et al., 2025). Community-scale treatment units or point-of-use filtration decentralized systems afford practical solutions for regions lacking centralized wastewater infrastructure. These systems need less capital investment and can be modified to local socio-environmental conditions. Additionally, integrating such technologies with participatory education, governance, and capacity-building initiatives can enable communities to achieve long-term sustainability and enhance technology acceptance (Obeidat et al., 2024; Ventura et al., 2024). Adaptive financing models, such as microfinancing and coupling technology adoption with inclusive policy frameworks, international collaboration, and results-based subsidies, are crucial for addressing these socio-economic disparities. Aligning these strategies with the United Nations Sustainable Development Goals, particularly SDG 6 (Clean Water and Sanitation), SDG 9 (Industry, Innovation and Infrastructure), and SDG 10 (Reduced Inequalities), and leveraging the innovations from Industry 4.0, which include and AI-enabled decision-making and remote monitoring can make equitable and resilient heavy metal management schemes across both developed and developing regions (Murtaza et al., 2024; Hesong Wang et al., 2024).

7.8 CONCLUSION AND OUTLOOK

To ensure environmental safety and resource sustainability, managing heavy metal-laden adsorbents is vital. Landfilling, incineration, and recovery-oriented technologies, including electrochemical regeneration, chemical leaching, and other emerging extraction methods, are discussed in this chapter, which examines essential disposal methods. Although these methods raise concerns about long-term stability and environmental impact, traditional techniques provide short-term containment.

Hybrid leaching systems, electrochemical regeneration, and advanced stabilization employing low-carbon binders are among the potentially effective and scalable recovery approaches. These methods minimize environmental risks and simplify the recovery of valuable metals, hence supporting circular economy principles. Membrane-assisted recovery, waste-to-energy conversion, and smart adsorbents are all innovations that further improve process efficiency and sustainability.

New technologies tailored to local economic and environmental frameworks, developed by combining established methods, are anticipated to be the most impactful advancements. For decentralized treatment systems, renewable energy-driven processes, cost-effective, and low-toxicity reagents, continued research will be indispensable. To ensure effective heavy metal management, policy support, industrial collaboration, and public awareness will play crucial roles in enabling the widespread adoption of a cleaner, more resource-efficient future.

ACKNOWLEDGMENTS

The authors recognize the Department of Science and Technology (DST), Government of India (File No. DST/TMD/EWO/WTI/DM/2021/83) and the Department of Science & Technology and Biotechnology, Government of West Bengal [Memo No.: 49 (Sanc.)/ST/P/S & T/15G-6/2018] for their support, which contributed to the broader research environment. The Ministry of Education (formerly the Ministry of Human Resource Development, or MHRD), a government agency in India, is also recognized for providing fellowship support.

REFERENCES

Adu, J. T. & Aneke, F. I. (2025). Evaluation of heavy metal contamination in landfills from e-waste disposal and its potential as a pollution source for surface water bodies. *Results in Engineering*, *25*(September 2024), 104431. https://doi.org/10.1016/j.rineng.2025.104431

Agarwal, M. & Chaudhry, K. (2015). Heavy metal sources impacts & removal technologies. *International Journal of Engineering Research & Technology (IJERT)*, *3*(03), 1–7.

Ahmaruzzaman, M. (2011). Industrial wastes as low-cost potential adsorbents for the treatment of wastewater laden with heavy metals. In *Advances in Colloid and Interface Science* (Vol. 166, Issues 1–2, pp. 36–59). https://doi.org/10.1016/j.cis.2011.04.005

Ajorloo, M., Ghodrat, M., Scott, J. & Strezov, V. (2022). Heavy metals removal/stabilization from municipal solid waste incineration fly ash: A review and recent trends. *Journal of Material Cycles and Waste Management*, *24*(5), 1693–1717. https://doi.org/10.1007/s10163-022-01459-w

Akar, T. & Tunali, S. (2006). Biosorption characteristics of Aspergillus flavus biomass for removal of Pb(II) and Cu(II) ions from an aqueous solution. *Bioresource Technology, 97*(15), 1780–1787. https://doi.org/10.1016/j.biortech.2005.09.009

Akhtar, M., Sarfraz, M., Ahmad, M., Raza, N. & Zhang, L. (2025). Use of low-cost adsorbent for waste water treatment: Recent progress, new trend and future perspectives. *Desalination and Water Treatment, 321*(September 2024), 100914. https://doi.org/10.1016/j.dwt.2024.100914

Ali, M. U., Gulzar, M. Z., Sattar, B., Sehar, S., Abbas, Q., Adnan, M., Sun, J., Luo, Z., Hu, G., Yu, R. & Wong, M. H. (2025). Silent threats of lead-based paints in toys and households to children's health and development. *Journal of Hazardous Materials, 486*(December 2024). https://doi.org/10.1016/j.jhazmat.2024.136984

Alidokht, L., Anastopoulos, I., Ntarlagiannis, D., Soupios, P., Tawabini, B., Kalderis, D. & Khataee, A. (2021). Recent advances in the application of nanomaterials for the remediation of arsenic-contaminated water and soil. *Journal of Environmental Chemical Engineering, 9*(4), 105533. https://doi.org/10.1016/j.jece.2021.105533

Appels, L., Baeyens, J., Degrève, J. & Dewil, R. (2008). Principles and potential of the anaerobic digestion of waste-activated sludge. *Progress in Energy and Combustion Science, 34*(6), 755–781. https://doi.org/10.1016/j.pecs.2008.06.002

Aucott, M. (2006). *The fate of heavy metals in landfills: A review.* Trenton, NJ, USA: New Jersey Department of Environmental Protection.

Babel, S. & del Mundo Dacera, D. (2006). Heavy metal removal from contaminated sludge for land application: A review. *Waste Management, 26*(9), 988–1004. https://doi.org/10.1016/j.wasman.2005.09.017

Bagotia, N. (2025). Regeneration strategies for exhausted adsorbents used in water treatment—A critical review. *Journal of Water Process Engineering, 69*(September 2024), 106560. https://doi.org/10.1016/j.jwpe.2024.106560

Barakat, M. A. (2011). New trends in removing heavy metals from industrial wastewater. In *Arabian Journal of Chemistry* (Vol. 4, Issue 4, pp. 361–377). https://doi.org/10.1016/j.arabjc.2010.07.019

Bhatnagar, A. & Minocha, A. K. (2009). Utilization of industrial waste for cadmium removal from water and immobilization in cement. *Chemical Engineering Journal, 150*(1), 145–151. https://doi.org/10.1016/j.cej.2008.12.013

Chen, T., Zhang, Y., Wang, H., Lu, W., Zhou, Z., Zhang, Y. & Ren, L. (2014). Influence of pyrolysis temperature on characteristics and heavy metal adsorptive performance of biochar derived from municipal sewage sludge. *Bioresource Technology, 164*, 47–54. https://doi.org/10.1016/j.biortech.2014.04.048

Choumane, R. & Peulon, S. (2022). Innovative electrochemical process for a total removal and/or separation of soluble heavy metals. *Journal of Environmental Chemical Engineering, 10*(6). https://doi.org/10.1016/j.jece.2022.108607

Coronas, J. & Santamaria, J. (2004). The use of zeolite films in small-scale and micro-scale applications. *Chemical Engineering Science, 59*(22–23), 4879–4885. https://doi.org/10.1016/j.ces.2004.08.001

Dawam, M., Gobara, M., Oraby, H., Y. Zorainy, M. & Nabil, I. M. (2025). Advances in Membrane Technologies for Heavy Metal Removal from Polluted Water: A Comprehensive Review. In *Water, Air, and Soil Pollution* (Vol. 236, Issue 7). Springer Nature. https://doi.org/10.1007/s11270-025-08035-6

Deng, W., Fu, P., Fang, G., Zhu, W., Li, S., Wang, X., Xue, T. & Chen, Y. (2022). Solidification/Stabilization of MSWI Fly Ash Using a Novel Metallurgical Slag-Based Cementitious Material. *Minerals, 12*(5). https://doi.org/10.3390/min12050599

Dhokpande, S. R., Deshmukh, S. M., Khandekar, A. & Sankhe, A. (2024). A review outlook on methods for removal of heavy metal ions from wastewater. *Separation and Purification Technology, 350*(May), 127868. https://doi.org/10.1016/j.seppur.2024.127868

Du, M., Liu, H., Hu, D., Huang, J., Liu, Z. & Fang, Y. (2020). The leaching mechanism of heavy metals (Ni, Cd, As) in a gasification slag during acidification. *Waste Management, 114*, 17–24. https://doi.org/10.1016/j.wasman.2020.06.029

El-Nadi, Y. A. (2017). Solvent extraction and its applications on ore processing and recovery of metals: Classical approach. In *Separation and Purification Reviews* (Vol. 46, Issue 3, pp. 195–215). Taylor and Francis Inc. https://doi.org/10.1080/15422119.2016.1240085

Firincă, C., Zamfir, L. G., Constantin, M., Răut, I., Jecu, M. L., Doni, M., Gurban, A. M. & Şesan, T. E. (2025). Innovative approaches and evolving strategies in heavy metal bioremediation: Current limitations and future opportunities. *Journal of Xenobiotics, 15*(3), 1–41. https://doi.org/10.3390/jox15030063

Florek, J., Negoro, M., Hu, Y., Kanamori, K., Nakanishi, K. & Kleitz, F. (2025). The role of nanoporous adsorbents in the circular economy—closing the loop of critical materials recovery. In *Advanced Functional Materials* (Vol. 35, Issue 1). John Wiley and Sons Inc. https://doi.org/10.1002/adfm.202409462

Fouda-Mbanga, B. G., Velempini, T., Pillay, K. & Tywabi-Ngeva, Z. (2024). Heavy metals removals from wastewater and reuse of the metal loaded adsorbents in various applications: A review. In *Hybrid Advances* (Vol. 6). Elsevier B.V. https://doi.org/10.1016/j.hybadv.2024.100193

Fu, F. & Wang, Q. (2011). Removal of heavy metal ions from wastewaters: A review. In *Journal of Environmental Management* (Vol. 92, Issue 3, pp. 407–418). Academic Press. https://doi.org/10.1016/j.jenvman.2010.11.011

Gebeyehu, K. B., Chen, L., Fan, L., Chao, Y. & Zhu, W. (2025). Recovery of high-value metals from the cathode of spent lithium-ion batteries via acid leaching: A review. In *Journal of Environmental Chemical Engineering* (Vol. 13, Issue 3). Elsevier Ltd. https://doi.org/10.1016/j.jece.2025.116174

Gode, F. & Pehlivan, E. (2006). Removal of chromium(III) from aqueous solutions using Lewatit S 100: The effect of pH, time, metal concentration and temperature. *Journal of Hazardous Materials, 136*(2), 330–337. https://doi.org/10.1016/j.jhazmat.2005.12.021

González Fernández, L. A., Navarro Frómeta, A. E., Carranza Álvarez, C., Flores Ramírez, R., Díaz Flores, P. E., Castillo Ramos, V., Sánchez Polo, M., Carrasco Marín, F. & Medellín Castillo, N. A. (2023). Valorization of sargassum biomass as potential material for the remediation of heavy-metals-contaminated waters. *International Journal of Environmental Research and Public Health, 20*(3), 1–21. https://doi.org/10.3390/ijerph20032559

Gulliani, S., Volpe, M., Messineo, A. & Volpe, R. (2023). Recovery of metals and valuable chemicals from waste electric and electronic materials: A critical review of existing technologies. *RSC Sustainability, 1*(5), 1085–1108. https://doi.org/10.1039/d3su00034f

Gusiatin, M. Z. & Rouhani, A. (2023). Application of selected methods to modify pyrolyzed biochar for the immobilization of metals in soil: A review. In *Materials* (Vol. 16, Issue 23). Multidisciplinary Digital Publishing Institute (MDPI). https://doi.org/10.3390/ma16237342

Hagelüken, C. & Goldmann, D. (2022). Recycling and circular economy—towards a closed loop for metals in emerging clean technologies. *Mineral Economics, 35*(3–4), 539–562. https://doi.org/10.1007/s13563-022-00319-1

Hariyani, D., Hariyani, P. & Mishra, S. (2025). Digital technologies for the sustainable development goals. *Green Technologies and Sustainability, 3*(3), 100202. https://doi.org/10.1016/j.grets.2025.100202

Hassan, G. S. & Abbar, A. H. (2025). Electrocoagulation process for cobalt removal from industrial wastewater: Optimization and kinetic study. *South African Journal of Chemical Engineering*, *53*(April), 233–241. https://doi.org/10.1016/j.sajce.2025.05.001

Hegazy, B., Khodeir, L. & Fathy, F. (2025). Achieving socio-economic resilience in neighborhood through nature-based solutions: A systematic review. *Results in Engineering*, *25*. https://doi.org/10.1016/j.rineng.2025.104266

Hejna, M., Moscatelli, A., Stroppa, N., Onelli, E., Pilu, S., Baldi, A. & Rossi, L. (2020). Bioaccumulation of heavy metals from wastewater through a Typha latifolia and Thelypteris palustris phytoremediation system. *Chemosphere*, *241*. https://doi.org/10.1016/j.chemosphere.2019.125018

Hwang, S. K. & Jho, E. H. (2018). Heavy metal and sulfate removal from sulfate-rich synthetic mine drainages using sulfate reducing bacteria. In *Science of the Total Environment* (Vol. 635, pp. 1308–1316). Elsevier B.V. https://doi.org/10.1016/j.scitotenv.2018.04.231

Islam, N. F., Gogoi, B., Saikia, R., Yousaf, B., Narayan, M. & Sarma, H. (2024). Encouraging circular economy and sustainable environmental practices by addressing waste management and biomass energy production. *Regional Sustainability*, *5*(4), 100174. https://doi.org/10.1016/j.regsus.2024.100174

Iyer, S., Deshmukh, S. M. & Tapre, R. W. (2024). Review on removal of heavy metals from industrial effluents by adsorption. *Reviews in Inorganic Chemistry*, 1–18. https://doi.org/10.1515/revic-2024-0079

Jain, A., Kumari, S., Agarwal, S. & Khan, S. (2021). Water purification via novel nano-adsorbents and their regeneration strategies. *Process Safety and Environmental Protection*, *152*, 441–454. https://doi.org/10.1016/j.psep.2021.06.031

Kazakis, N. A. (2025). Green approaches to heavy metal removal from wastewater: Microalgae solutions in a circular economy framework. *Societal Impacts*, *5*, 100103. https://doi.org/10.1016/j.socimp.2025.100103

Khoshyomn, S., Heidari, A., Farzam, M., Shariatmadari, Z. & Karimian, Z. (2024). Integrated approaches for heavy metal–contaminated soil remediation: Harnessing the potential of Paulownia elongata S. Y. Hu, Oscillatoria sp., arbuscular mycorrhizal fungi (Glomus mosseae and Glomus intraradices), and iron nanoparticles. *Environmental Science and Pollution Research*, *31*(13), 19595–19614. https://doi.org/10.1007/s11356-024-32380-4

Kim, H. M., Choi, T. Y., Park, M. J. & Jeong, D. W. (2022). Heavy metal removal using an advanced removal method to obtain recyclable paper incineration ash. *Scientific Reports*, *12*(1). https://doi.org/10.1038/s41598-022-16486-8

Kinhikar, V. R. (2021). Recovery and reuse of adsorbed heavy metal ions by adsorption technology. *Journal of Scientific Research*, *65*(08), 101–105. https://doi.org/10.37398/jsr.2021.650821

Klimek-Szczykutowicz, M., Szopa, A., Blicharska, E., Dziurka, M., Komsta, Ł. & Ekiert, H. (2019). Bioaccumulation of selected macro- and microelements and their impact on antioxidant properties and accumulation of glucosinolates and phenolic acids in in vitro cultures of Nasturtium officinale (watercress) microshoots. *Food Chemistry*, *300*(July), 125184. https://doi.org/10.1016/j.foodchem.2019.125184

Kurniawan, T. A., Lo, W., Liang, X., Goh, H. H., Othman, M. H. D., Chong, K. K. & Chew, K. W. (2023). Remediation technologies for contaminated groundwater due to arsenic (As), mercury (Hg), and/or fluoride (F): A critical review and way forward to contribute to carbon neutrality. *Separation and Purification Technology*, *314*(January), 123474. https://doi.org/10.1016/j.seppur.2023.123474

Kuyucak, N. & Akcil, A. (2013). Cyanide and removal options from effluents in gold mining and metallurgical processes. *Minerals Engineering*, *50–51*, 13–29. https://doi.org/10.1016/j.mineng.2013.05.027

Lata, S., Singh, P. K. & Samadder, S. R. (2015). Regeneration of adsorbents and recovery of heavy metals: A review. In *International Journal of Environmental Science and Technology* (Vol. 12, Issue 4, pp. 1461–1478). Center for Environmental and Energy Research and Studies. https://doi.org/10.1007/s13762-014-0714-9

Liang, M., Lu, L., He, H., Li, J., Zhu, Z. & Zhu, Y. (2021). Applications of biochar and modified biochar in heavy metal contaminated soil: A descriptive review. In *Sustainability (Switzerland)* (Vol. 13, Issue 24). MDPI. https://doi.org/10.3390/su132414041

Lim, S. S., Fontmorin, J. M., Pham, H. T., Milner, E., Abdul, P. M., Scott, K., Head, I. & Yu, E. H. (2021). Zinc removal and recovery from industrial wastewater with a microbial fuel cell: Experimental investigation and theoretical prediction. *Science of the Total Environment*, *776*, 145934. https://doi.org/10.1016/j.scitotenv.2021.145934

Mahmoud, M. E., Osman, M. M., Hafez, O. F., Hegazi, A. H. & Elmelegy, E. (2010). Removal and preconcentration of lead (II) and other heavy metals from water by alumina adsorbents developed by surface-adsorbed-dithizone. *Desalination*, *251*(1–3). https://doi.org/10.1016/j.desal.2009.08.008

Maitra, S. (2018). In situ bioremediation—an overview | scienceDirect topics. *Research Journal of Life Sciences, Bioinformatics, Pharmaceutical and Chemical Sciences*, *4*(6), 576–598. https://doi.org/10.26479/2018.0406.45

Misganaw, A., Akenaw, B. & Getu, S. (2024). Determination of the level of chromium (III) and comparison of chemical precipitating agents to recover and reuse it from tannery waste water. *Desalination and Water Treatment*, *317*(February), 100150. https://doi.org/10.1016/j.dwt.2024.100150

Mohd Salim, R., Khan Chowdhury, A. J., Rayathulhan, R., Yunus, K. & Sarkar, M. Z. I. (2016). Biosorption of Pb and Cu from aqueous solution using banana peel powder. *Desalination and Water Treatment*, *57*(1), 303–314. https://doi.org/10.1080/19443994.2015.1091613

Murtaza, G., Ahmed, Z., Valipour, M., Ali, I., Usman, M., Iqbal, R., Zulfiqar, U., Rizwan, M., Mahmood, S., Ullah, A., Arslan, M., Rehman, M. H. ur, Ditta, A. & Tariq, A. (2024). Recent trends and economic significance of modified/functionalized biochars for remediation of environmental pollutants. In *Scientific Reports* (Vol. 14, Issue 1). Nature Publishing Group UK. https://doi.org/10.1038/s41598-023-50623-1

Najafi, M., Behnajady, B. & Karimi, S. (2025). Potential dissolvability of Ni, Cd, and Zn in different choline chloride-based green deep eutectic solvents on cold filter cake leaching and molecular dynamic simulation. *Journal of Sustainable Metallurgy*, 0123456789. https://doi.org/10.1007/s40831-025-01138-8

Obeidat, N., Shatanawi, K., Kassab, G. & Halalsheh, M. (2024). Performance of decentralized wastewater treatment system employing Upflow anaerobic sludge blanket and vertical flow constructed wetland. *Case Studies in Chemical and Environmental Engineering*, *9*(January), 100695. https://doi.org/10.1016/j.cscee.2024.100695

Ojha, A. & Tiwary, D. (2021). Various remediation measures for groundwater contamination. *Groundwater Geochemistry: Pollution and Remediation Methods*, *1*, 326–351. https://doi.org/10.1002/9781119709732.ch16

Okuyelu, O. & Adaji, O. (2024). AI-driven real-time quality monitoring and process optimization for enhanced manufacturing performance. *Journal of Advances in Mathematics and Computer Science*, *39*(4), 81–89. https://doi.org/10.9734/jamcs/2024/v39i41883

Oliveira, R. C., Hammer, P., Guibal, E., Taulemesse, J. M. & Garcia, O. (2014). Characterization of metal-biomass interactions in the lanthanum(III) biosorption on Sargassum sp. using SEM/EDX, FTIR, and XPS: Preliminary studies. *Chemical Engineering Journal*, *239*, 381–391. https://doi.org/10.1016/j.cej.2013.11.042

Pandey, A., Kalamdhad, A. & Chandra Sharma, Y. (2023). Recent advances of nanocellulose as biobased adsorbent for heavy metal ions removal: A sustainable approach integrating with waste management. *Environmental Nanotechnology, Monitoring and Management*, *20*(January), 100791. https://doi.org/10.1016/j.enmm.2023.100791

Peng, C., Jin, R., Li, G., Li, F. & Gu, Q. (2014). Recovery of nickel and water from wastewater with electrochemical combination process. *Separation and Purification Technology*, *136*, 42–49. https://doi.org/10.1016/j.seppur.2014.08.025

Pet, I., Sanad, M. N., Farouz, M., ElFaham, M. M., El-Hussein, A., El-sadek, M. S. A., Althobiti, R. A. & Ioanid, A. (2024). Review: Recent developments in the implementation of activated carbon as heavy metal removal management. In *Water Conservation Science and Engineering* (Vol. 9, Issue 2). Springer Nature. https://doi.org/10.1007/s41101-024-00287-3

Qasem, N. A. A., Mohammed, R. H. & Lawal, D. U. (2021). Removal of heavy metal ions from wastewater: A comprehensive and critical review. *NPJ Clean Water*, *4*(1). https://doi.org/10.1038/s41545-021-00127-0

Qomariyah, L., Ira Kumalasari, P., Suprapto, S., Fajar Puspita, N., Agustiani, E., Altway, S. & Hardi, H. (2022). Bioremediation of heavy metals in petroleum sludge through bacterial mixtures. *Materials Today: Proceedings*, *63*, S140–S142. https://doi.org/10.1016/j.matpr.2022.02.054

Rai, V., Liu, D., Xia, D., Jayaraman, Y. & Gabriel, J. C. P. (2021). Electrochemical approaches for the recovery of metals from electronic waste: A critical review. In *Recycling* (Vol. 6, Issue 3). MDPI AG. https://doi.org/10.3390/recycling6030053

Raji, Z., Karim, A., Karam, A. & Khalloufi, S. (2023). Adsorption of heavy metals: Mechanisms, kinetics, and applications of various adsorbents in wastewater remediation—A review. *Waste*, *1*(3), 775–805. https://doi.org/10.3390/waste1030046

Razzak, S. A., Faruque, M. O., Alsheikh, Z., Alsheikhmohamad, L., Alkuroud, D., Alfayez, A., Hossain, S. M. Z. & Hossain, M. M. (2022). A comprehensive review on conventional and biological-driven heavy metals removal from industrial wastewater. *Environmental Advances*, *7*. https://doi.org/10.1016/j.envadv.2022.100168

Renu, Agarwal, M. & Singh, K. (2017). Heavy metal removal from wastewater using various adsorbents: A review. *Journal of Water Reuse and Desalination*, *7*(4), 387–419. https://doi.org/10.2166/wrd.2016.104

Rodrigues, I. R., Deferm, C., Binnemans, K. & Riaño, S. (2022). Separation of cobalt and nickel via solvent extraction with Cyanex-272: Batch experiments and comparison of mixer-settlers and an agitated column as contactors for continuous countercurrent extraction. *Separation and Purification Technology*, *296*(April). https://doi.org/10.1016/j.seppur.2022.121326

Sadehlari, F., Cassol, G. O. & Dubljevic, S. (2025). Screening of parameters and optimization of nickel extraction by green emulsion liquid membrane using statistical experimental design. *Canadian Journal of Chemical Engineering*, *June 2024*, 4106–4124. https://doi.org/10.1002/cjce.25657

Saravanan, A., Kumar, P. S., Duc, P. A. & Rangasamy, G. (2023). Strategies for microbial bioremediation of environmental pollutants from industrial wastewater: A sustainable approach. *Chemosphere*, *313*. https://doi.org/10.1016/j.chemosphere.2022.137323

Shabbirahmed, A. M., Jacob, A., Dey, P., Somu, P. & Haldar, D. (2025). Biomass as eco-friendly adsorbents for the removal of emerging pollutants from wastewater: A review. *Discover Applied Sciences*, *7*(7). https://doi.org/10.1007/s42452-025-07463-7

Sharma, B. & Shukla, P. (2021). A comparative analysis of heavy metal bioaccumulation and functional gene annotation towards multiple metal resistant potential by Ochrobactrum intermedium BPS-20 and Ochrobactrum ciceri BPS–26. *Bioresource Technology*, *320*. https://doi.org/10.1016/j.biortech.2020.124330

Shen, H., Lou, B., Liu, B., Zhang, J., Zhang, X., Liu, J., Zhang, R., Chen, M. & Zhang, S. (2024). Reduction of heavy metals in municipal solid waste incineration fly ash followed by making transparent glass. *Waste Management*, *183*, 184–190. https://doi.org/10.1016/j.wasman.2024.05.008

Shih, Y. J., Chien, S. K., Jhang, S. R. & Lin, Y. C. (2019). Chemical leaching, precipitation and solvent extraction for sequential separation of valuable metals in cathode material of spent lithium ion batteries. *Journal of the Taiwan Institute of Chemical Engineers*, *100*, 151–159. https://doi.org/10.1016/j.jtice.2019.04.017

Shrestha, R., Ban, S., Devkota, S., Sharma, S., Joshi, R., Tiwari, A. P., Kim, H. Y. & Joshi, M. K. (2021). Technological trends in heavy metals removal from industrial wastewater: A review. In *Journal of Environmental Chemical Engineering* (Vol. 9, Issue 4). Elsevier Ltd. https://doi.org/10.1016/j.jece.2021.105688

Shuai, J., Liu, W., Rohani, S., Wang, Z., He, M., Ding, C. & Lv, X. (2025). Efficient extraction and separation of valuable elements from spent lithium-ion batteries by leaching and solvent extraction: A review. *Chemical Engineering Journal*, *503*(September 2024), 158114. https://doi.org/10.1016/j.cej.2024.158114

Shukla, S. R. & Pai, R. S. (2005). Adsorption of Cu(II), Ni(II) and Zn(II) on dye loaded groundnut shells and sawdust. *Separation and Purification Technology*, *43*(1), 1–8. https://doi.org/10.1016/j.seppur.2004.09.003

Song, S., Li, Q., Leslie, G. & Shen, Y. (2024). Water treatment methods in heavy metals removal during photovoltaic modules recycling: A review. In *Resources, Conservation and Recycling* (Vol. 208). Elsevier B.V. https://doi.org/10.1016/j.resconrec.2024.107701

Sonone, S. S., Jadhav, S., Sankhla, M. S. & Kumar, R. (2021). Water contamination by heavy metals and their toxic effect on aquaculture and human health through food chain. *Letters in Applied NanoBioScience*, *10*(2), 2148–2166. https://doi.org/10.33263/LIANBS102.21482166

Syeda, H. I., Sultan, I., Razavi, K. S. & Yap, P. S. (2022). Biosorption of heavy metals from aqueous solution by various chemically modified agricultural wastes: A review. *Journal of Water Process Engineering*, *46*(January), 102446. https://doi.org/10.1016/j.jwpe.2021.102446

Tang, J., Su, M., Wu, Q., Wei, L., Wang, N., Xiao, E., Zhang, H., Wei, Y., Liu, Y., Ekberg, C., Steenari, B. M. & Xiao, T. (2019). Highly efficient recovery and clean-up of four heavy metals from MSWI fly ash by integrating leaching, selective extraction and adsorption. *Journal of Cleaner Production*, *234*, 139–149. https://doi.org/10.1016/j.jclepro.2019.06.198

Tumanov, V. V., Storozhenko, P. A., Grachev, A. A. & Fedin, O. I. (2023). Group extraction of the rare-earth elements by alkyl phosphine oxides of the hexyl–octyl series. *Russian Journal of Inorganic Chemistry*, *68*(4), 474–478. https://doi.org/10.1134/S0036023622602902

Twizerimana, P. & Wu, Y. (2024). Overview of integrated electrocoagulation-adsorption strategies for the removal of heavy metal pollutants from wastewater. *Discover Chemical Engineering*, *4*(1). https://doi.org/10.1007/s43938-024-00053-w

Van Wendel De Joode, B., Barbeau, B., Bouchard, M. F., Mora, A. M., Skytt, Å., Córdoba, L., Quesada, R., Lundh, T., Lindh, C. H. & Mergler, D. (2016). Manganese concentrations in drinking water from villages near banana plantations with aerial mancozeb spraying in Costa Rica: Results from the Infants' Environmental Health Study (ISA). *Environmental Pollution*, *215*, 247–257. https://doi.org/10.1016/j.envpol.2016.04.015

Vardhan, K. H., Kumar, P. S. & Panda, R. C. (2019). A review on heavy metal pollution, toxicity and remedial measures: Current trends and future perspectives. *Journal of Molecular Liquids*, *290*, 111197. https://doi.org/10.1016/j.molliq.2019.111197

Velempini, T., Ahamed, M. E. H. & Pillay, K. (2023). Heavy-metal spent adsorbents reuse in catalytic, energy and forensic applications- a new approach in reducing secondary pollution associated with adsorption. In *Results in Chemistry* (Vol. 5). Elsevier B.V. https://doi.org/10.1016/j.rechem.2023.100901

Ventura, J. R. S., Tulipan, J. U., Banawa, A., Umali, K. D. C. & Villanueva, J. A. L. (2024). Advancements and challenges in decentralized wastewater treatment: A comprehensive review. In *Desalination and Water Treatment* (Vol. 320). Elsevier B.V. https://doi.org/10.1016/j.dwt.2024.100830

Verma, R., Vinoda, K. S., Papireddy, M. & Gowda, A. N. S. (2016). Toxic pollutants from plastic waste- A review. *Procedia Environmental Sciences*, *35*, 701–708. https://doi.org/10.1016/j.proenv.2016.07.069

Wagner, T. P. & Raymond, T. (2015). Landfill mining: Case study of a successful metals recovery project. *Waste Management*, *45*, 448–457. https://doi.org/10.1016/j.wasman.2015.06.034

Wang, B., Lan, J., Bo, C., Gong, B. & Ou, J. (2023). Adsorption of heavy metal onto biomass-derived activated carbon: Review. In *RSC Advances* (Vol. 13, Issue 7, pp. 4275–4302). Royal Society of Chemistry. https://doi.org/10.1039/d2ra07911a

Wang, H., Yang, J., Zhang, H., Zhao, J., Liu, H., Wang, J., Li, G. & Liang, H. (2024). Membrane-based technology in water and resources recovery from the perspective of water social circulation: A review. *Science of the Total Environment*, *908*(August 2023), 168277. https://doi.org/10.1016/j.scitotenv.2023.168277

Wang, H., Lv, Z., Wang, B., Wang, Y. N., Sun, Y., Tsang, Y. F., Zhao, J. & Zhan, M. (2019). Effective stabilization of antimony in waste-to-energy fly ash with recycled laboratory iron–rich residuals. *Journal of Cleaner Production*, *230*, 685–693. https://doi.org/10.1016/j.jclepro.2019.05.128

Wong, K. Y., Ng, J. H., Chong, C. T., Lam, S. S. & Chong, W. T. (2019). Biodiesel process intensification through catalytic enhancement and emerging reactor designs: A critical review. In *Renewable and Sustainable Energy Reviews* (Vol. 116). Elsevier Ltd. https://doi.org/10.1016/j.rser.2019.109399

Xu, Y., Chen, Y. & Feng, Y. (2013). Stabilization treatment of the heavy metals in fly ash from municipal solid waste incineration using diisopropyl dithiophosphate potassium. *Environmental Technology (United Kingdom)*, *34*(11), 1411–1419. https://doi.org/10.1080/09593330.2012.752871

Xue, Y. & Liu, X. (2021). Detoxification, solidification and recycling of municipal solid waste incineration fly ash: A review. *Chemical Engineering Journal*, *420*(P3), 130349. https://doi.org/10.1016/j.cej.2021.130349

Yao, X., Zhu, J., Li, E., Xiong, F. & Zhang, B. (2024). Comprehensive benefit analysis of zero-liquid discharge system for cold-rolling wastewater in iron and steel industry. *Journal of Water Process Engineering*, *63*(April), 105537. https://doi.org/10.1016/j.jwpe.2024.105537

Ye, G., Liu, H., Miao, Y., Zheng, H. & Zhang, G. (2025). Investigating the efficacy and mechanistic pathways of mildly alkaline activators in the immobilization and stabilization of high-arsenic metallurgical slag using slag-based cementitious matrices. *Process Safety and Environmental Protection*, *198*. https://doi.org/10.1016/j.psep.2025.107138

Yenduri, S., Hegde, B. M. & Koppuravuri, N. P. (2025). Biogenic nanomaterials in heavy metal pre-concentration: A green future. *Microchemical Journal*, *216*, 114694. https://doi.org/10.1016/j.microc.2025.114694

Yesil, H. & Tugtas, A. E. (2019). Removal of heavy metals from leaching effluents of sewage sludge via supported liquid membranes. *Science of the Total Environment*, *693*. https://doi.org/10.1016/j.scitotenv.2019.133608

Zhang, H., Zhang, L. L., Li, J., Chen, M. & An, R. D. (2020). Comparative study on the bioaccumulation of lead, cadmium and nickel and their toxic effects on the growth and enzyme defence strategies of a heavy metal accumulator, Hydrilla verticillata (L.f.) Royle. *Environmental Science and Pollution Research*, *27*(9), 9853–9865. https://doi.org/10.1007/s11356-019-06968-0

Zhang, Z., Yu, Y., Rao, Y., Wang, Y., Yu, C., Luo, Z., Zhao, H., Saffarzadeh, A., Wang, X., Wu, C. & Wang, Q. (2024). Stabilization of heavy metals in municipal solid waste incineration fly ash using organic chelating agents: Insight into risk assessment and function mechanism. *Science of the Total Environment, 914*. https://doi.org/10.1016/j.scitotenv.2024.169986

Zhang, Z., Cheng, L., Hu, Q. & Hu, Z. (2025). Influence of pyrolysis temperature on speciation, leaching and environmental risk assessment of heavy metals in biochar and bio-oil from pyrolysis of wet sewage sludge. *Biomass Conversion and Biorefinery*. https://doi.org/10.1007/s13399-025-06530-8

Zhou, B., Su, H., Liu, W., Zhu, Z., Wang, L. & Qi, T. (2025). Solvent extraction of metal ions from the leaching solutions of waste lithium-ion battery materials: A review. *Separation and Purification Technology*, *354*(P5), 129173. https://doi.org/10.1016/j.seppur.2024.129173

Zhou, B., Zhang, T. & Wang, F. (2023). Microbial-based heavy metal bioremediation: Toxicity and eco-friendly approaches to heavy metal decontamination. *Applied Sciences (Switzerland)*, *13*(14). https://doi.org/10.3390/app13148439

Zhou, W., Liu, X., Lyu, X., Gao, W., Su, H. & Li, C. (2022a). Extraction and separation of copper and iron from copper smelting slag: A review. *Journal of Cleaner Production*, *368*(July), 133095. https://doi.org/10.1016/j.jclepro.2022.133095

Zhou, W., Liu, X., Lyu, X., Gao, W., Su, H. & Li, C. (2022b). Extraction and separation of copper and iron from copper smelting slag: A review. In *Journal of Cleaner Production* (Vol. 368). Elsevier Ltd. https://doi.org/10.1016/j.jclepro.2022.133095

Index

Note: Page references in *italics* and **bold** indicate figures and tables on the corresponding page.

For Product Safety Concerns and Information please contact our EU representative GPSR@taylorandfrancis.com
Taylor & Francis Verlag GmbH, Kaufingerstraße 24, 80331 München, Germany

www.ingramcontent.com/pod-product-compliance
Ingram Content Group UK Ltd.
Pitfield, Milton Keynes, MK11 3LW, UK
UKHW020200200926
475702UK00004B/545
* 9 7 8 1 0 3 2 9 8 5 1 9 0 *